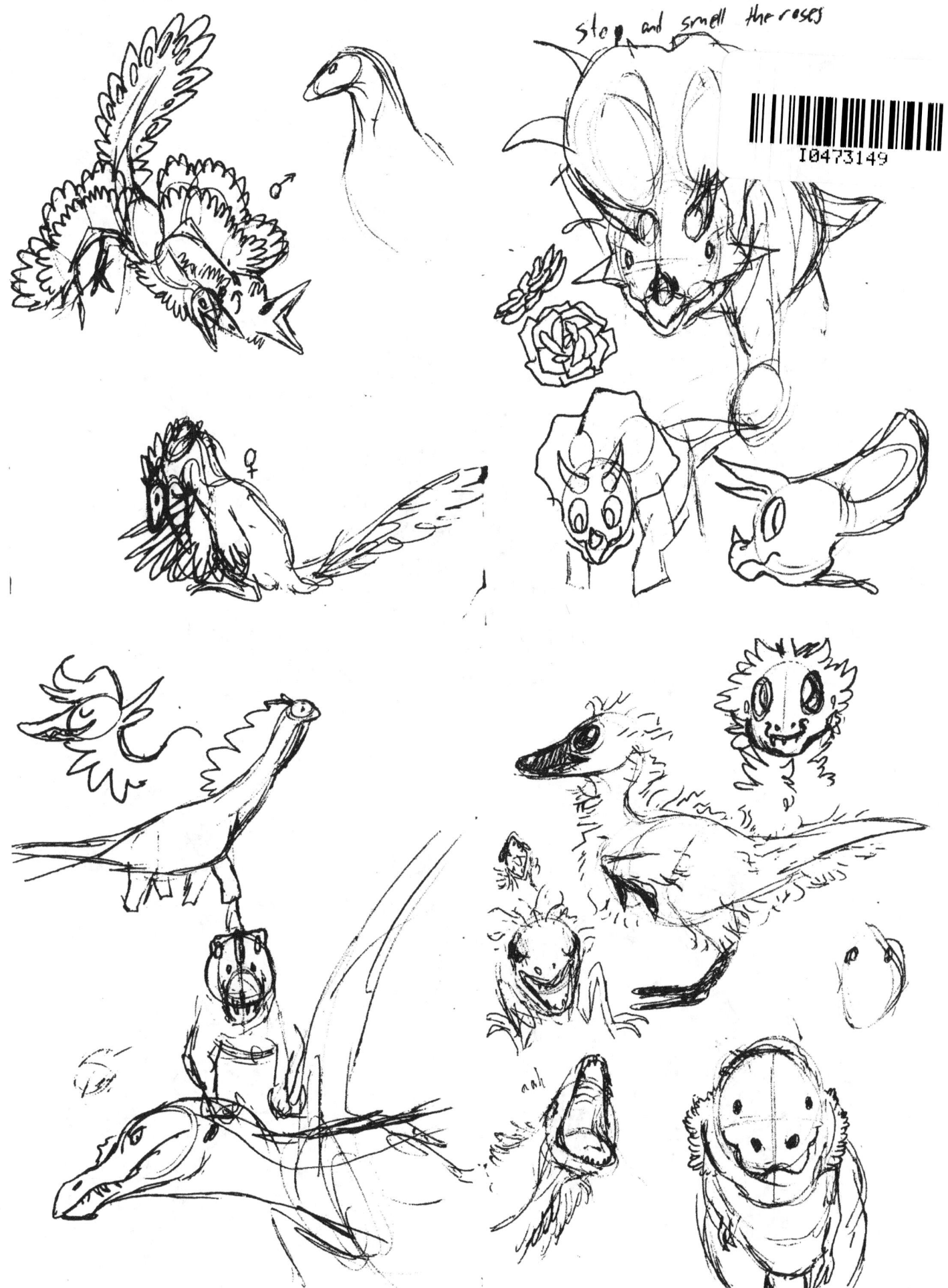

step and smell the roses

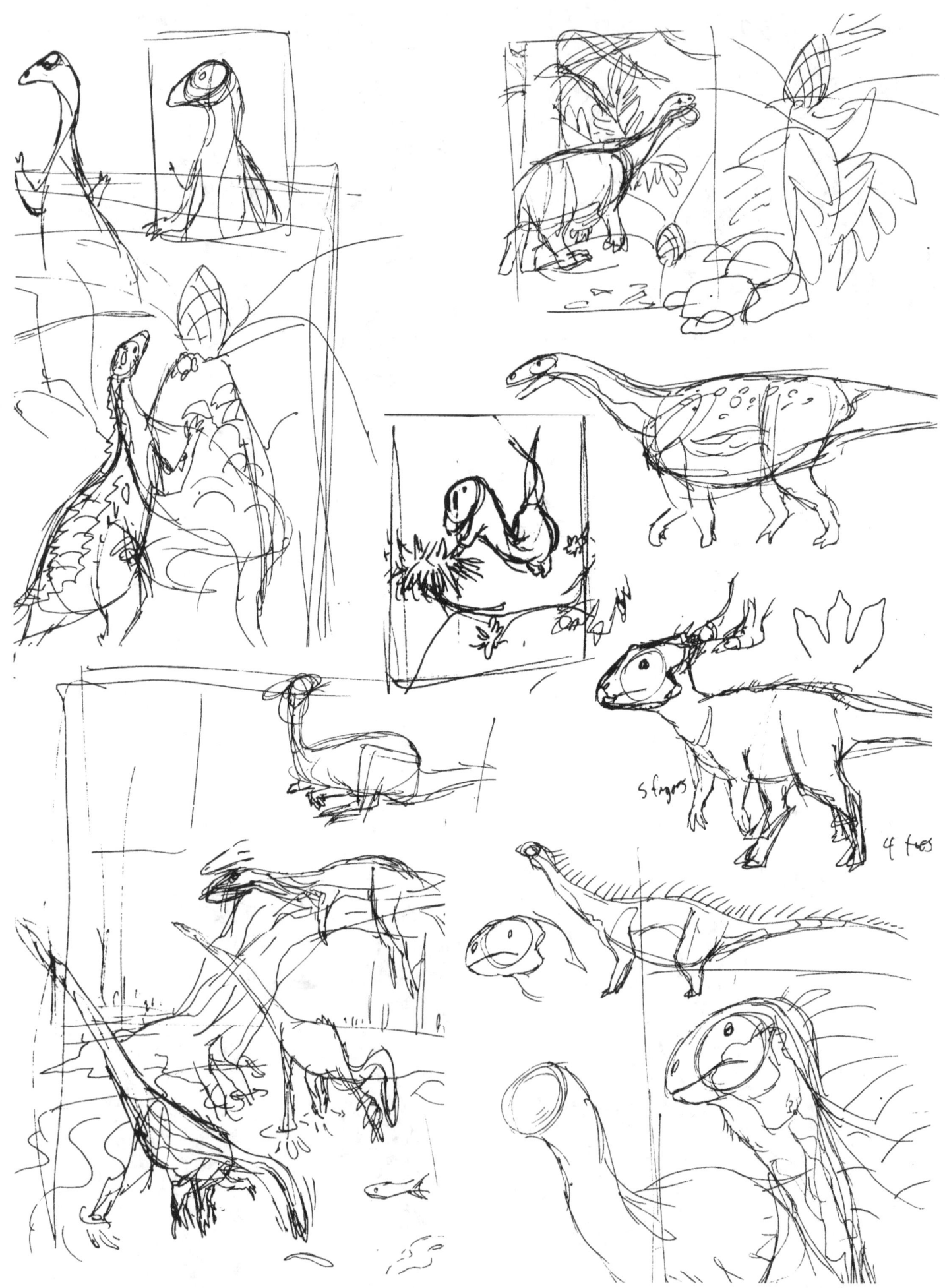

5 fingers

4 toes

Many of you may remember learning in school (or on your own) that one of the things we probably will never know about dinosaurs and other extinct animals is what color they were. Though paleontology has come a long way, and there are some specimens that have preserved their pigments by being buried in fine substrate, most of these animals' colors remain a mystery. That is where you come in!

This is a coloring book that strives to be scientifically minded, from following the latest developments in our knowledge of these creatures, to the research done in creating plausible situations and environments that illustrate how they might have lived their lives. Paleoart is an important way the public is able to access and learn about prehistoric life-forms, and it is important for these artists to not perpetuate outdated models. These were real animals that lived and thrived for hundreds of millions of years, not some movie monsters that look a certain way just to look scary.

The animals in this book are labelled by their genus names, and are listed in chronological order.
Because of new evidence and a little bit of speculation, many of the creatures featured in this book have been drawn with some kind of fluffy, feathery coating. After all, especially for those incredibly ancient specimens, softer-tissue is rarely preserved, and this includes both evidence of fuzz and the lack thereof. Current evidence points to the origins of hair-like structures being much further in the past than was previously thought.

Keep in mind that though much effort has been invested into keeping the content accurate, the author/illustrator is still human, and you may discover a few errors! Happy coloring!

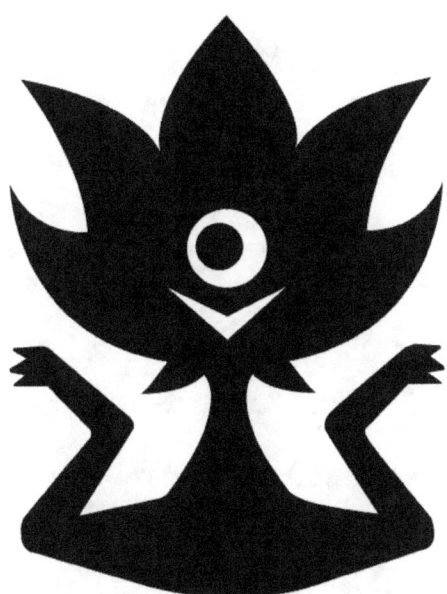

dianeramic@gmail.com
www.dramic.wixsite.com/home
dianeramic.tumblr.com

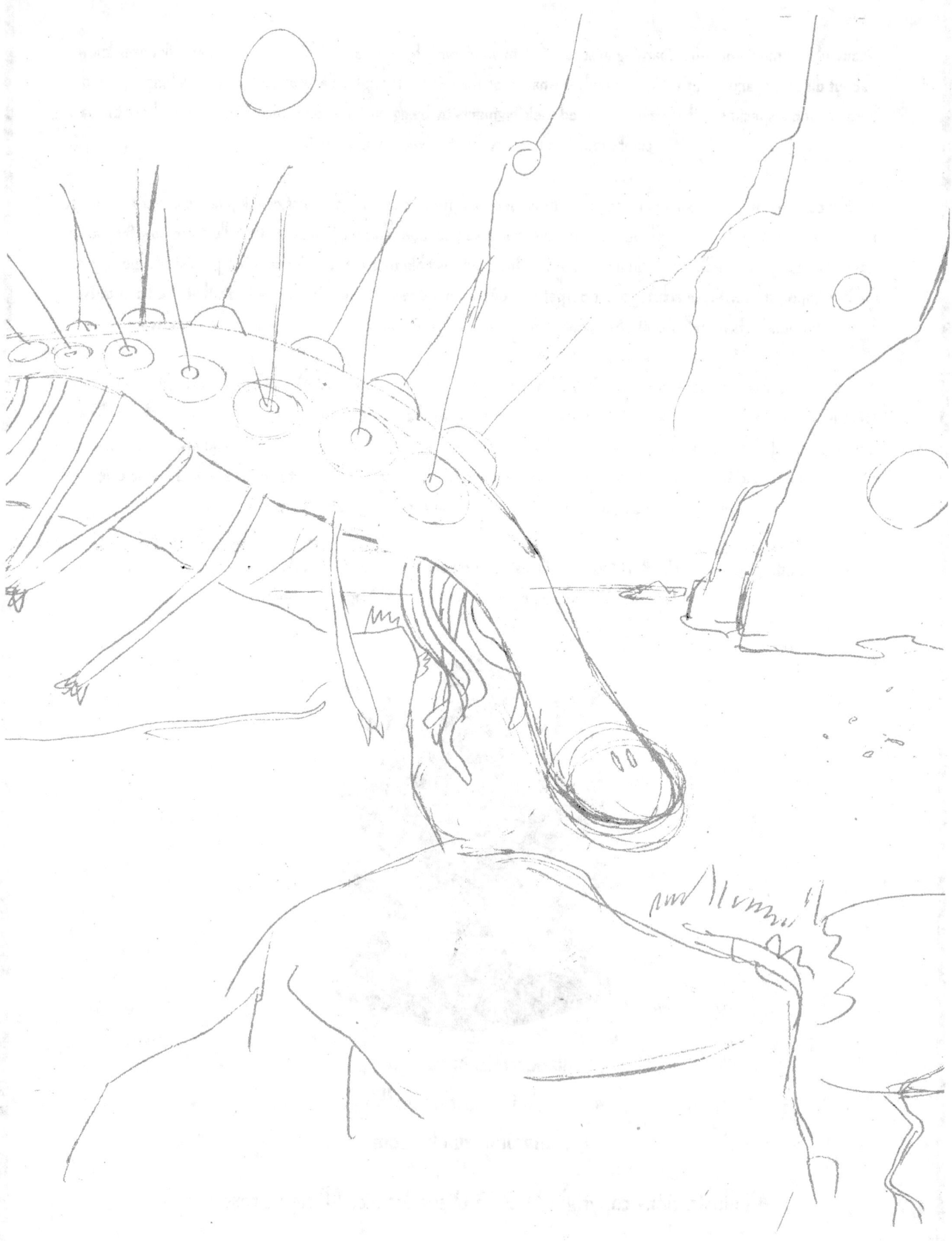

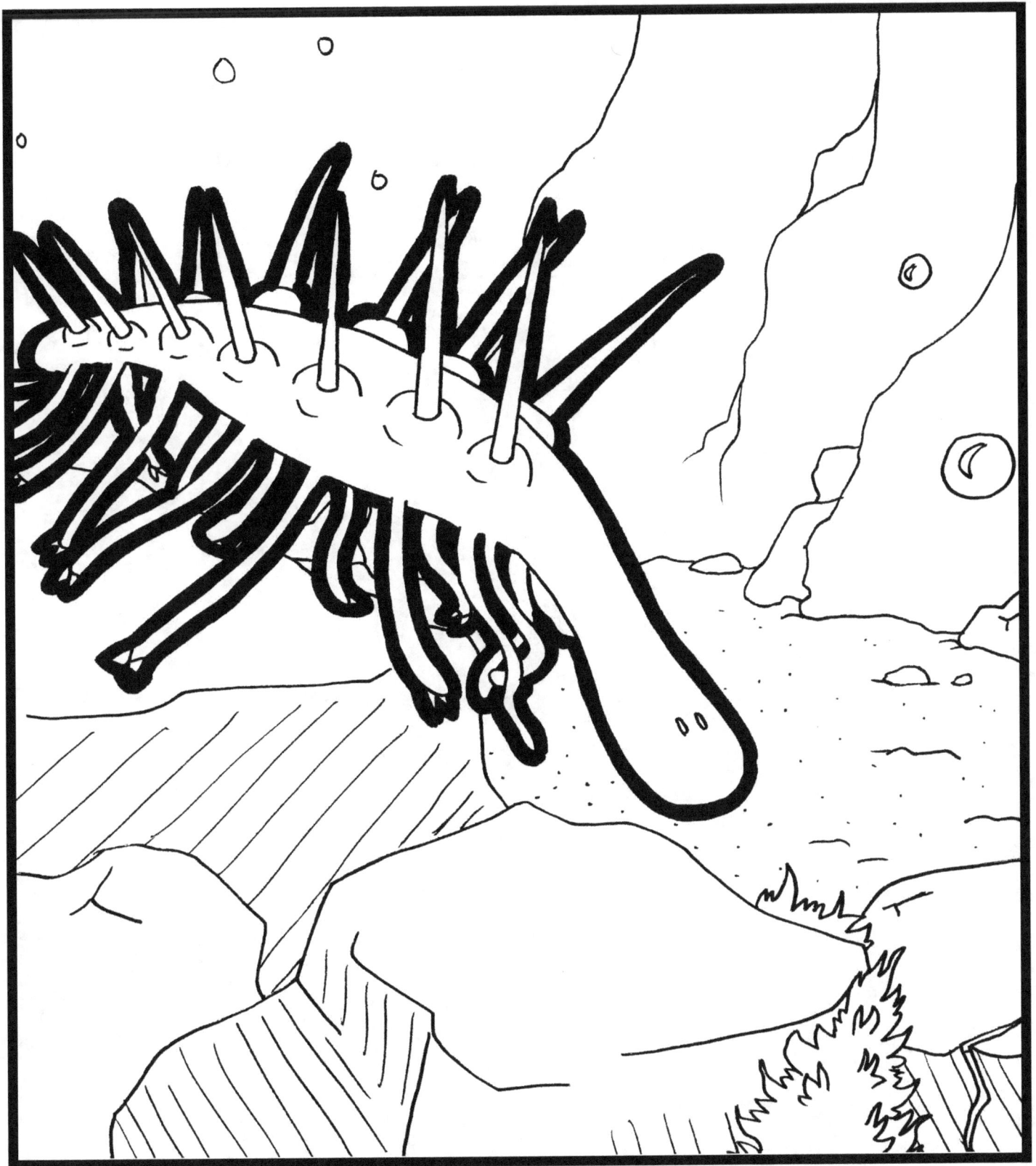

HALLUCIGENIA

520 MYA, Middle Cambrian, Canada, China

Upon discovery, this invertebrate baffled researchers, and many different reconstructions of this creature exist today. Some label it a velvet worm ancestor, some call it an arthropod, while others are content with it being a lobopod, a worm with legs. Early depictions of this animal were recontructed upside down and back to front, adding to its strange history since discovery.

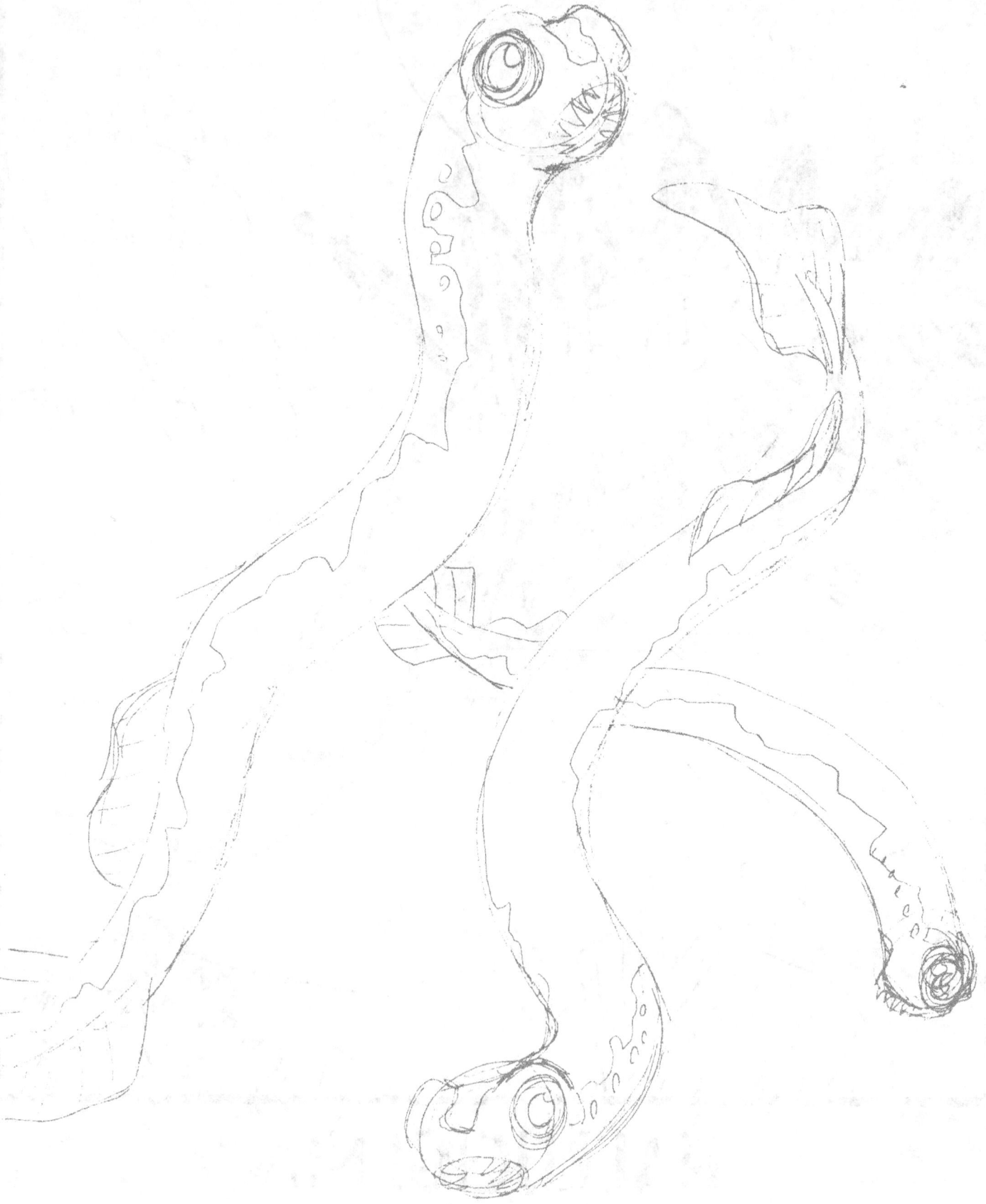

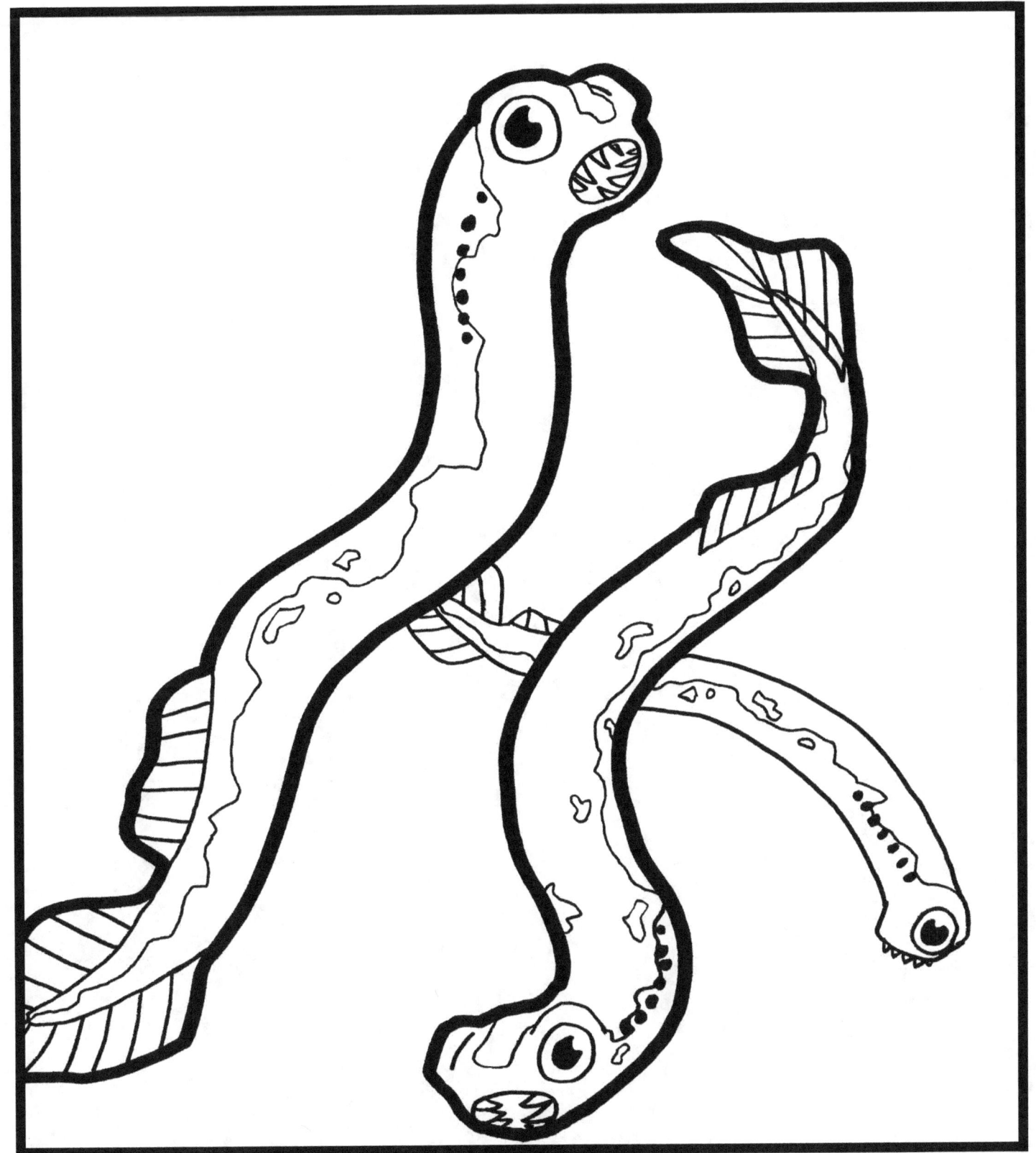

CONODONT

520 - 200 MYA, Late Cambrian to Late Triassic, South Africa, Canada, United States, Scotland

Despite the teeth being one of the only parts of the body that could be easily preserved, fossils of these early vertebrates have been found all over the world, and are commonly used to identify geologic periods. They were eel-shaped, and ranged in length from one to forty centimeters. Some species ate plankton, some had grasping teeth, but overall, they were not fit for high speed chases.

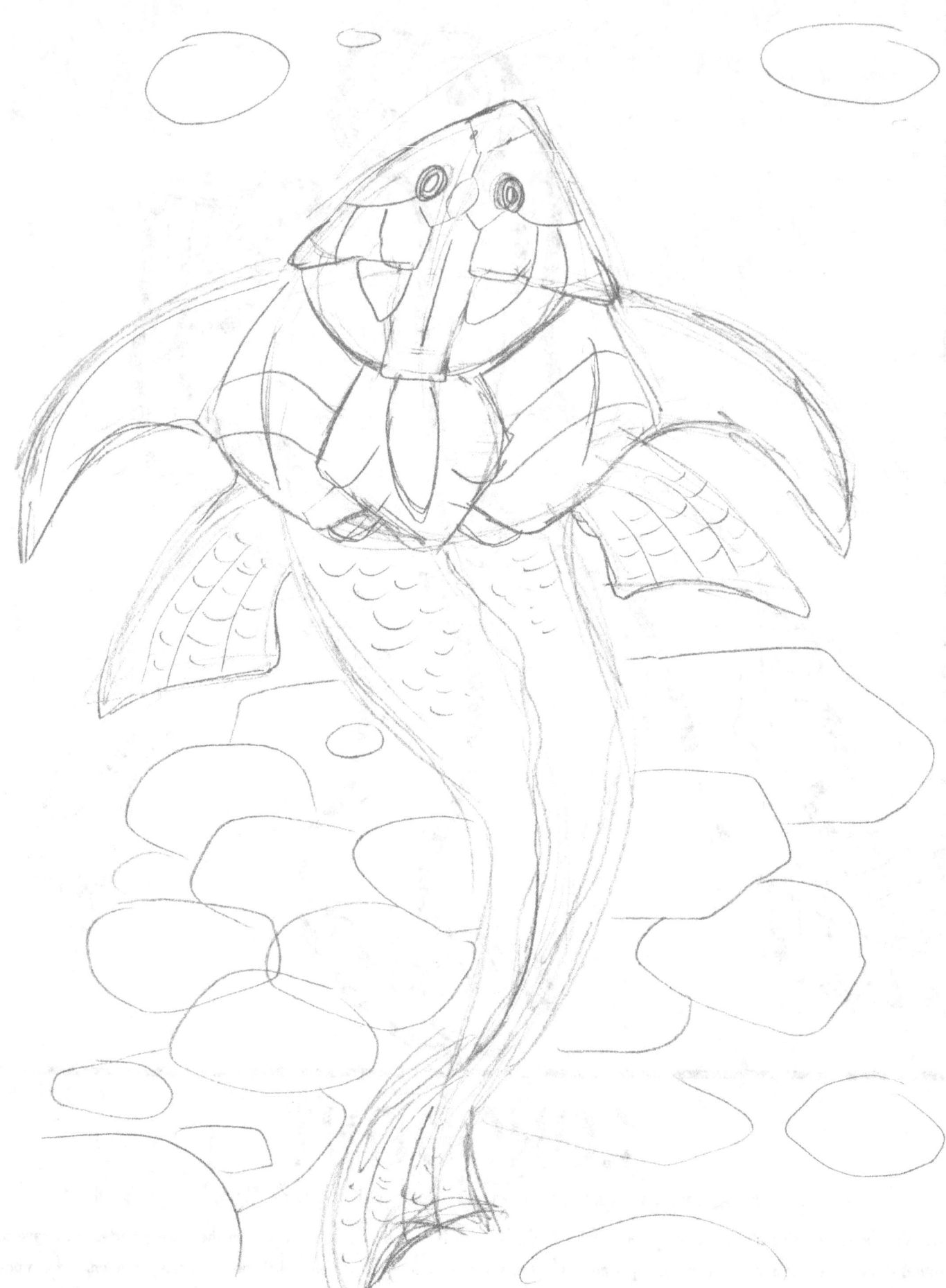

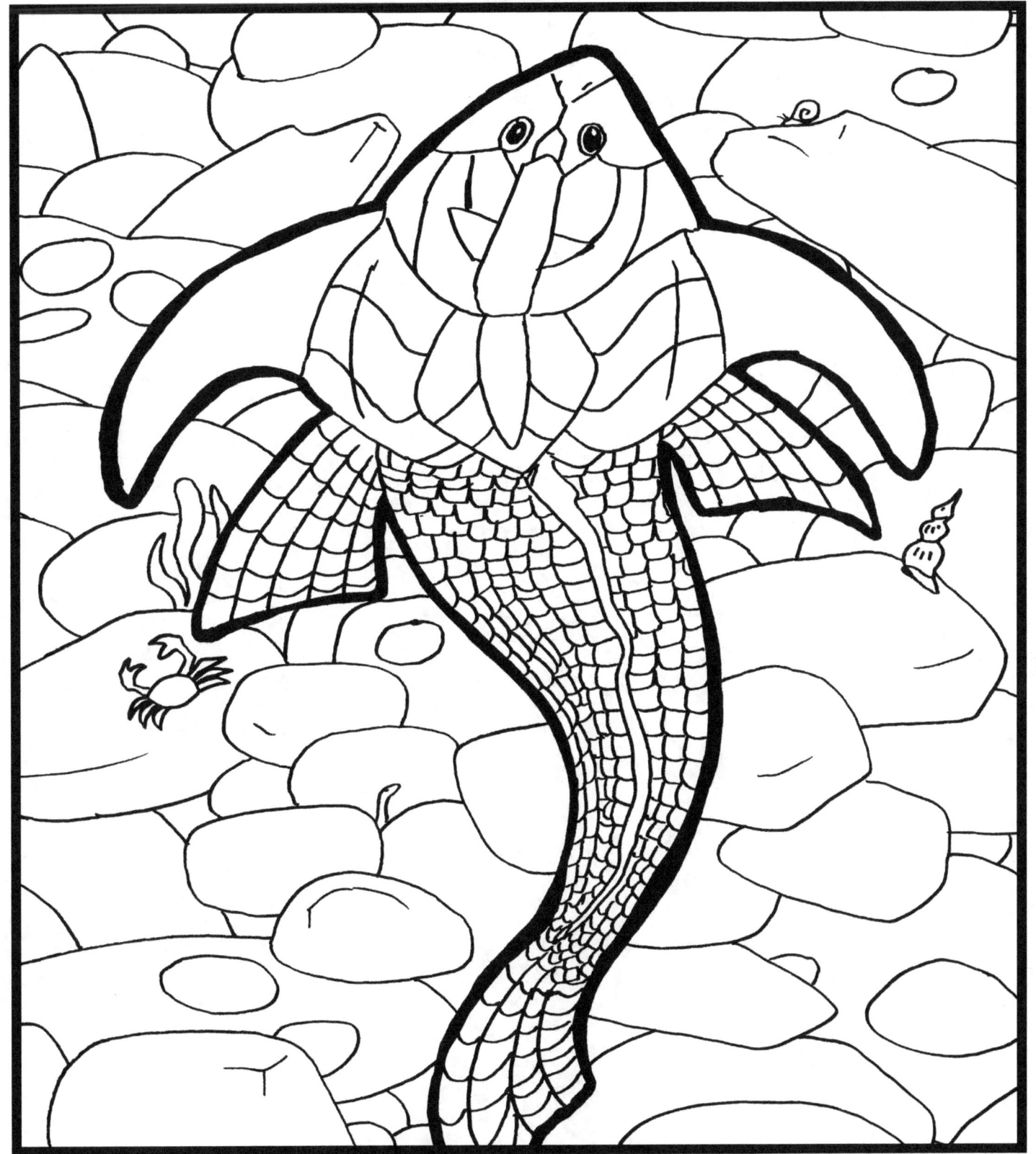

LUNASPIS

409 - 402.5 MYA, Early Devonian, Australia, China, Germany

This placoderm lived in shallow marine environments, and was given the name of "moon-shield" for the crescent shaped plates on either side of the head. These hunters were active swimmers, and had long, whiplike tails.

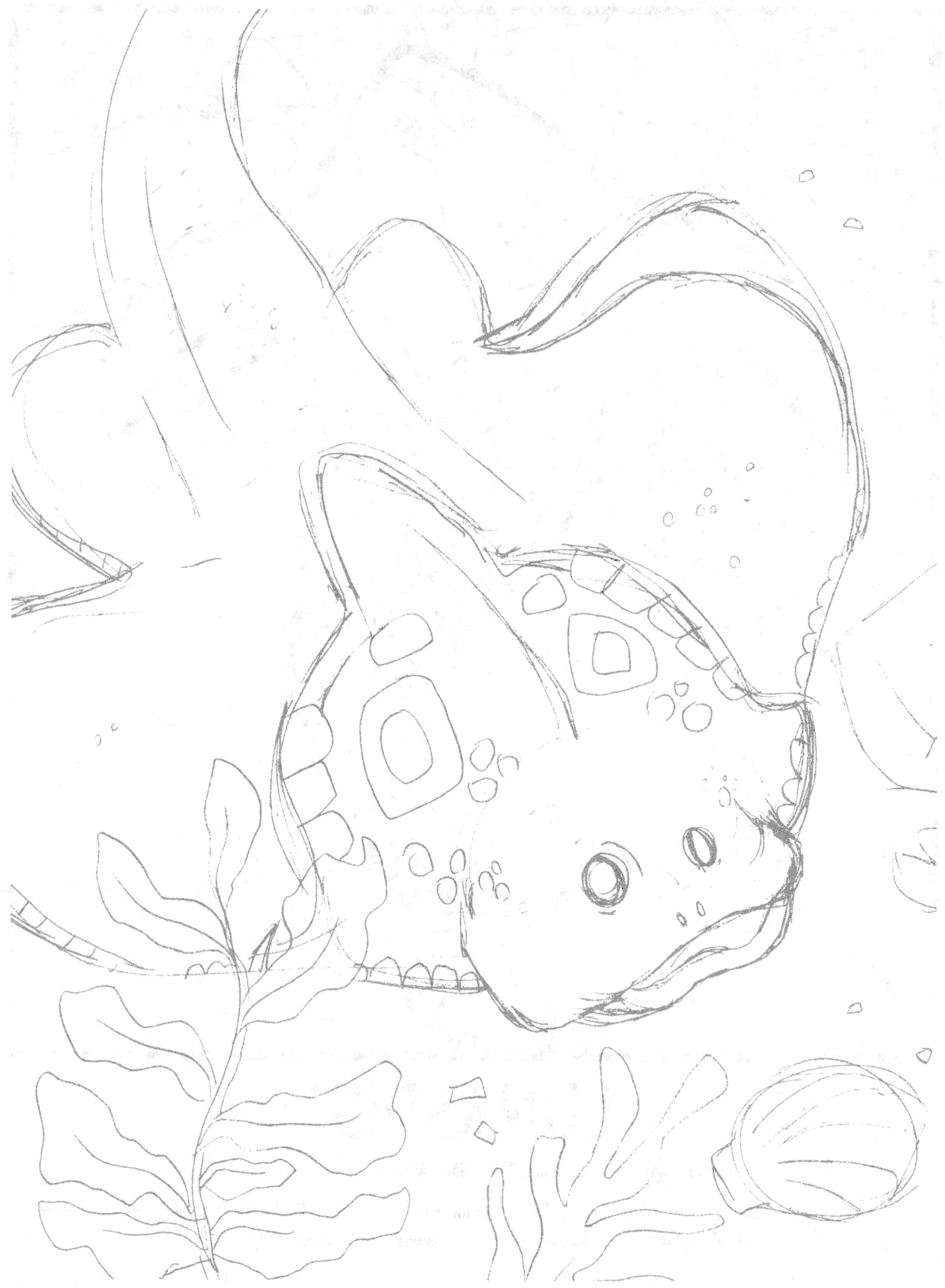

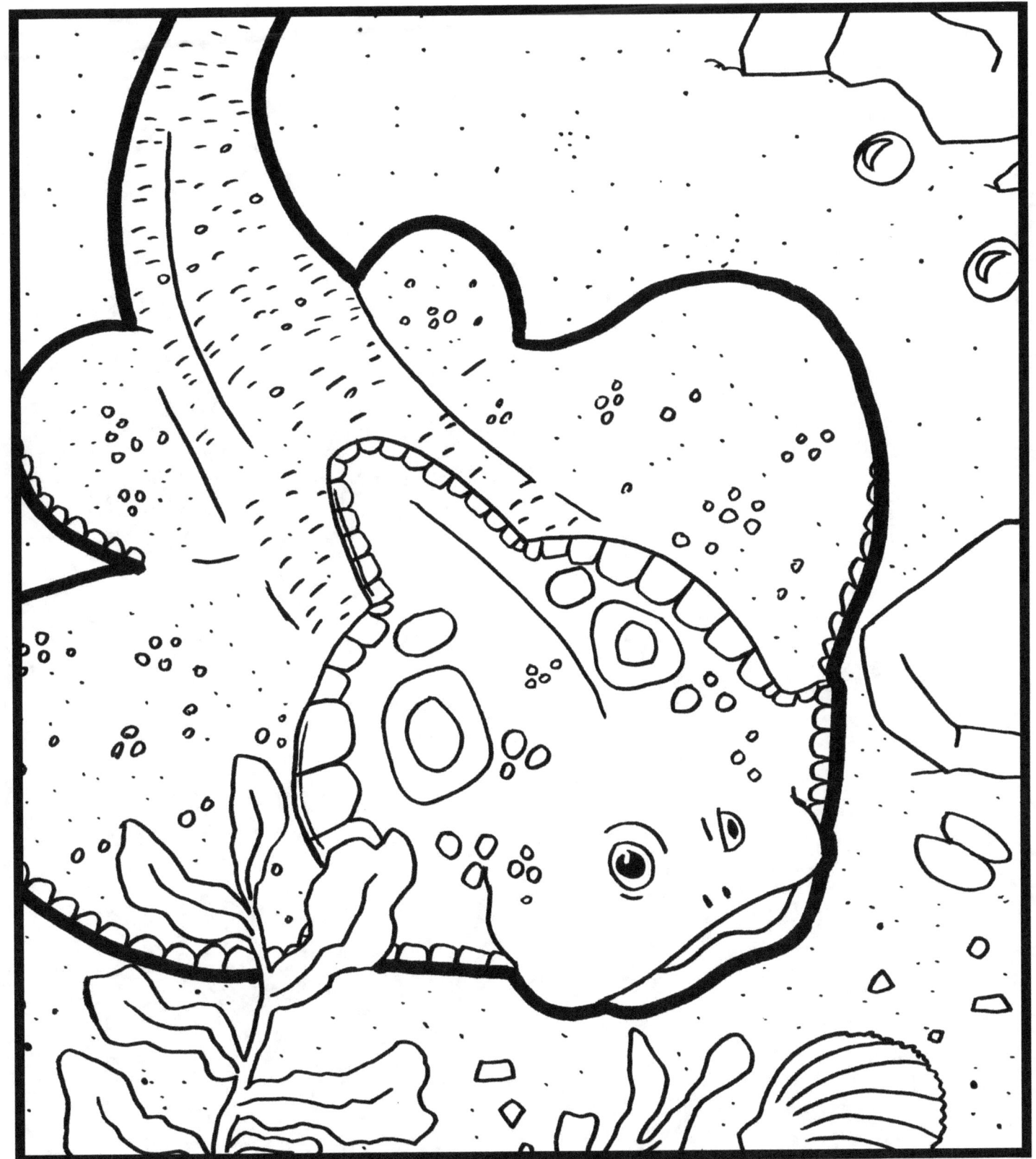

GEMUENDINA

409 - 402 MYA, Early Devonian, Germany

Gemuendina was a bone-plated placoderm fish with the outward appearance of a modern stingray. Growing up to a meter long, it dwelled at the bottom of the sea, suctioning up small prey that swam too close to its upturned mouth.

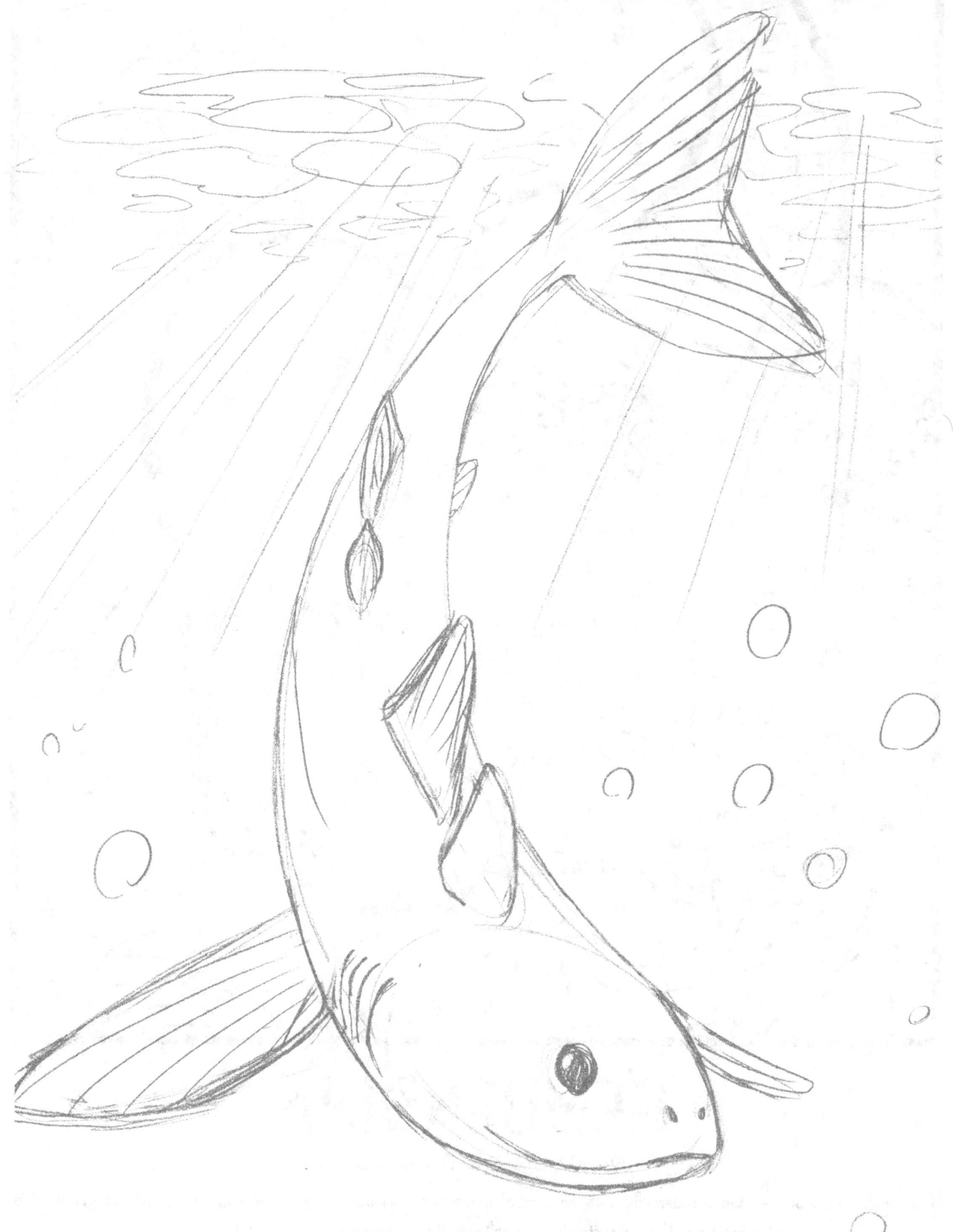

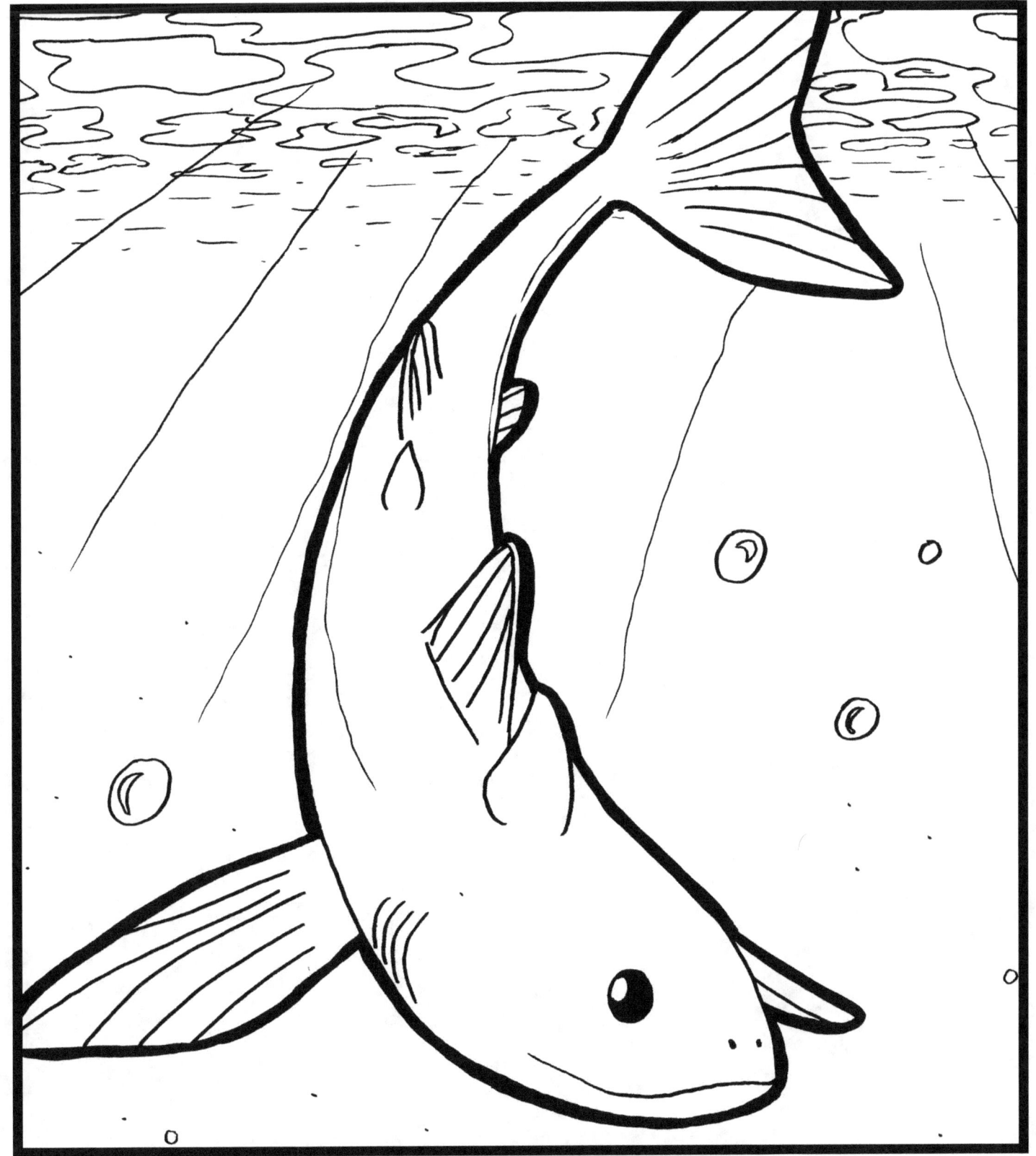

CLADOSELACHE

385 - 359 MYA, Late Devonian, United States and Canada

This was an early shark that lacked scales over most of its body, preserving them on the fins, mouth, and around the eyes. This shark was a fast and agile swimmer, and it had to be, for not only did it hunt fast prey, it was a target of another large predator: the armored fish Dunkleosteus.

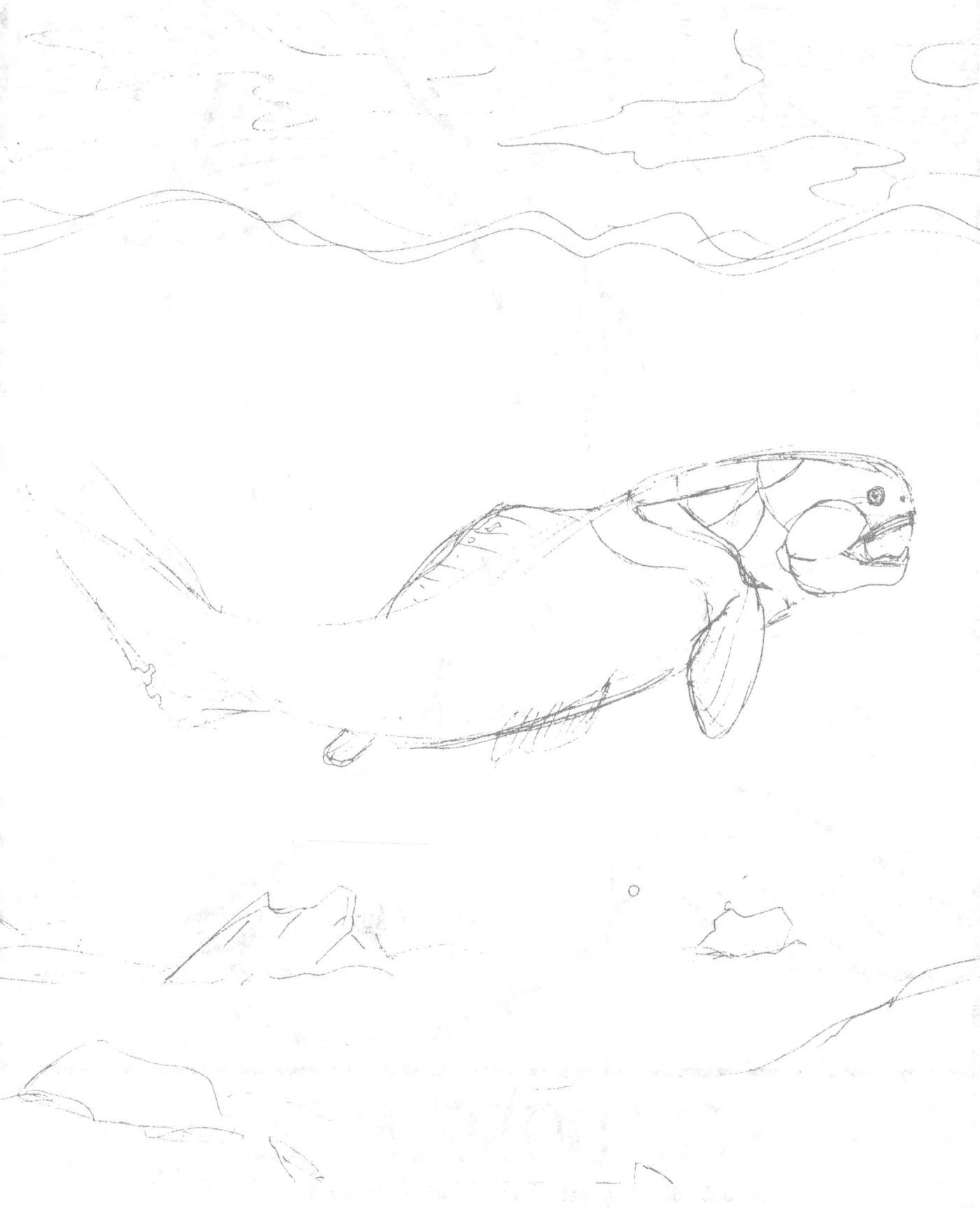

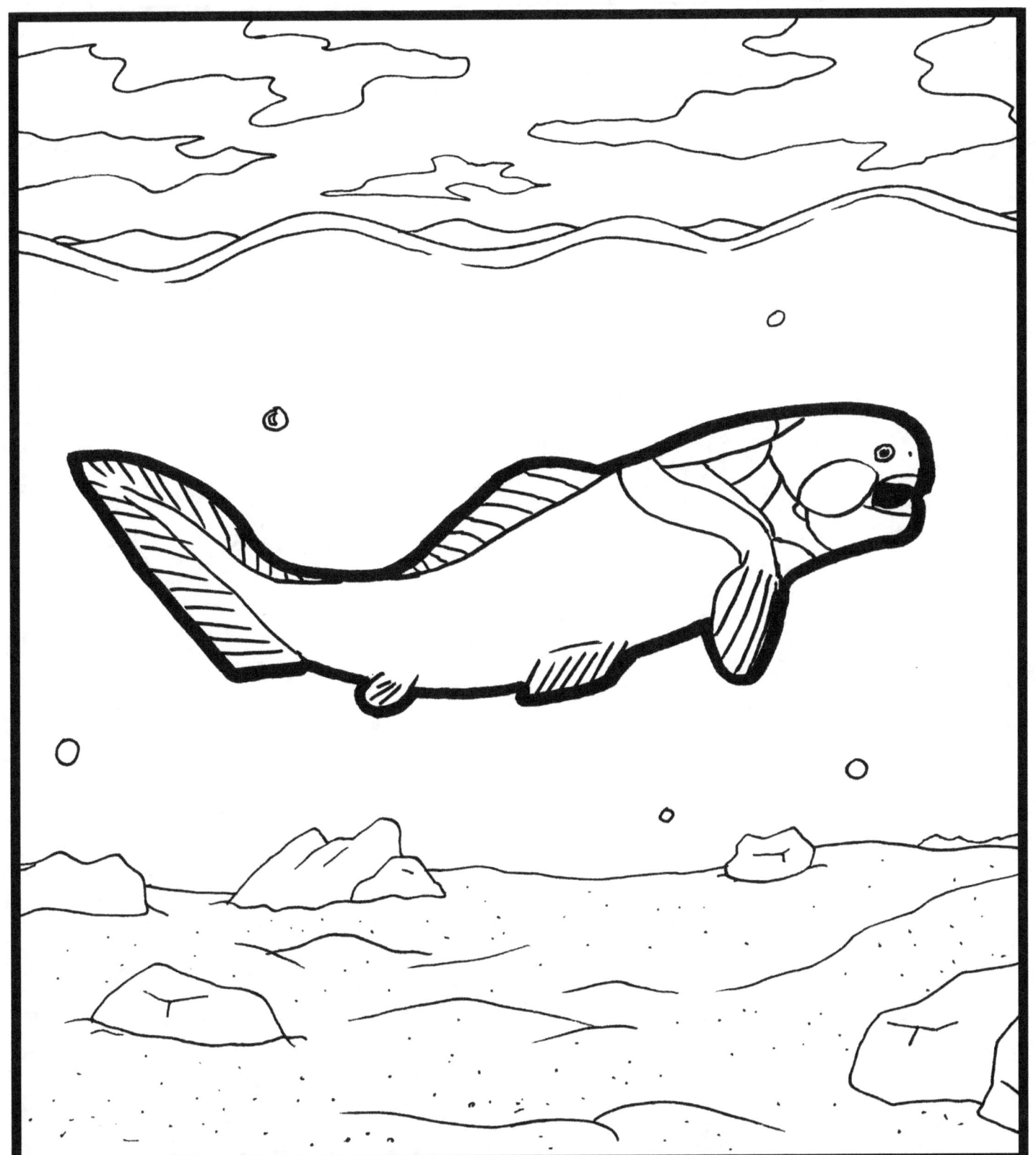

DUNKLEOSTEUS

382 - 358 MYA, Late Devonian, Morocco, Belgium, Poland, Canada, and United States

The largest member of this carnivorous genus an apex predator; it was 6 meters long and weighed up to a ton. As a placoderm, it was a fish with plenty of bony plates in its body, and instead of teeth, it had a pair of bony plates that resembled a beak.
It was a slow, but powerful swimmer.

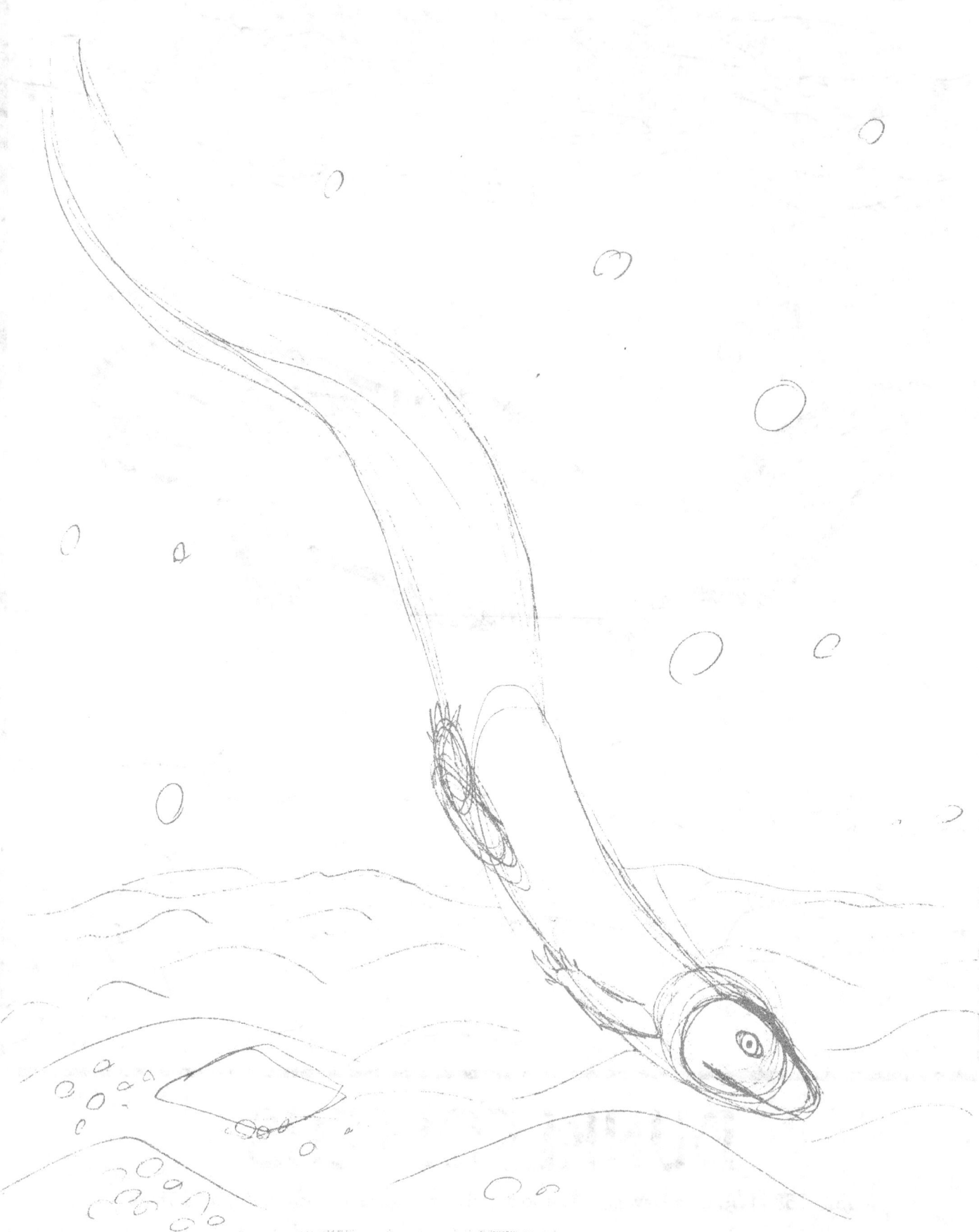

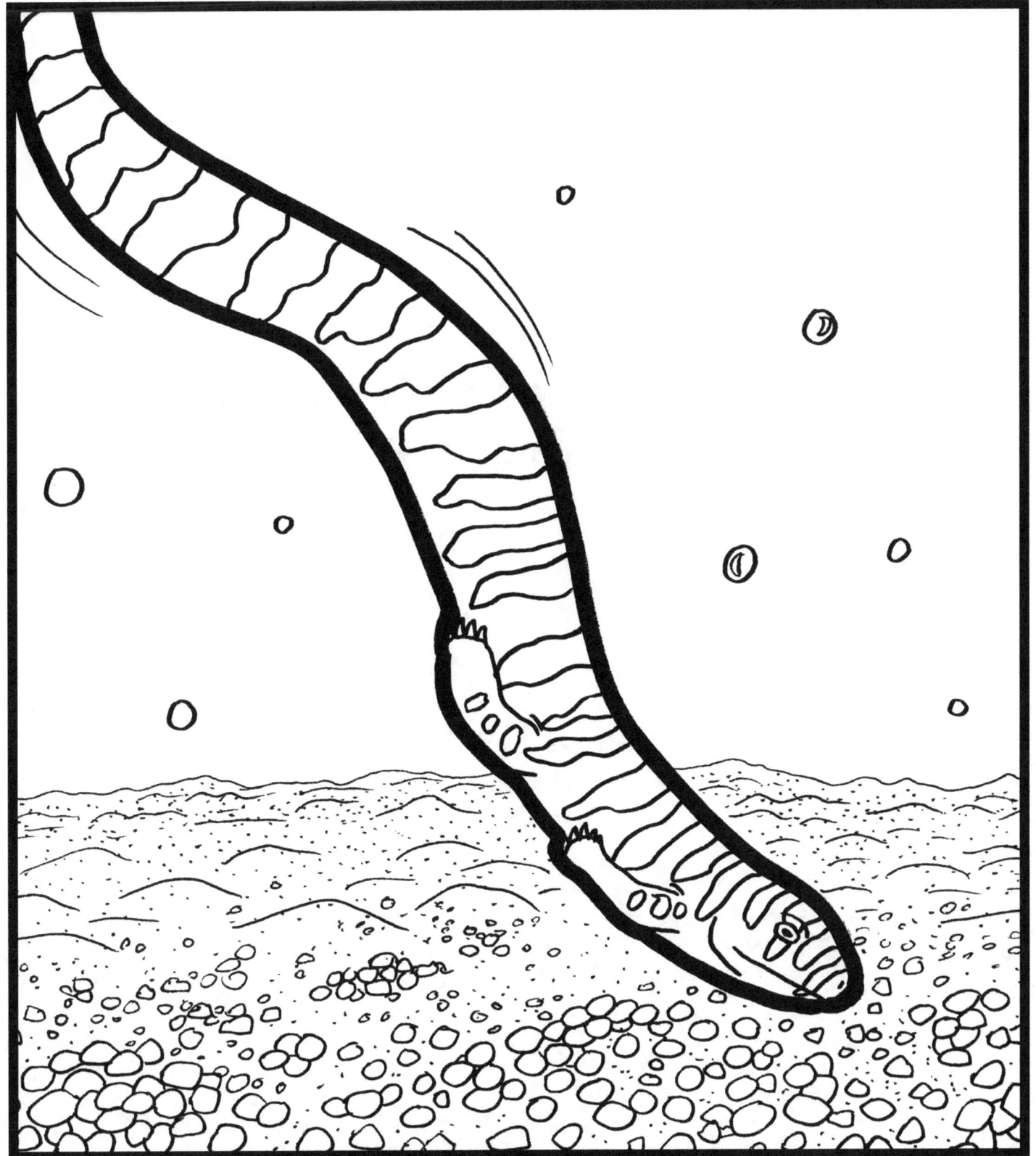

UROCORDYLUS

318 - 314 MYA, Late Carboniferous, Ireland

This was a small amphibian that could reach a length of about 15 centimeters, most of which was part of its tail.
It used this long, paddle-shaped tail to swim through the waters in which it lived, but probably ventured on land from time to time.

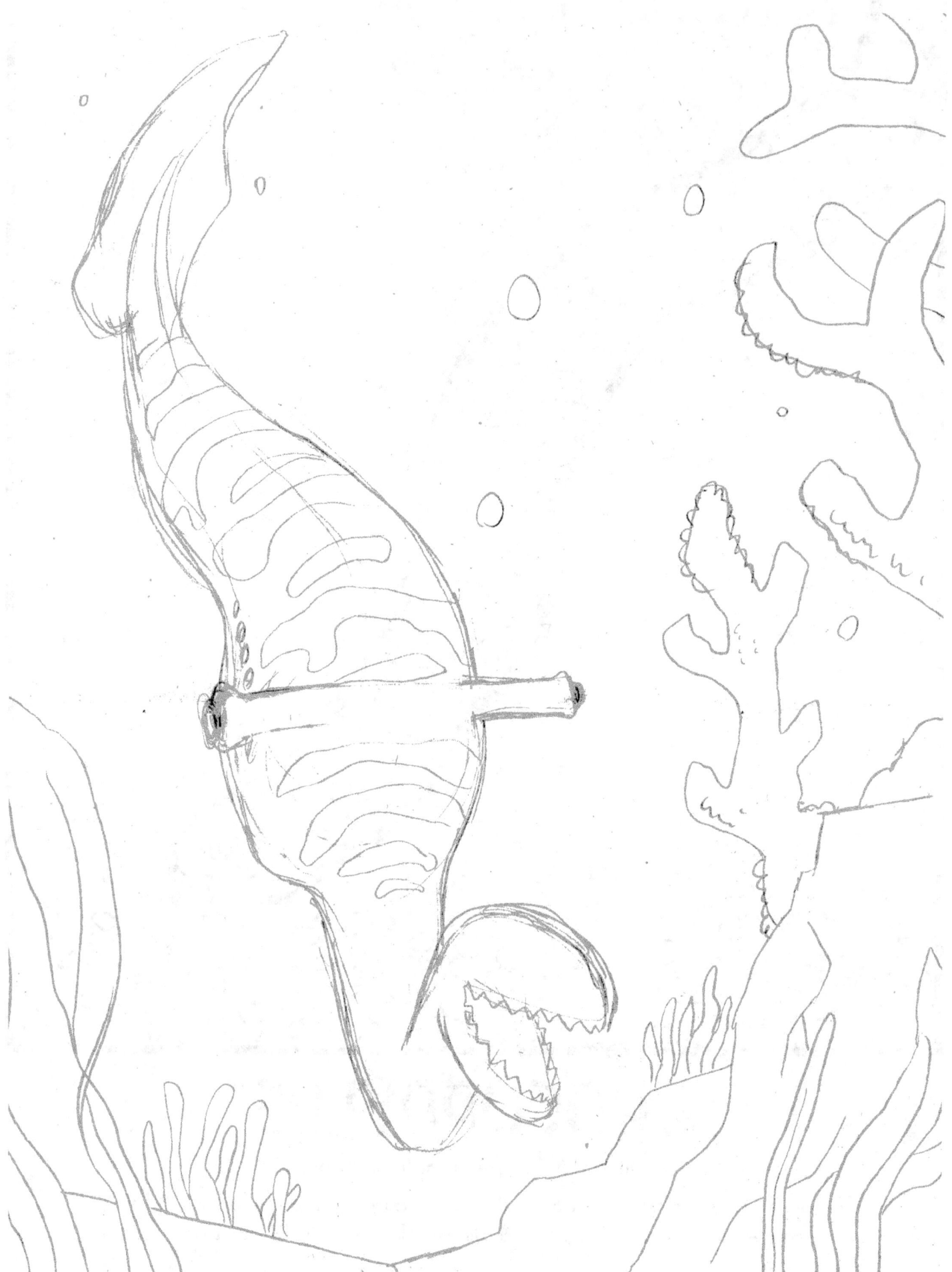

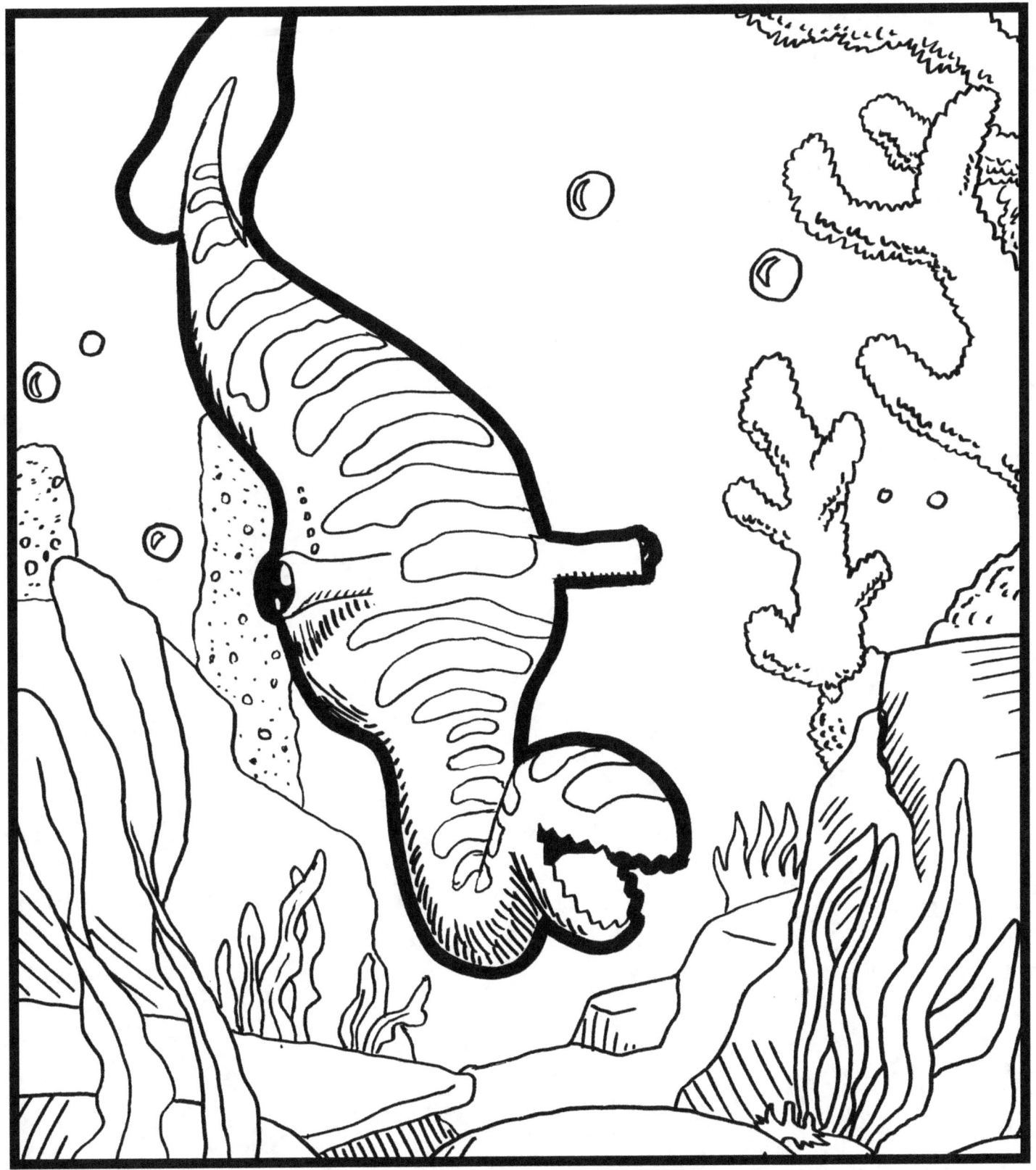

TULLIMONSTRUM

300 MYA, Late Carboniferous, United States

Though originally thought to have been some sort of marine invertebrate, new research has suggested that this stangely-built creature was actually a basal vertebrate related to lampreys. Fossil specimens have also preserved pigments on the animal, showing where darker patches would have been.

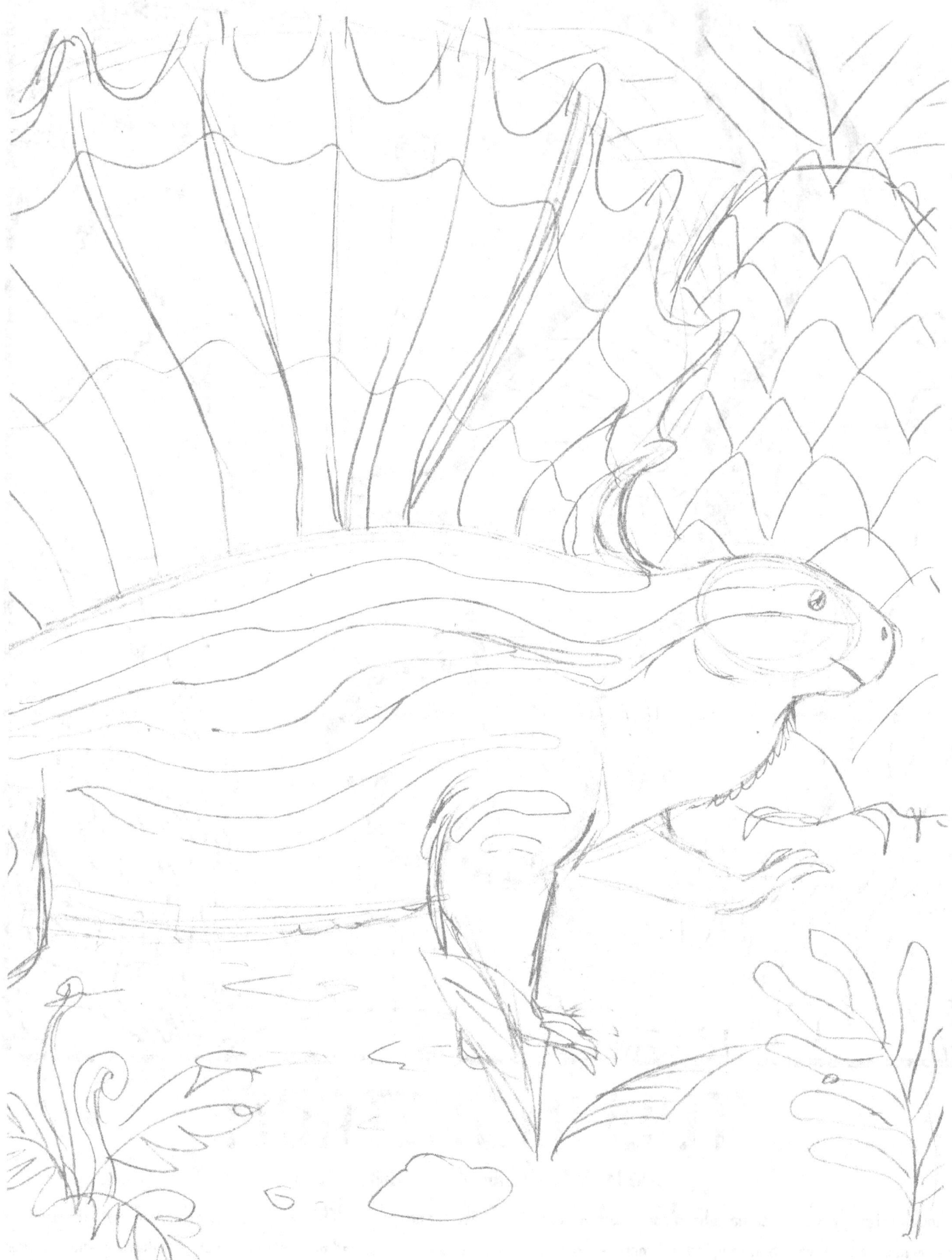

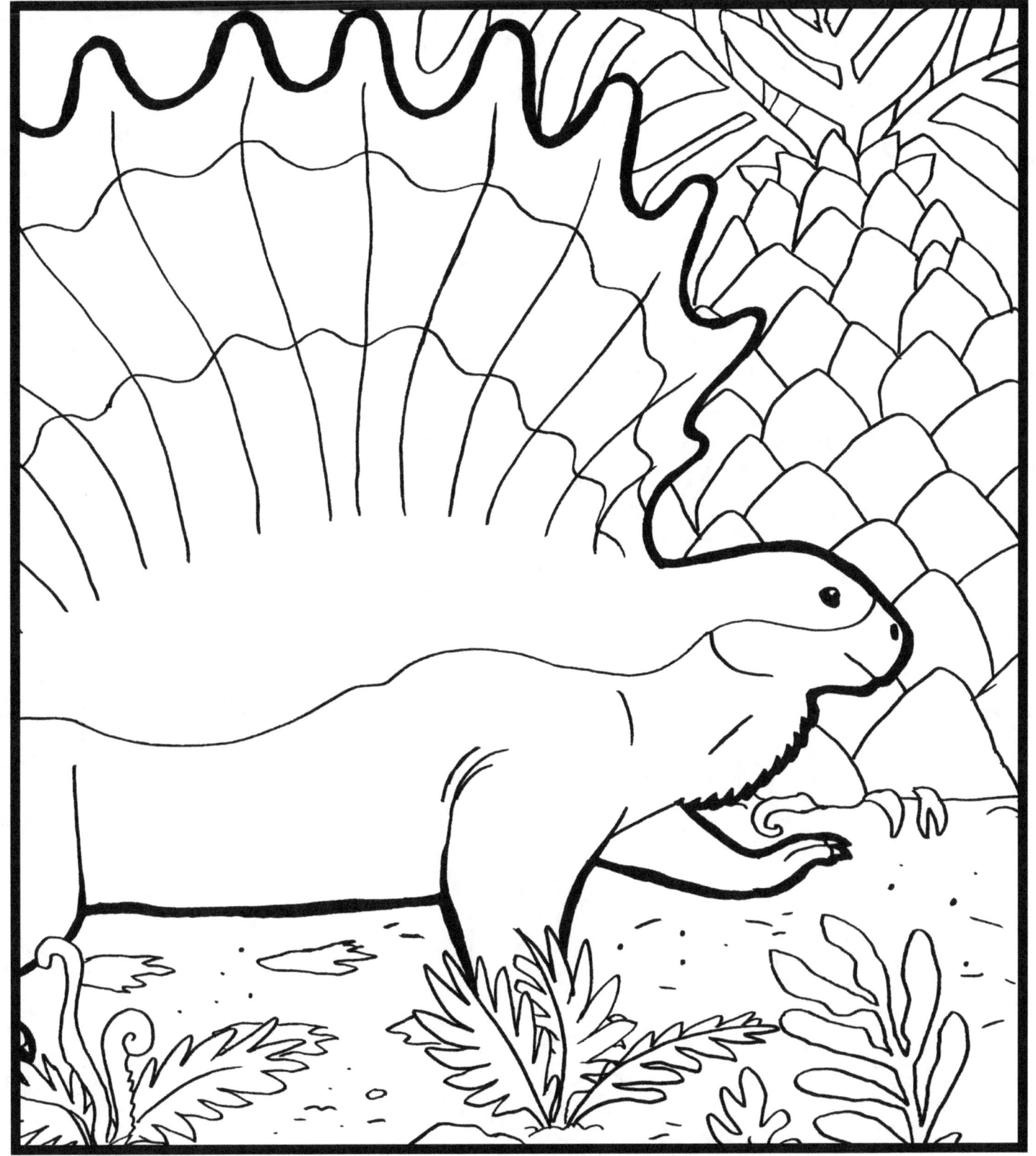

EDAPHOSAURUS

300 - 280 MYA, Late Carboniferous to Early Permian, Germany, Czech Republic, and United States

These animals were some of the earliest large, plant-eating amniote tetrapods. Like many other sail-backed animals, many different uses were proposed for the sail, including but not limited to: courtship, thermoregulation, fat storage, muscle support, camouflage, protection against predators, and species recognition. It's likely the sail had more than one purpose.

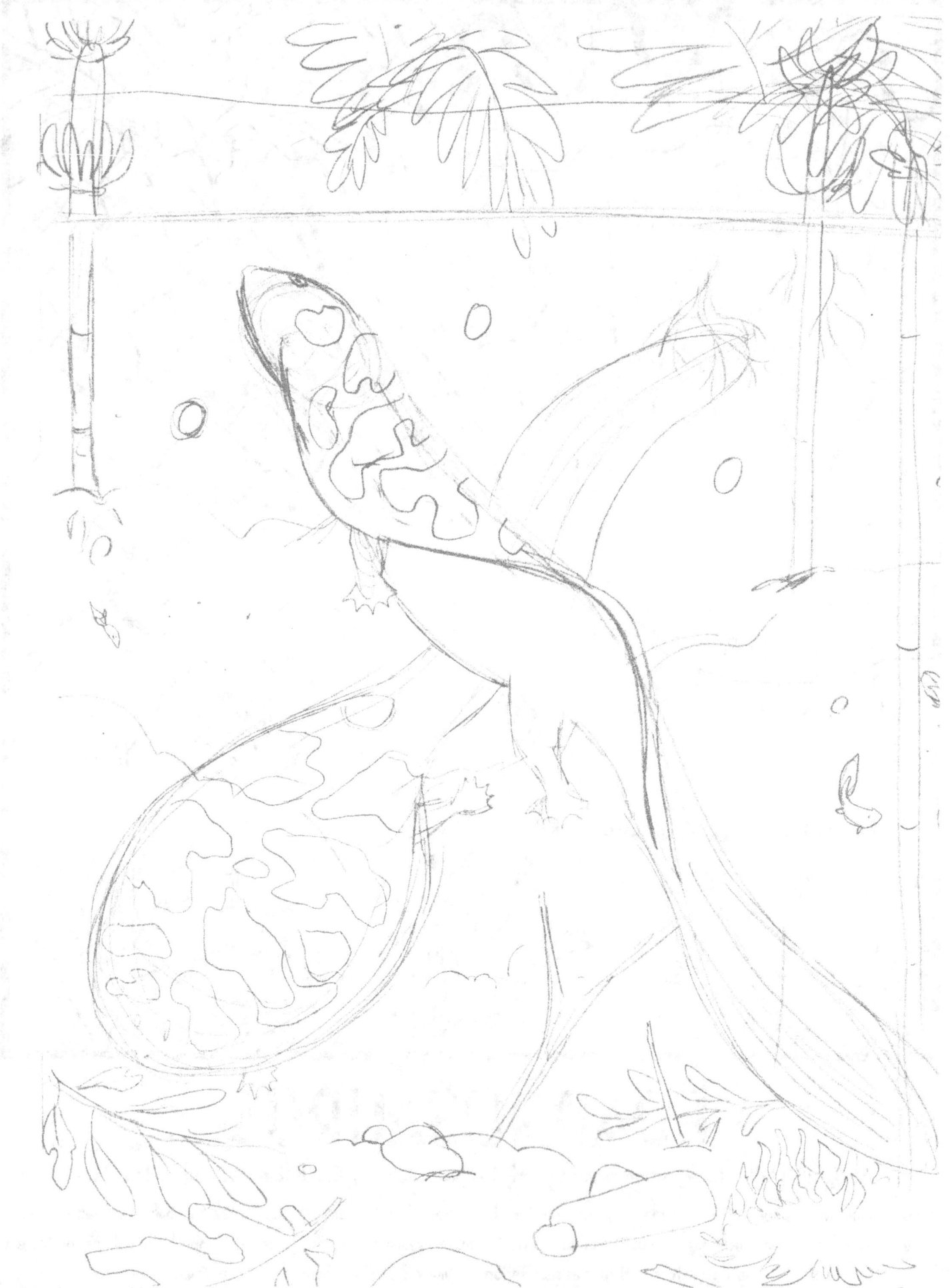

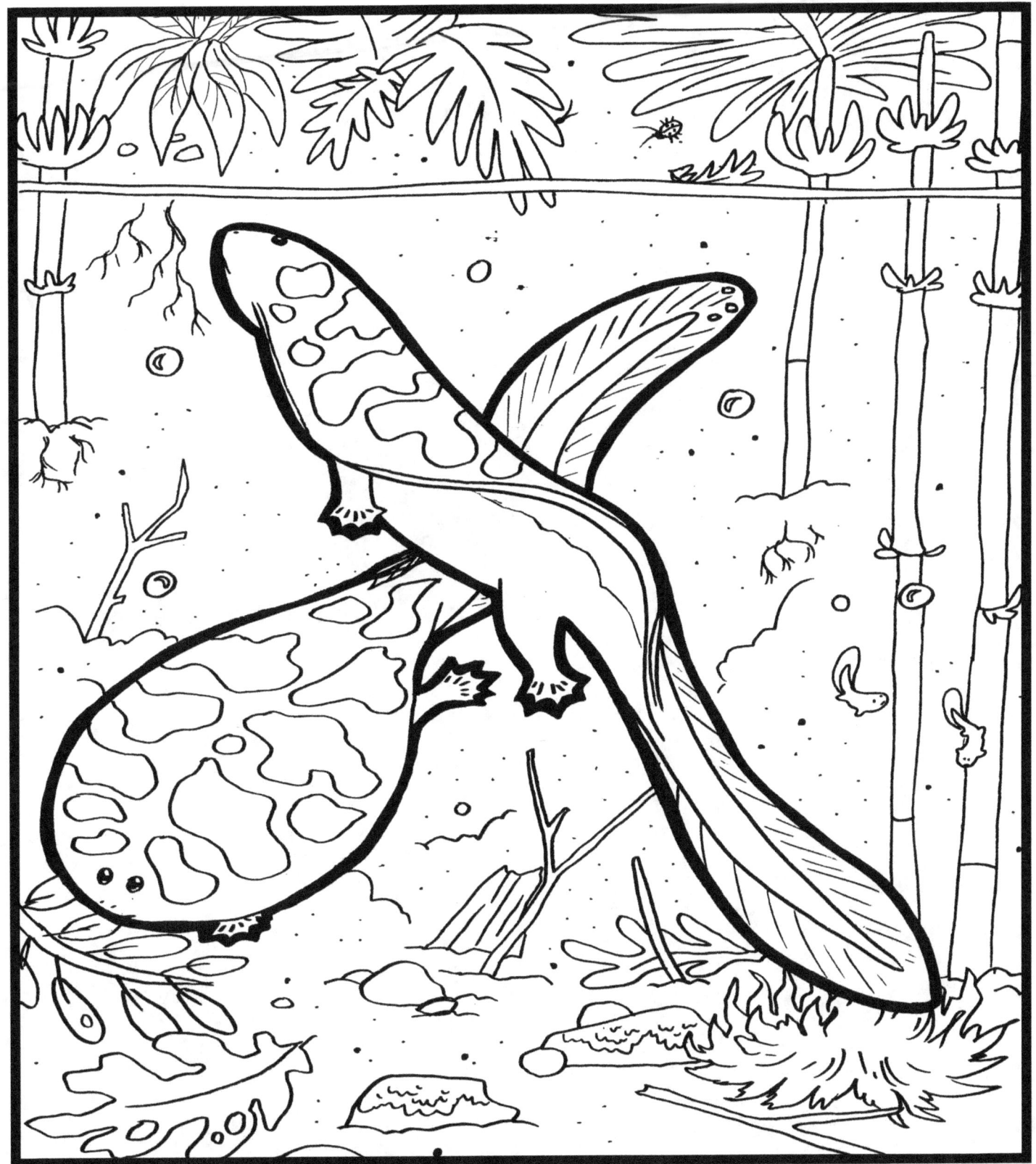

DIPLOCAULUS

299 - 251 MYA, Permian, North America

This long-lasting genus of amphibians were around for nearly the entire Permian. They have been known to dig burrows, which helped it cope with long periods of drought. The purpose of such an oddly-shaped skull has been thought to be several things: to make it more difficult to be eaten, as an attachment for a membrane, and/or to make the animal more hydrodynamic.

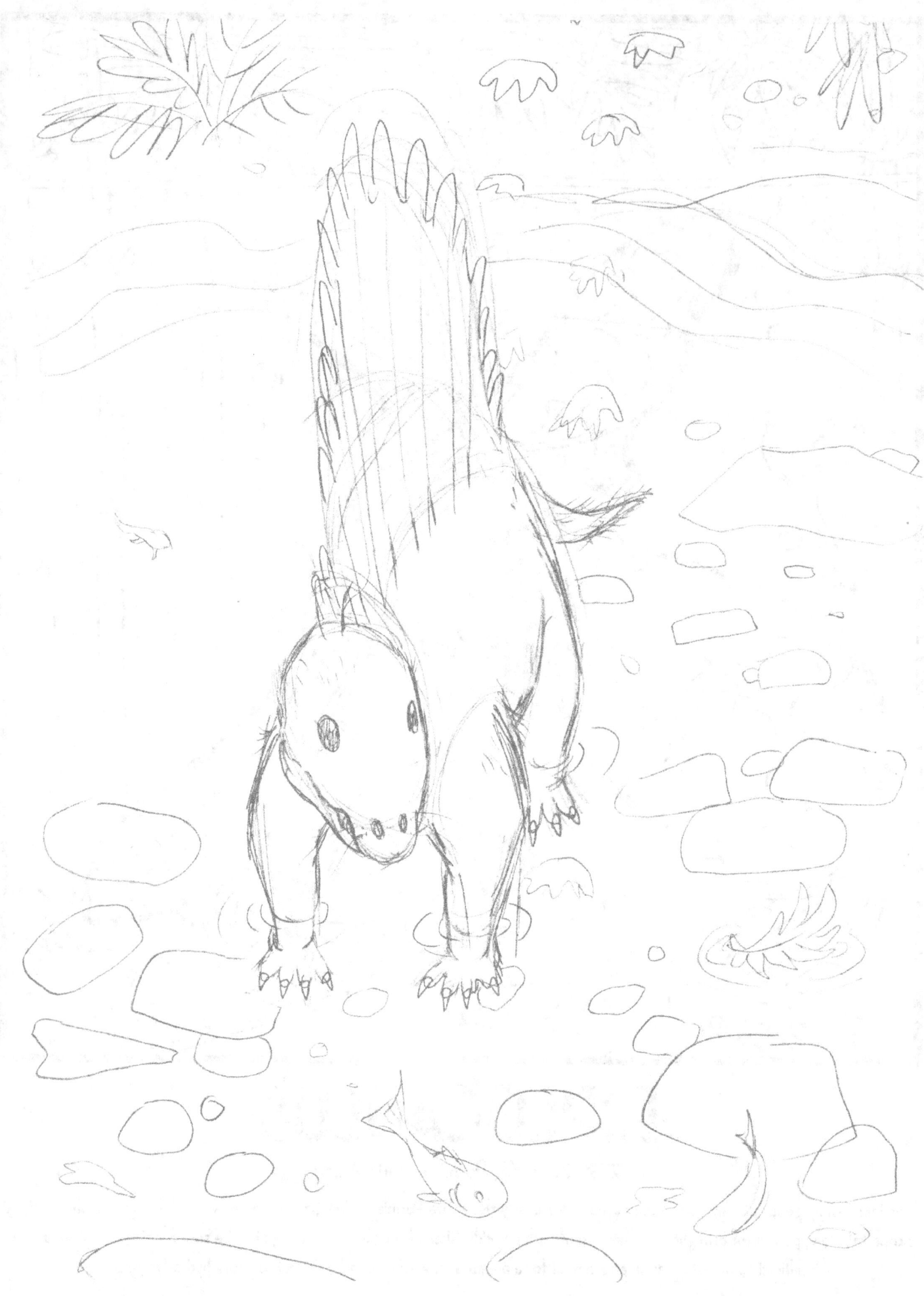

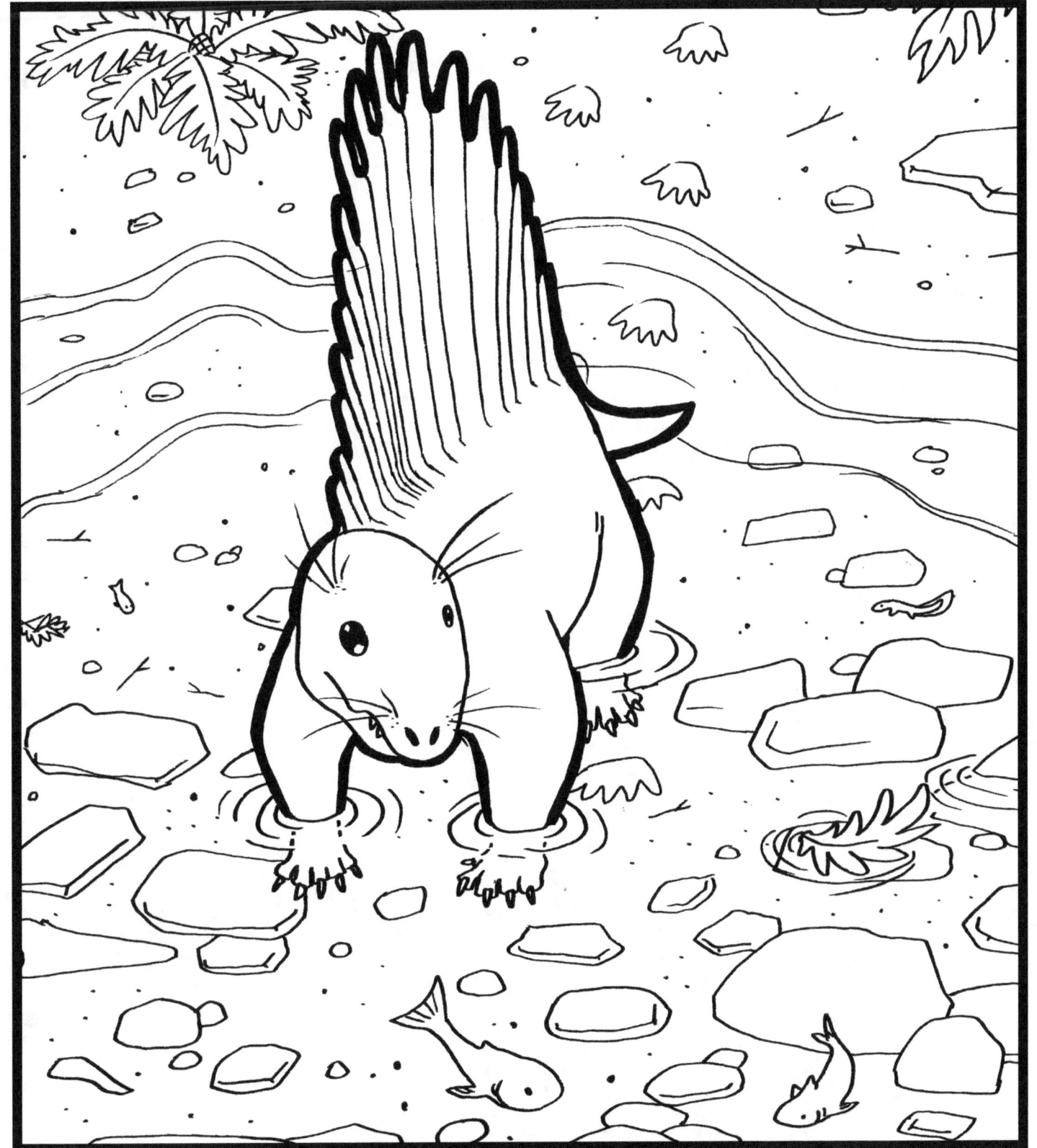

DIMETRODON

295 - 272 MYA, Early Permian, United States and Germany

Despite being lumped in the same category as dinosaurs in countless media, Dimetrodon were not only around 40 million years before any dinosaur, but were also more closely related to mammals. The group it's in, Synapsida, also contains mammals as we know them today, and is separate from Sauropsida, which include dinosaurs. The sail-back was likely used in courtship, but probably had other uses as well.

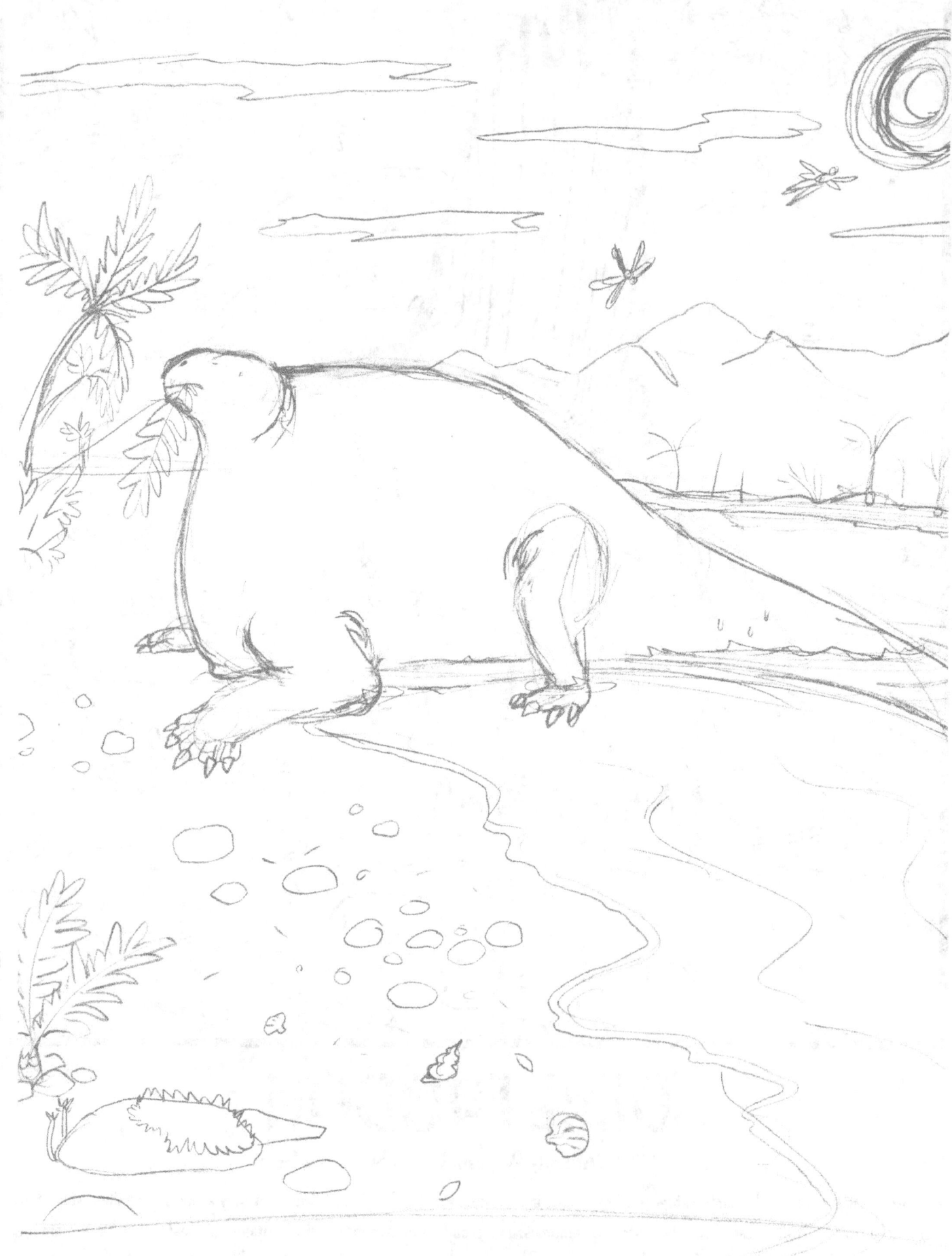

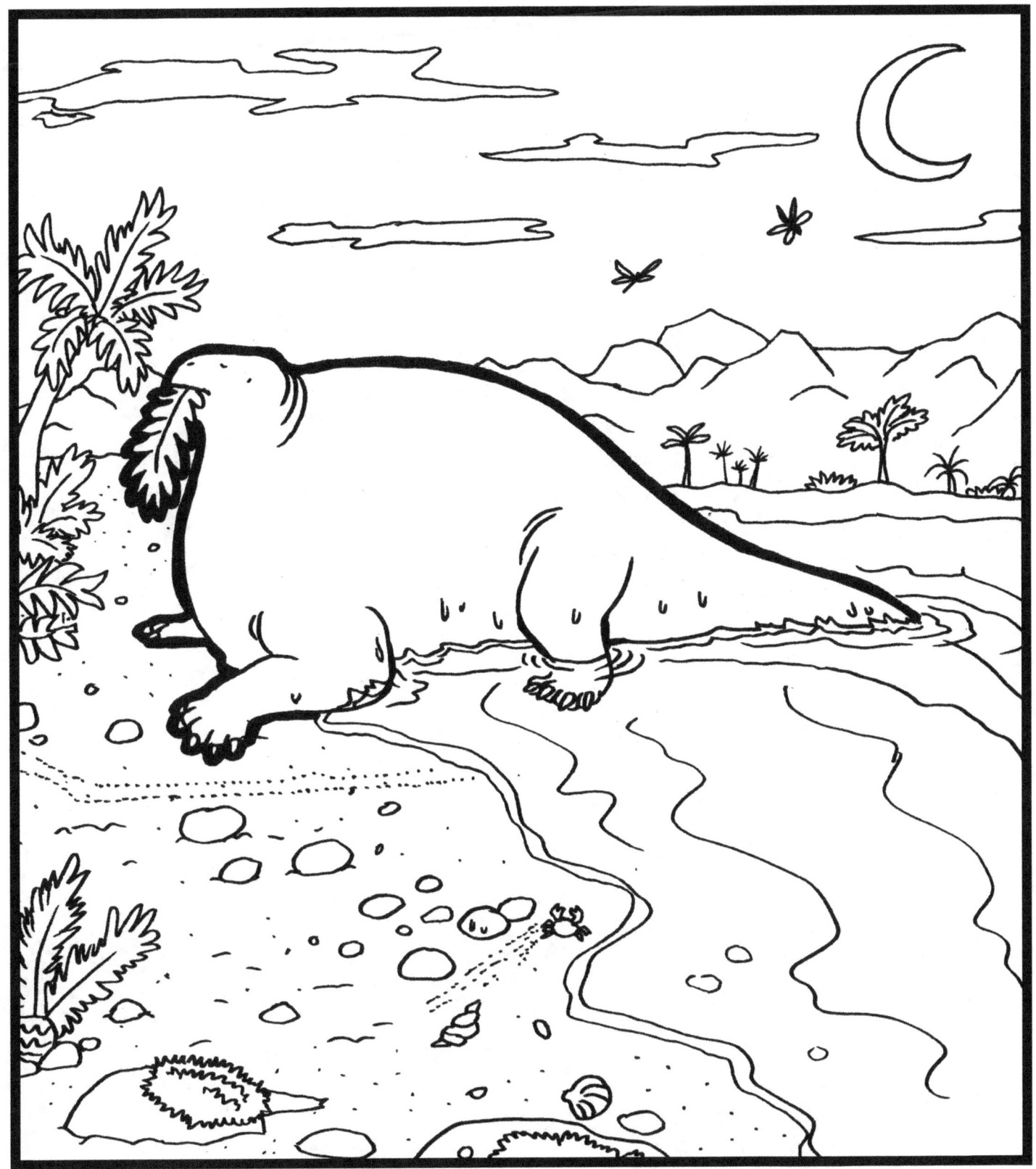

COTYLORHYNCHUS

279 - 272 MYA, Early Permian, United States

At 6 meters long, and with a body this massive, this was one of the largest synapsids of the early Permian, and might have been partly aquatic. The hands could have been used as paddles as the animal swam through water, like a turtle. Its teeth were similar to an iguana's, and probably needed to spend a lot of its time eating in order to sustain such a large form.

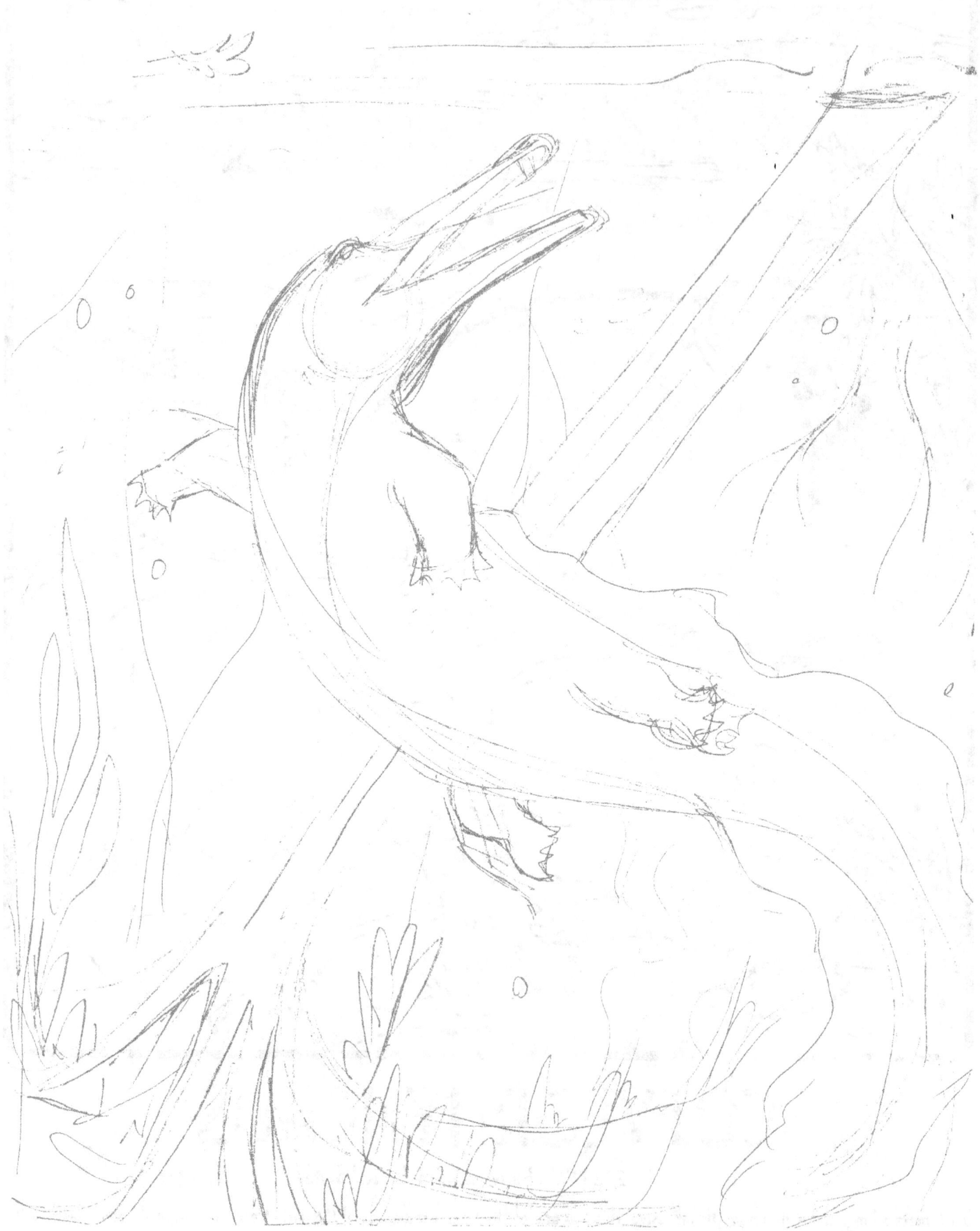

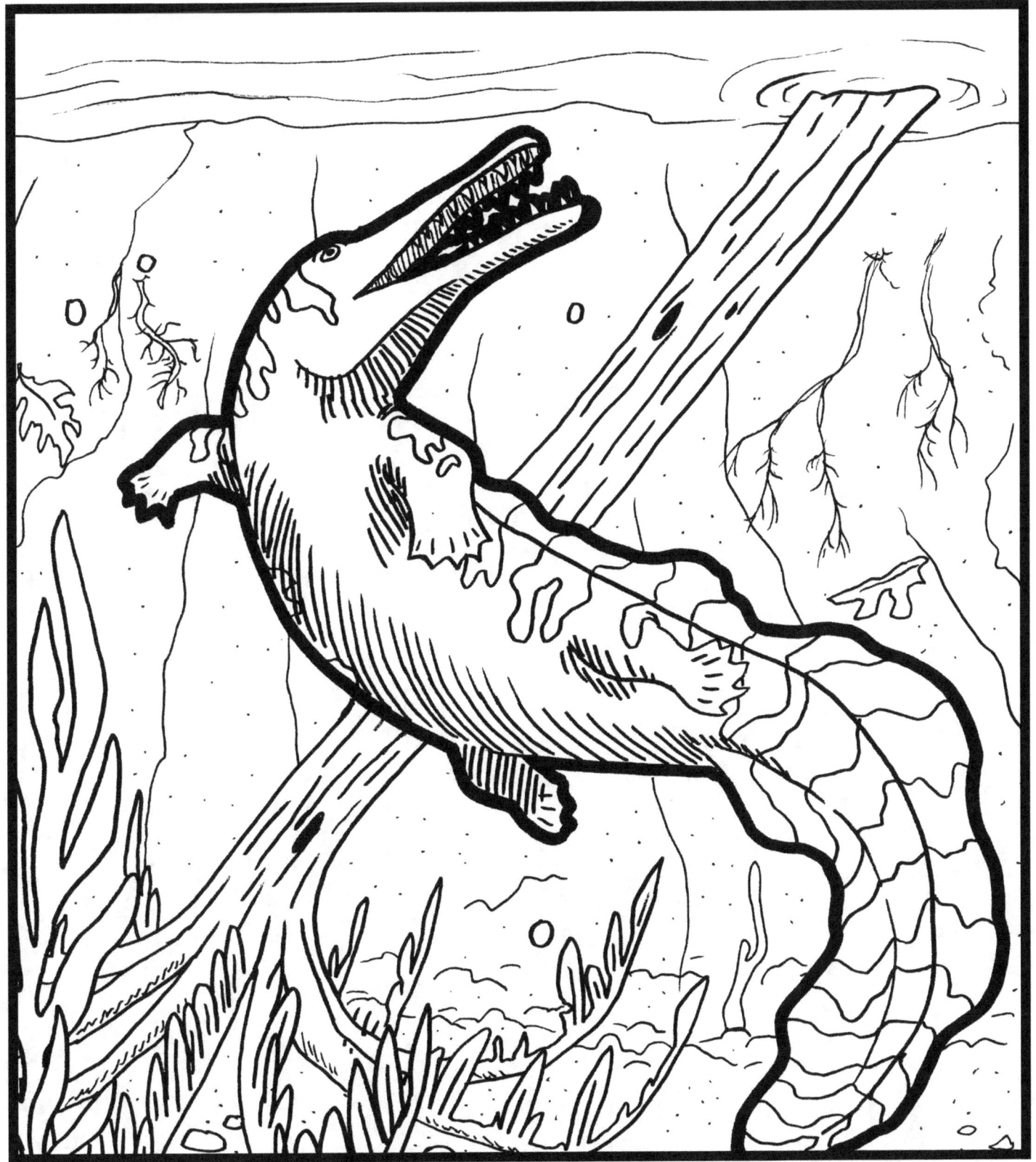

PRIONOSUCHUS

270 MYA, Middle Permian, Brazil

This gharial-like, piscivorous amphibian lived in a humid, tropical environment. Most of them have been found to be between 2 and 2.5 meters long, but there was one individual specimen that had such a huge skull, the body would have been 9 meters long. If indeed this large, it would been one of the largest predators of the Permian.

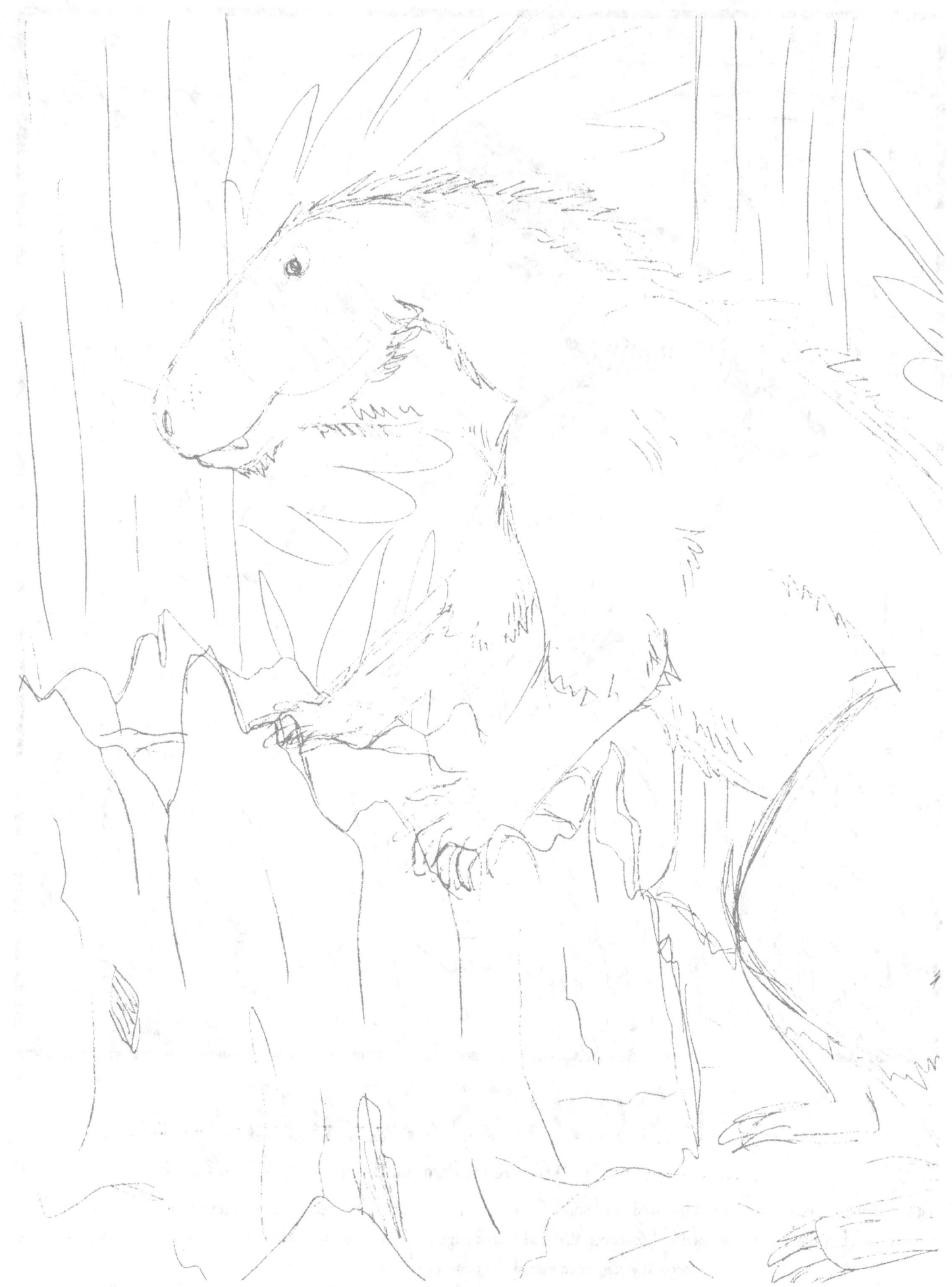

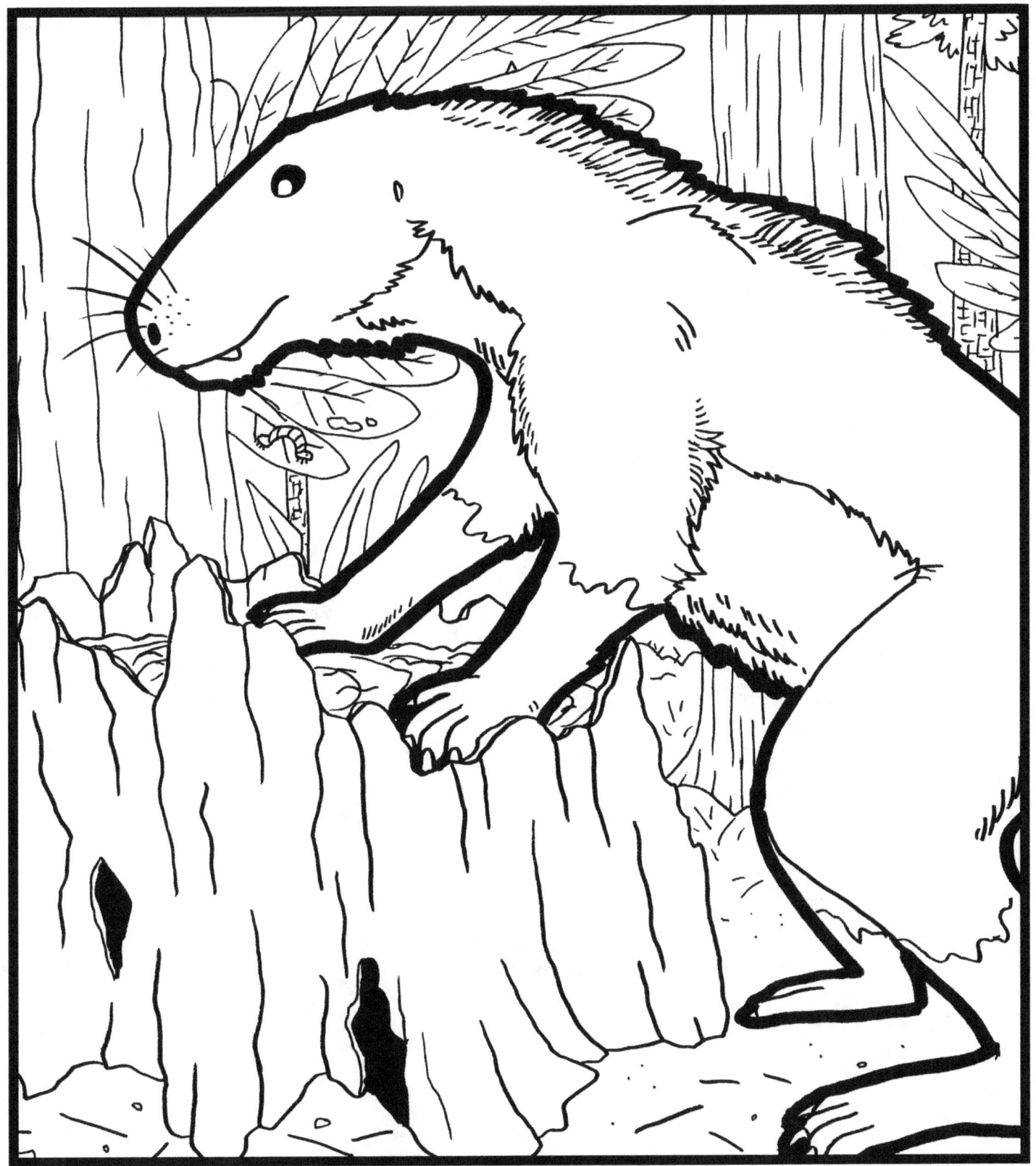

LYCAENOPS

270 - 251 MYA, Mid to Late Permian, South Africa

This "wolf-face" was a carnivorous therapsid, and were about a meter long, and weighed about 15 kilograms. Being in the same group as mammals, they were likely covered in fur, and could hold their legs under their body rather than splayed to the sides. This allowed them to outrun other animals.

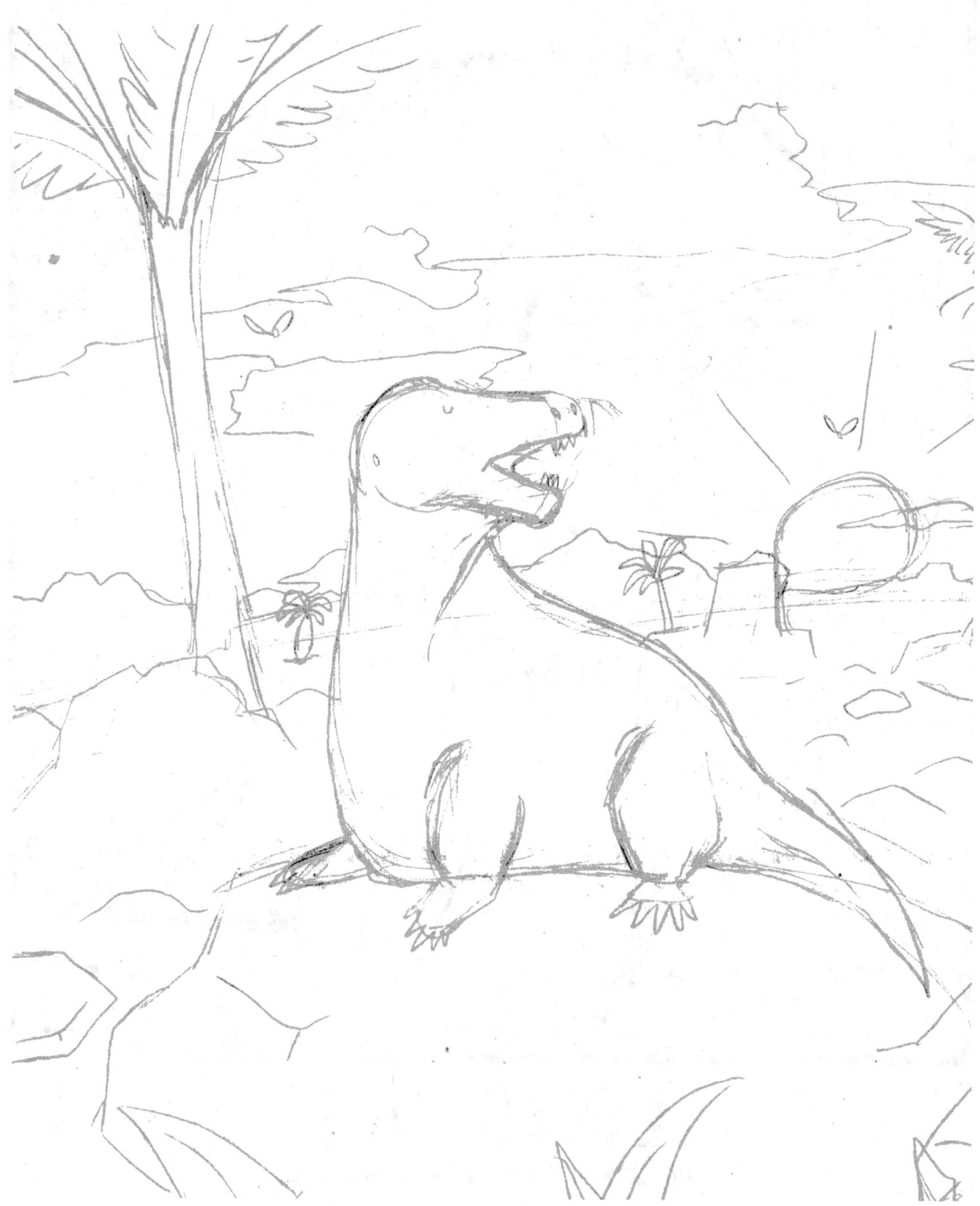

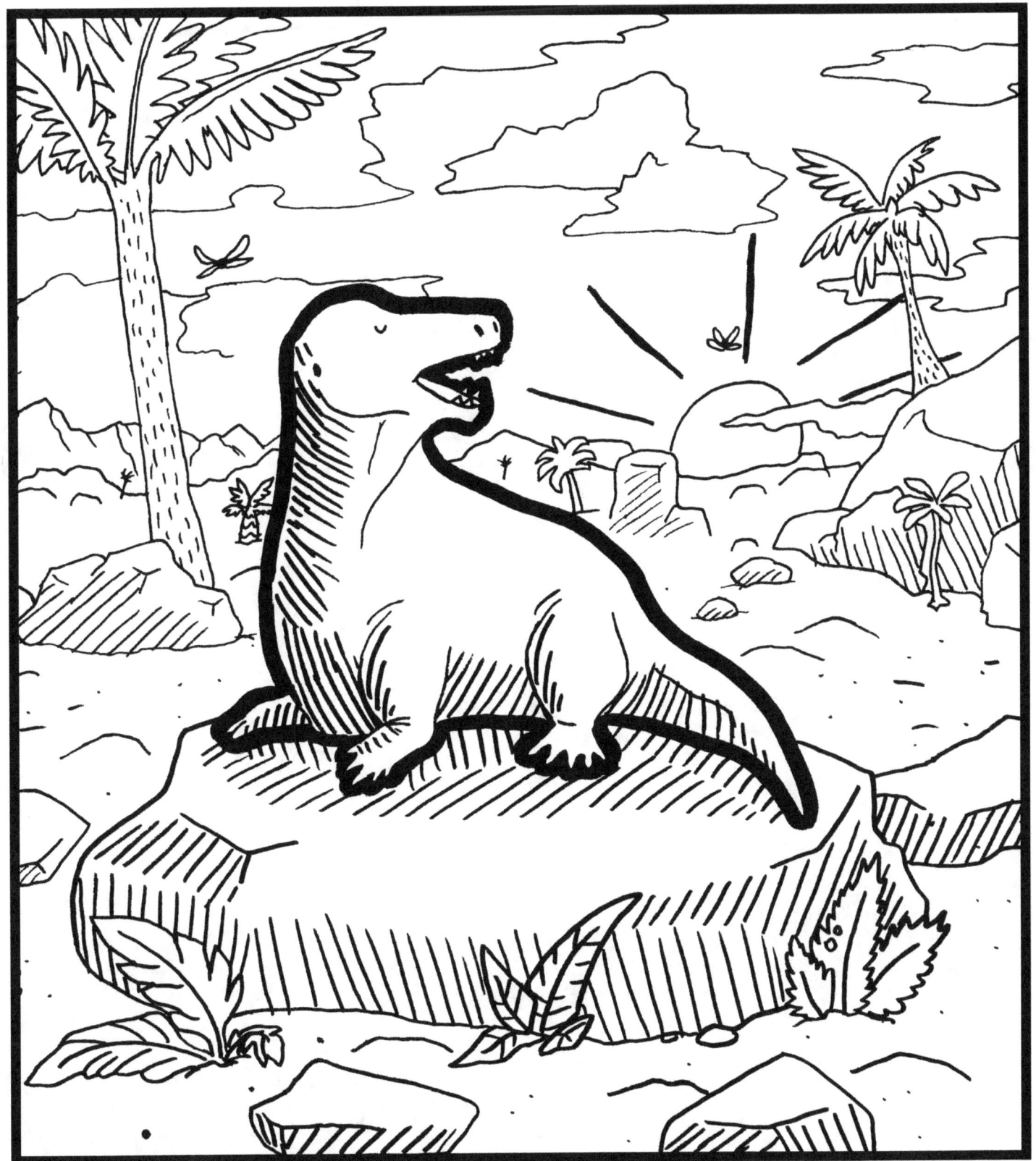

TITANOSUCHUS

265 MYA, Mid Permian, South Africa

Though usually depicted as a carnivore, it has been argued that this therapsid could just as well have used its mouth full of incisors to go at plants instead. It depends on how the fossils are interpreted, and a misinterpretation of what the teeth belonged to led to it being named a "titan crocodile." Though not related to any crocodiles, it did descend from the same group that contained animals like Dimetrodon.

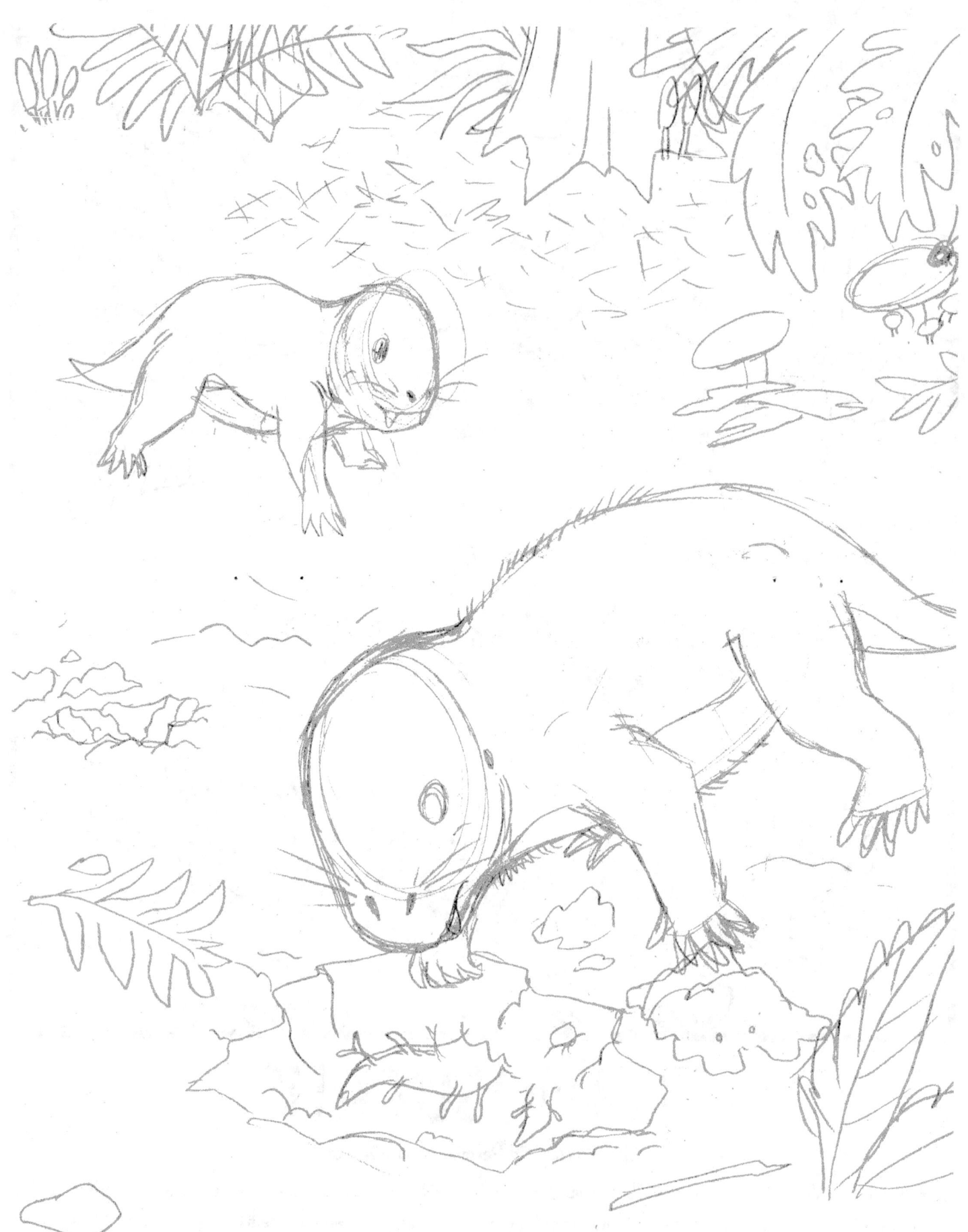

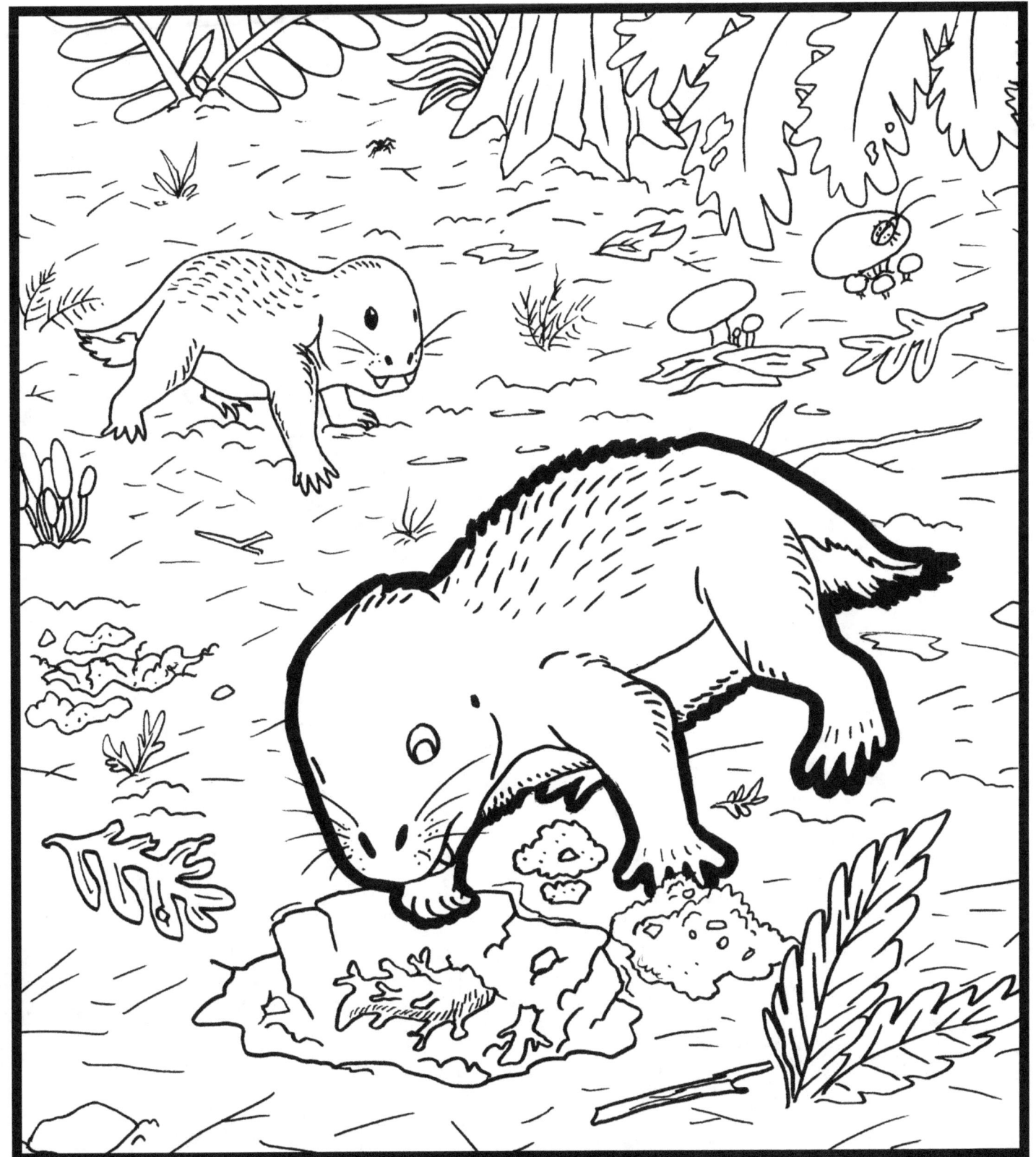

ROBERTIA

265 - 260 MYA, Middle to Late Permian, South Africa

This was a small, early dicynodont, a group of herbivorous synapsids that had a pair of tusks. The horny beak was capable of severing tough plant matter such as stems or twigs. It was only about 20 centimeters long, and probably burrowed in the ground.

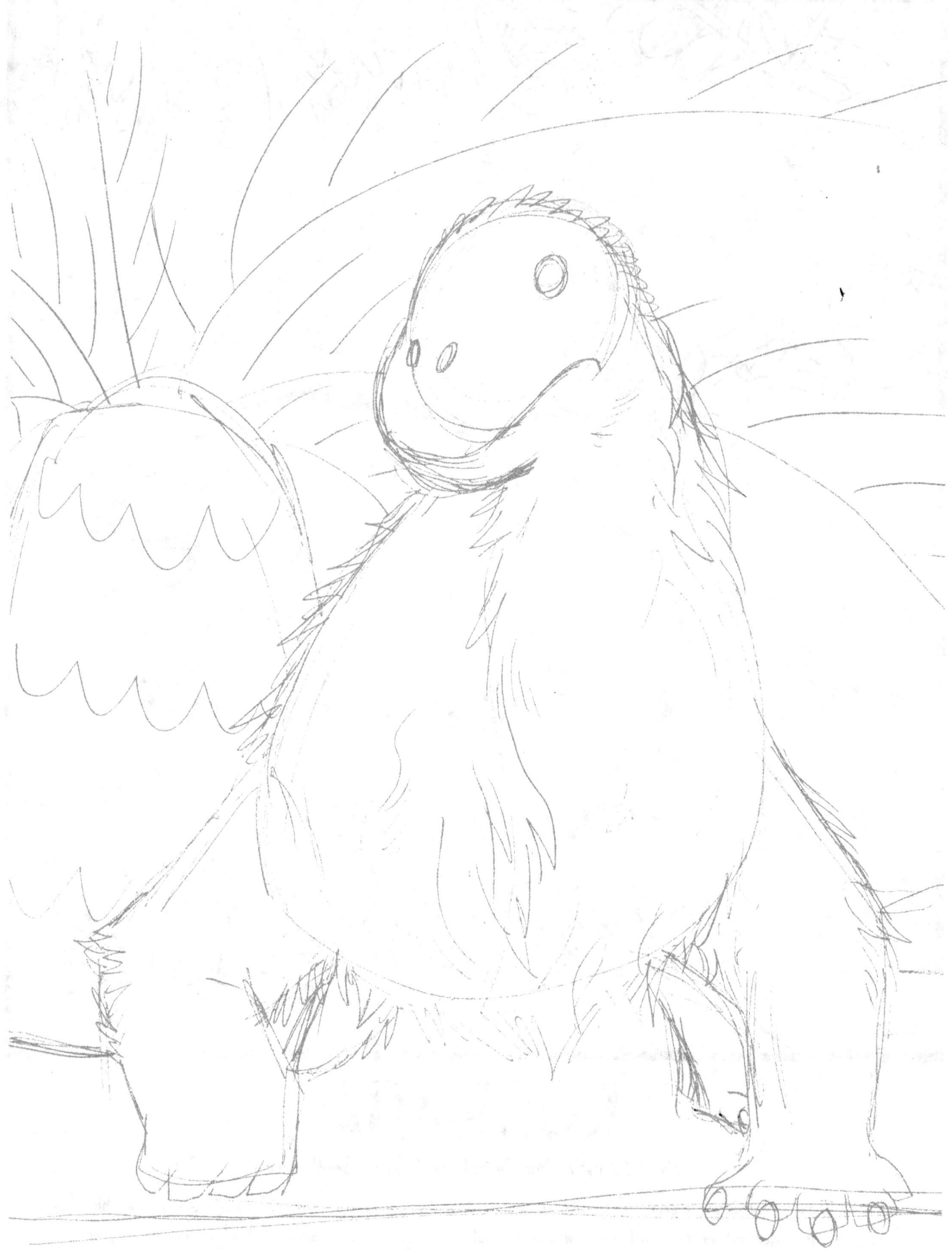

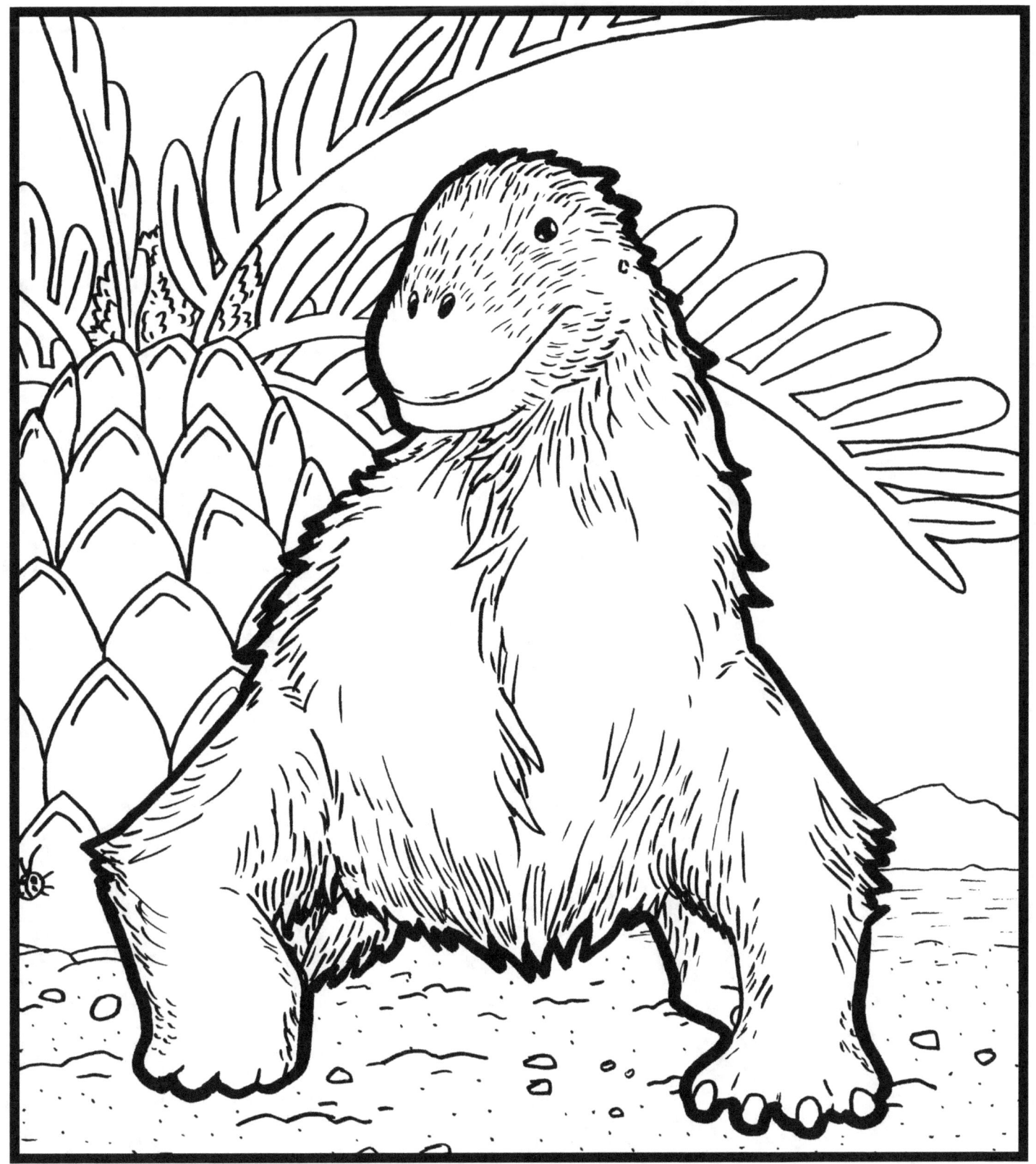

MOSCHOPS

265 - 260 MYA, Middle Permian, South Africa

This massive, slow-moving synapsid could grow up to 5 meters long, and fed mainly upon tough, nutrient-poor vegetation. As a result, it had to have spent most of its day eating. Its name also means "calf-face." It might have used its 10 centimeter-thick skull in dominace displays, where two individuals would push at one another with their heads.

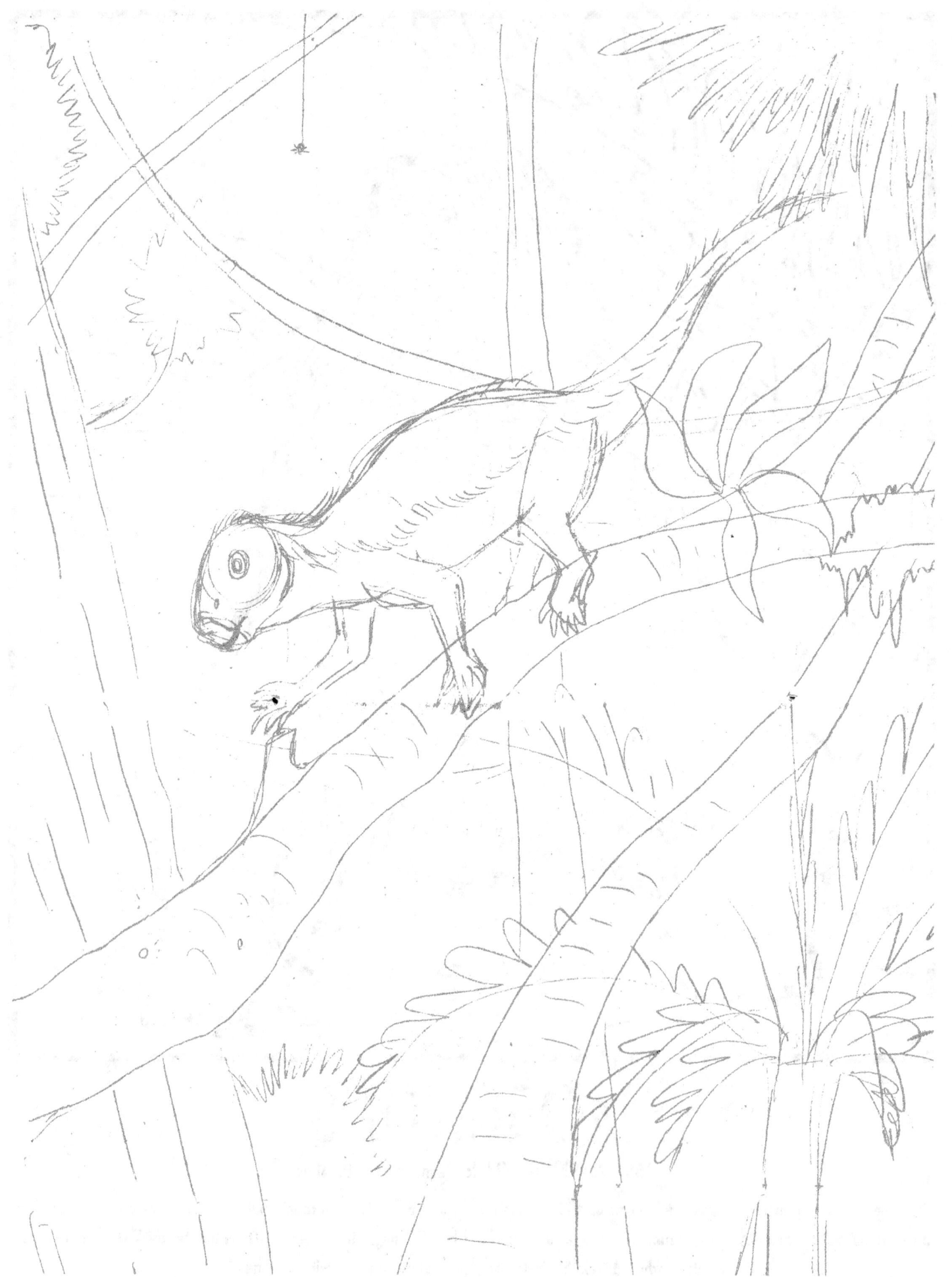

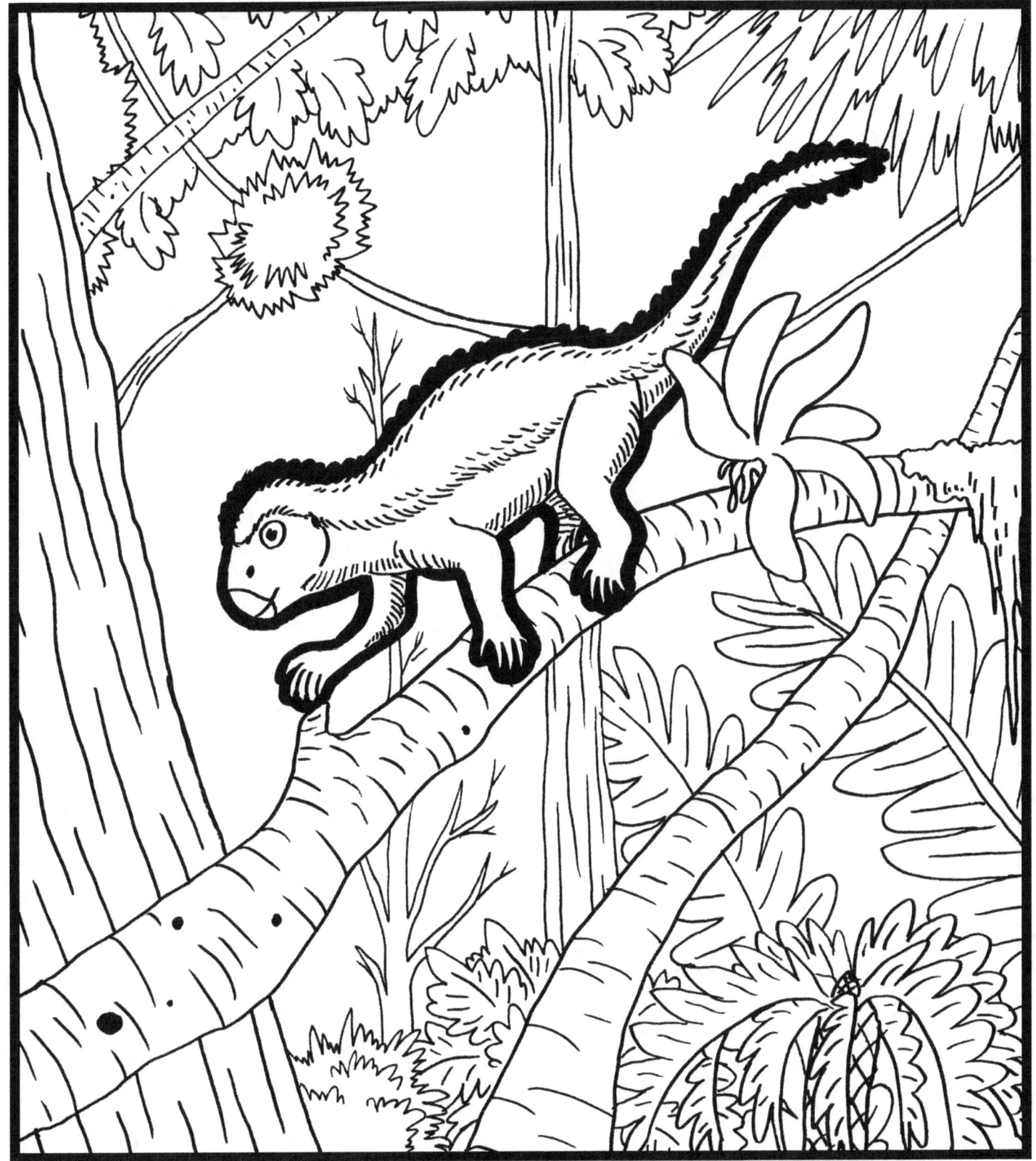

SUMINIA

260 MYA, Late Permian, Russia

It has been suggested that this small early synapsid was at least partly arboreal, owing to the hands being capable of grasping. If so, this would be the earliest arboreal vertebrate known so far. Heavy tooth-wear indicated that it was an herbivore, despite the teeth themselves being both large and sharp.

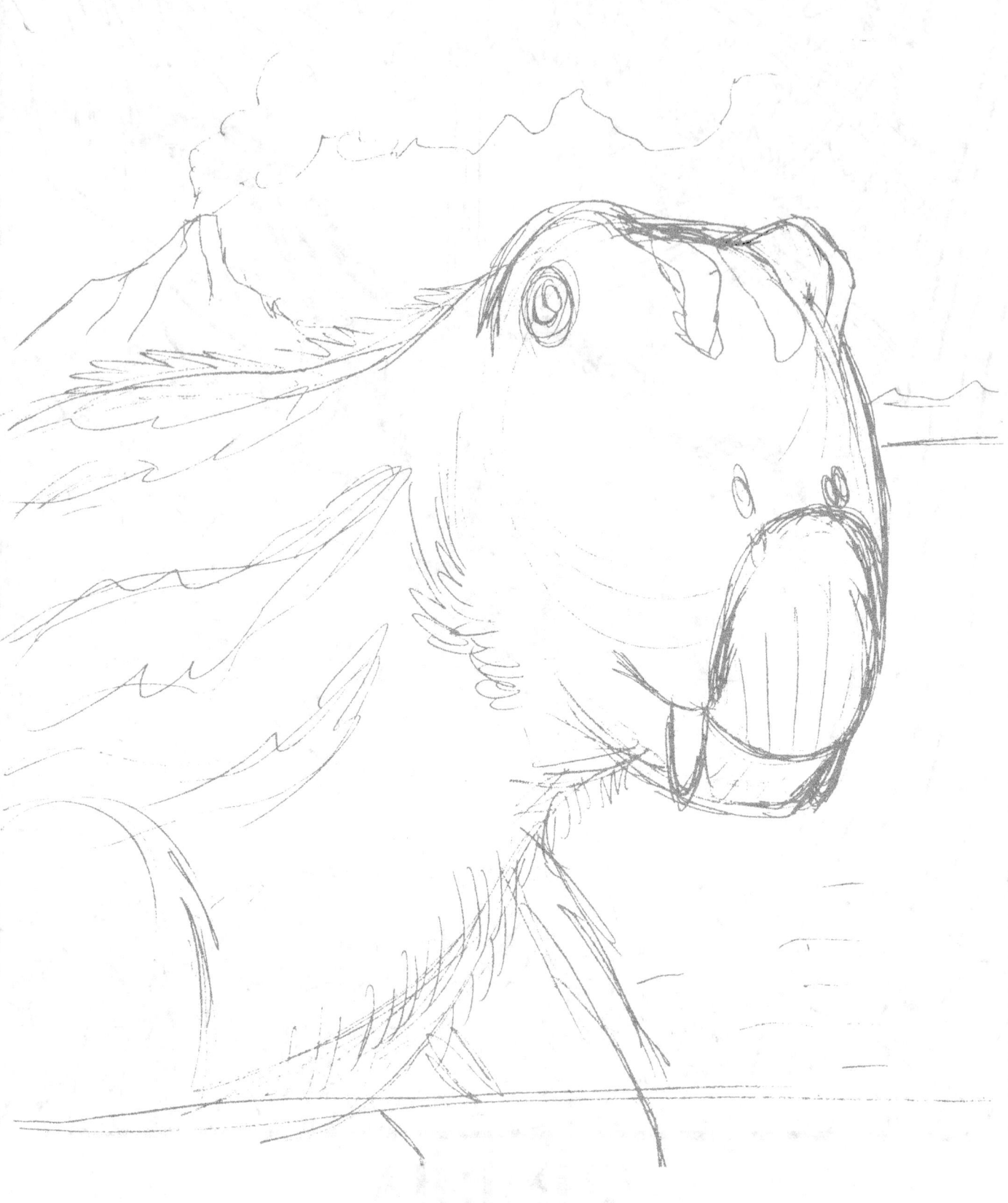

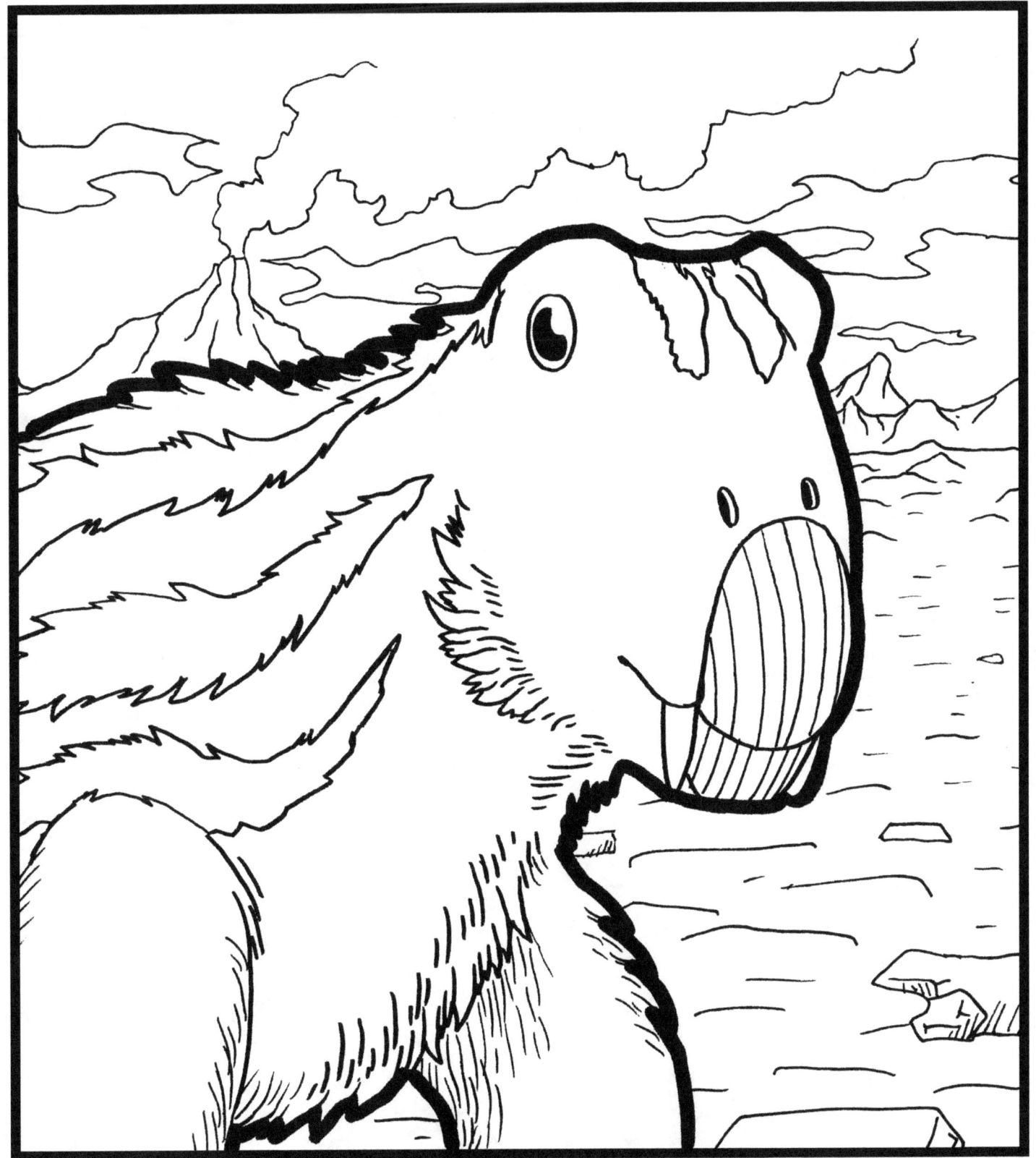

LYSTROSAURUS

255 - 250 MYA, Late Permian to Early Triassic, Antarctica, India, South Africa

An herbivorous therapsid, this "shovel-lizard" used its beak to cut away tough vegetation, and probably burrowed for food or shelter. During the Early Triassic they dominated the southern part of Pangaea, with up to 95% of all land-vertebrate fossil specimens belonging to this animal.

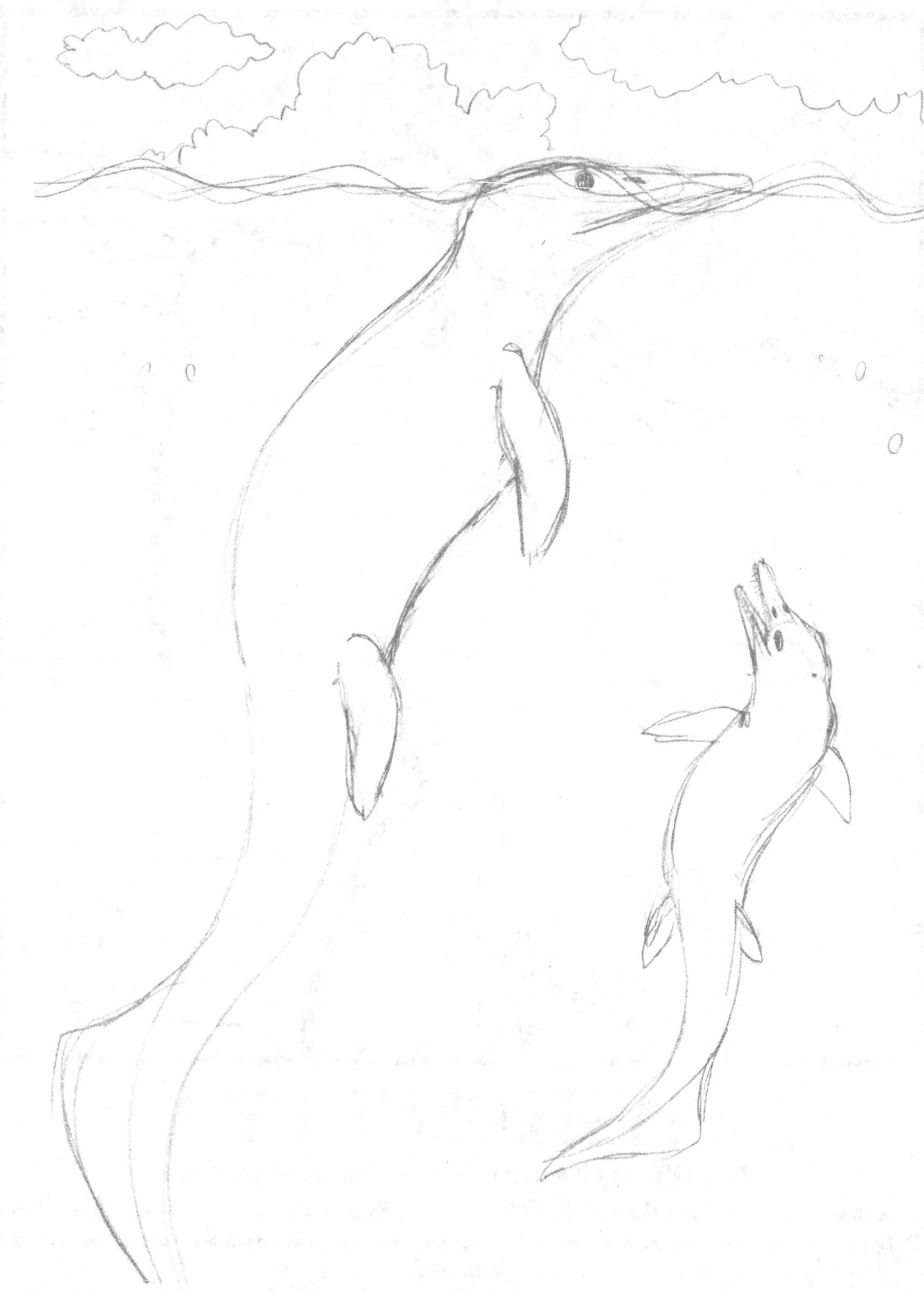

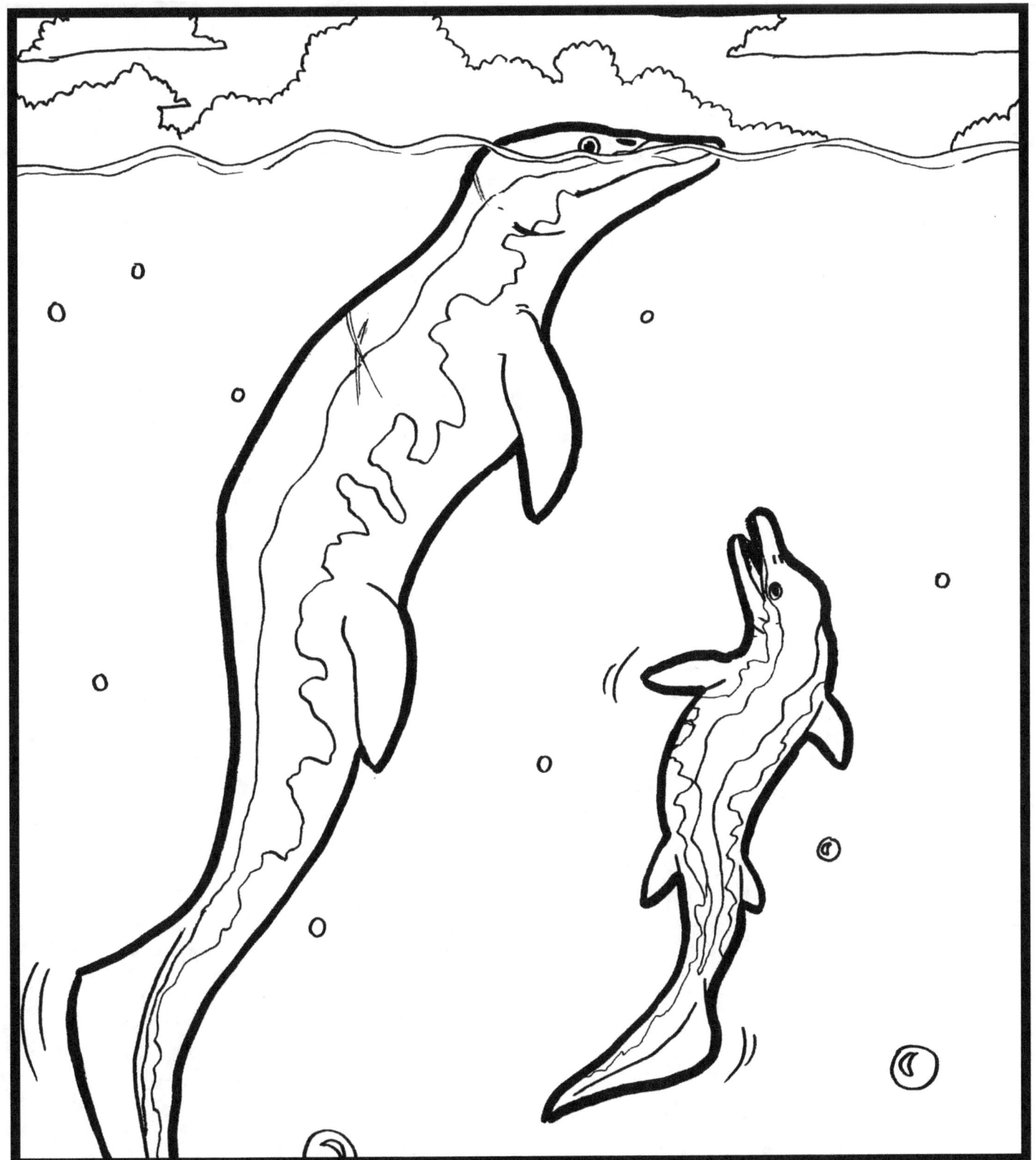

CHAOHUSAURUS

248 MYA, Early Triassic, China

This animal was an ancestor of the ichthyosaurs, marine reptiles with a dolphin-like shape. Compared to later species, the head was shorter, and the body more lizard-like. They did not lay eggs, but unlike later ichthyosaurs which are born tail-first to prevent suffocation, they were born head-first, which gave evidence for the terrestrial ancestor also bearing live young head-first.

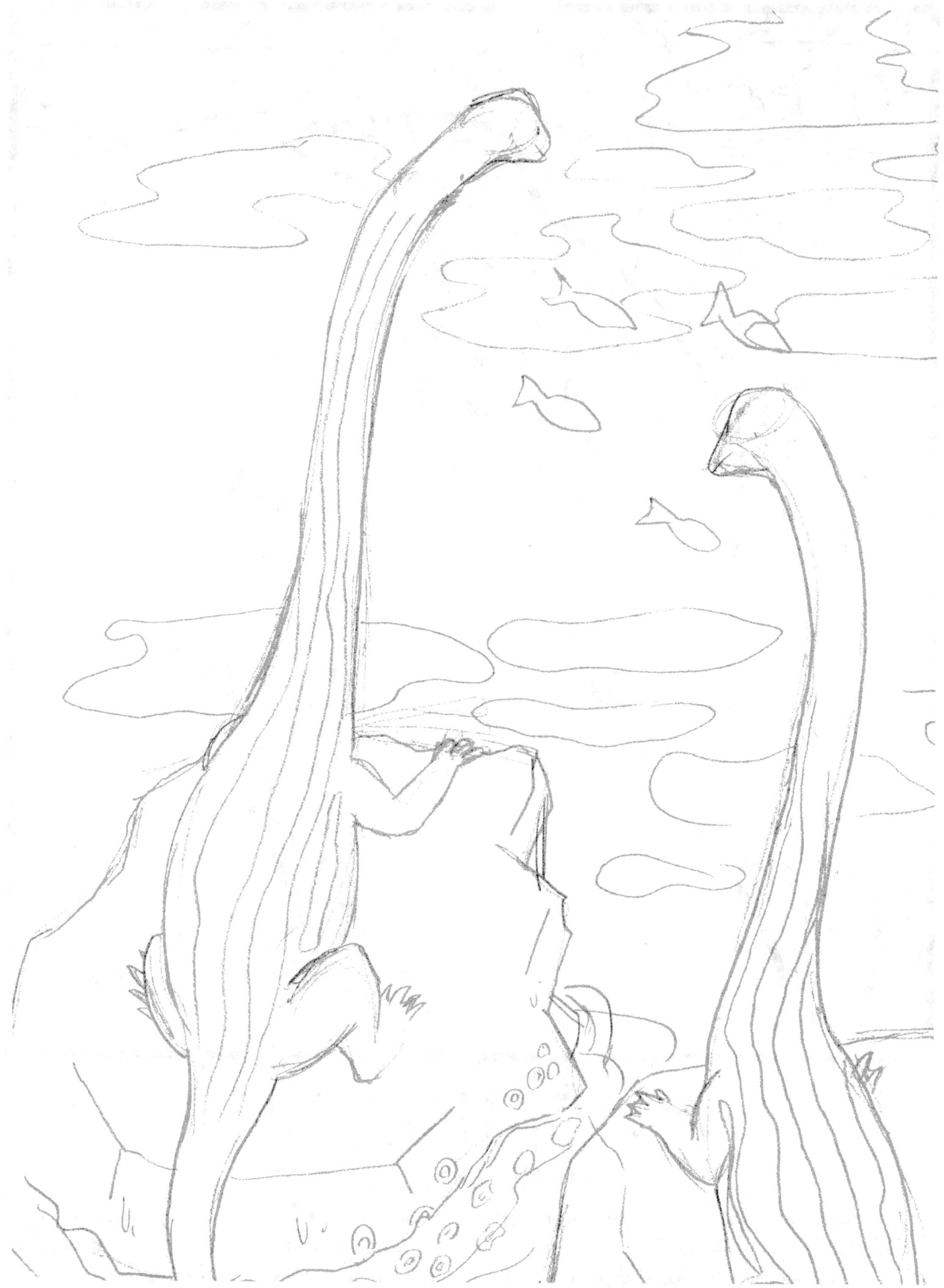

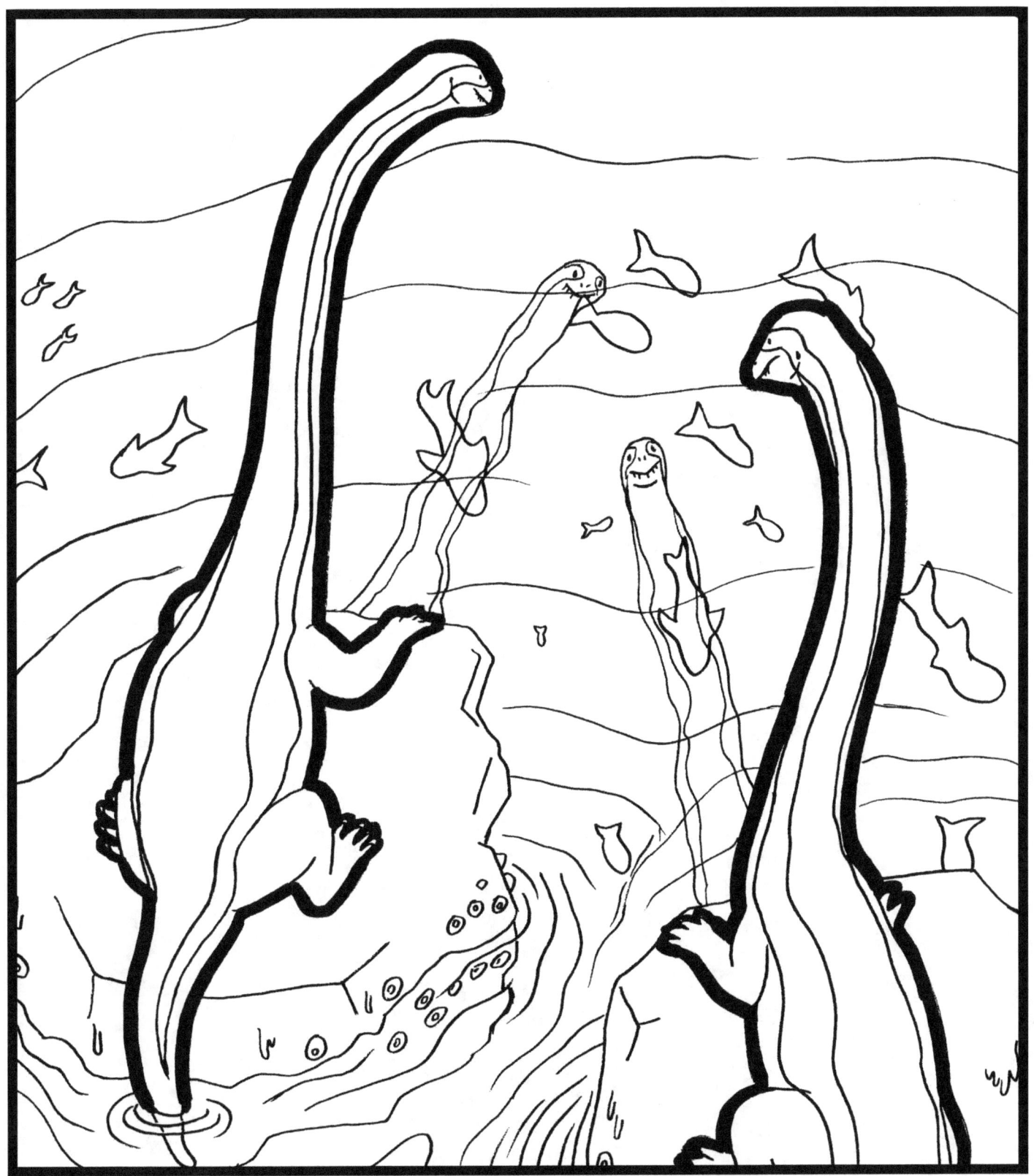

TANYSROPHEUS

245 - 228 MYA, Middle Triassic, Italy and China

This was a fish-eater with a neck that was longer than both its body and tail combined, giving a total length of around 6 meters. The neck was very stiff, made up of 12-13 extremely long vertebrae. It was also not very heavy, with most of the animal's mass being near the back, where the powerful hind limbs shifted the center of mass towards the back. This allowed it to easily swing its head at prey.

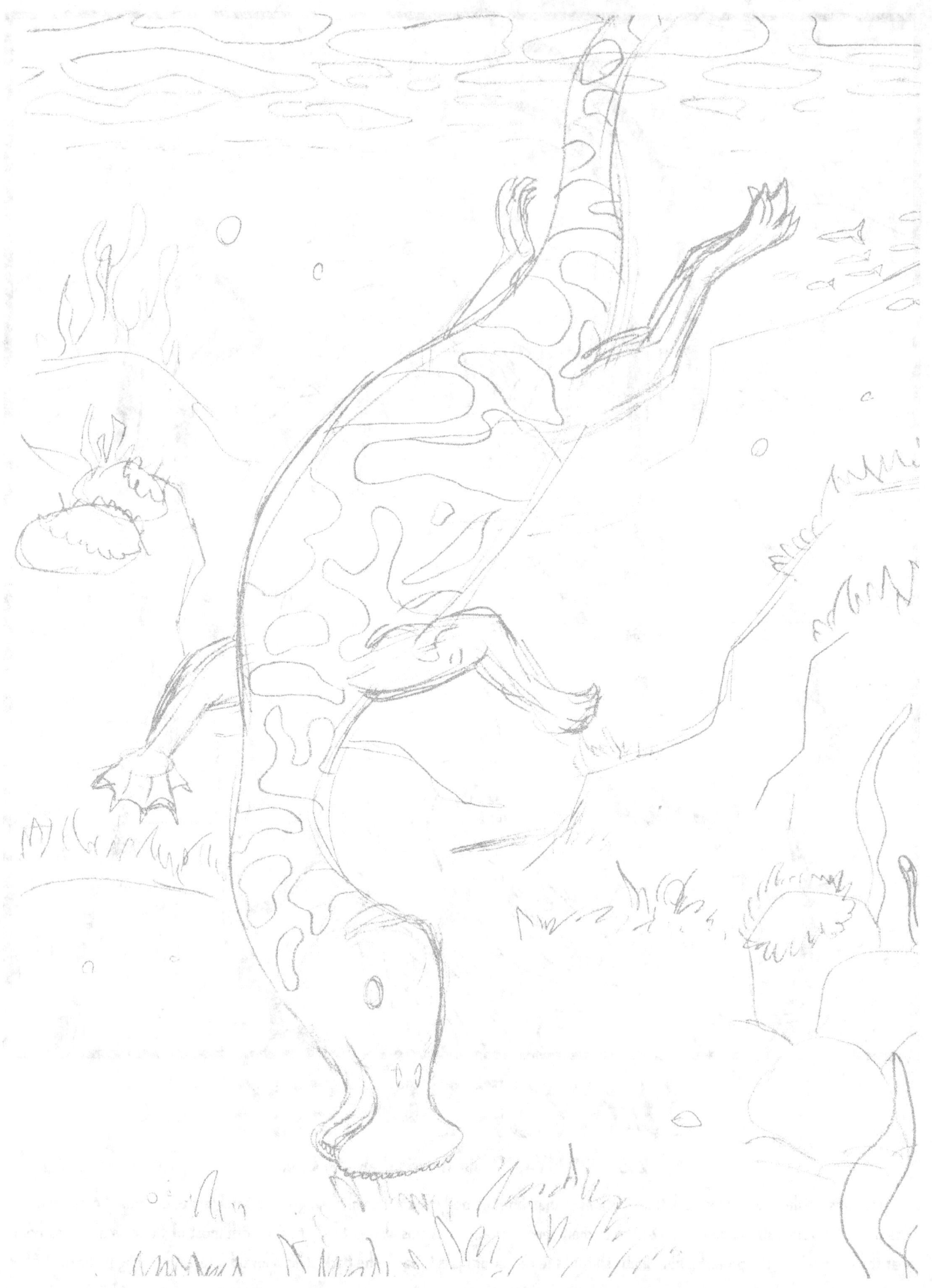

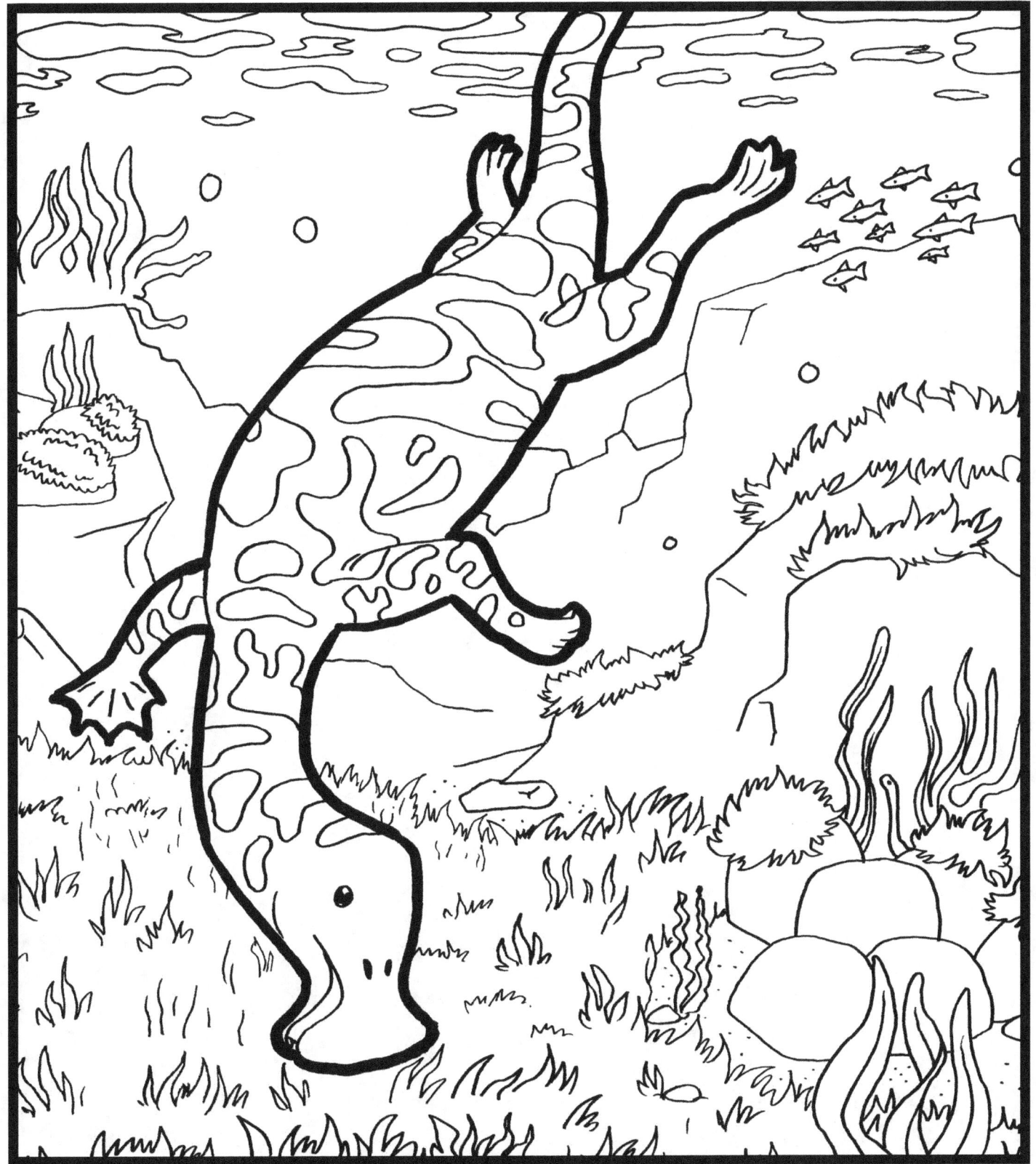

ATOPODENTATUS

240 MYA, Middle Triassic, Southwest China

This algae-eating marine reptile's odd, hammer-shaped head helped it root through the seafloor for food.
It is the earliest known herbivorous marine reptile, which is made even more unique when compared to most other marine reptiles, which tended to be omnivores or carnivores.

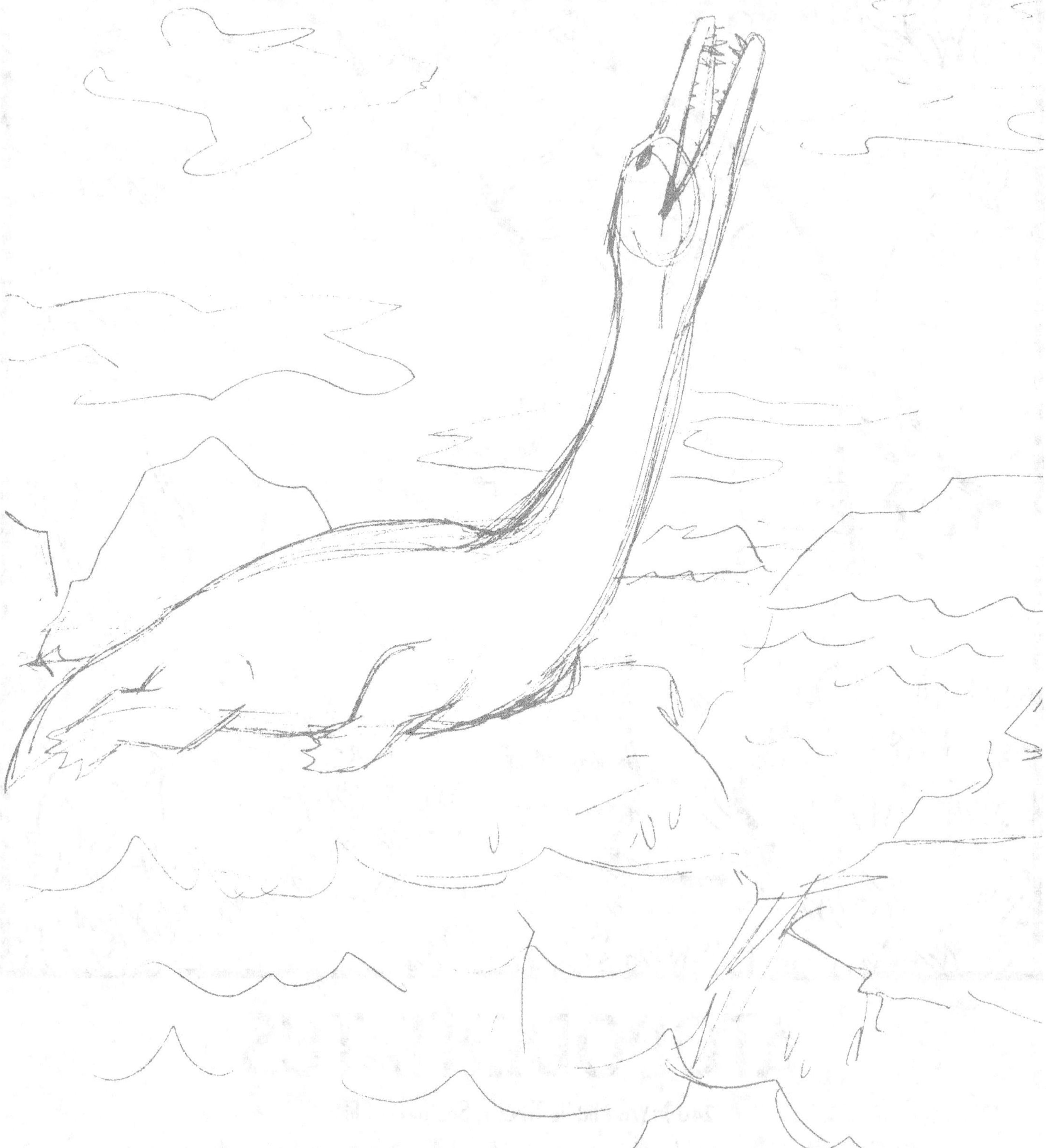

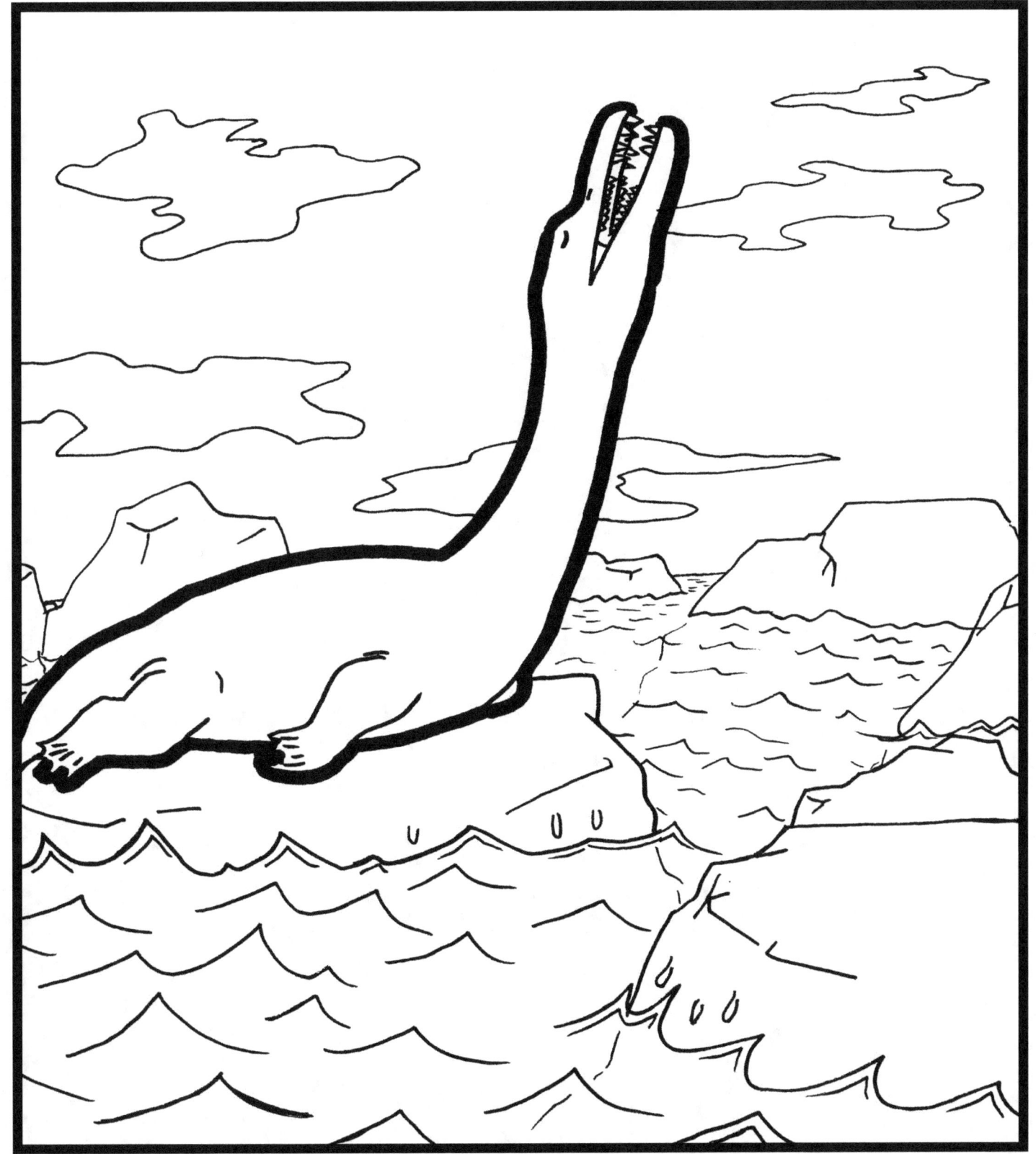

NOTHOSAURUS

240 - 210 MYA, Early to Late Triassic, Germany, Spain, China, Netherlands

Like modern-day seals, nothosaurus spent part of its time on land, and the other part in the sea. It is unknown if they gave birth to live young, or if they laid eggs, but there is evidence that suggests a common ancestor of both nothosaurs and ichthyosaurs was capable of live-births.

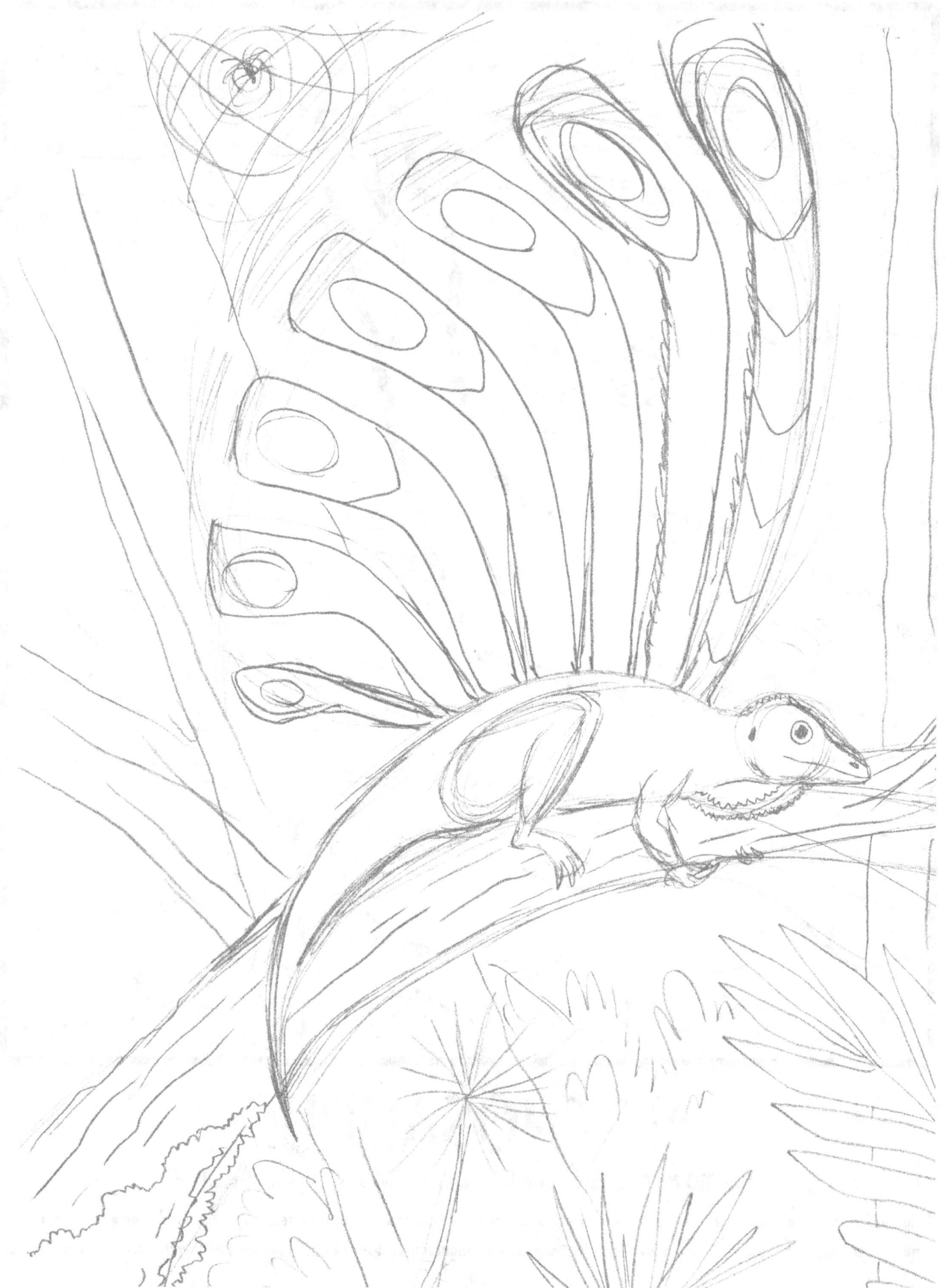

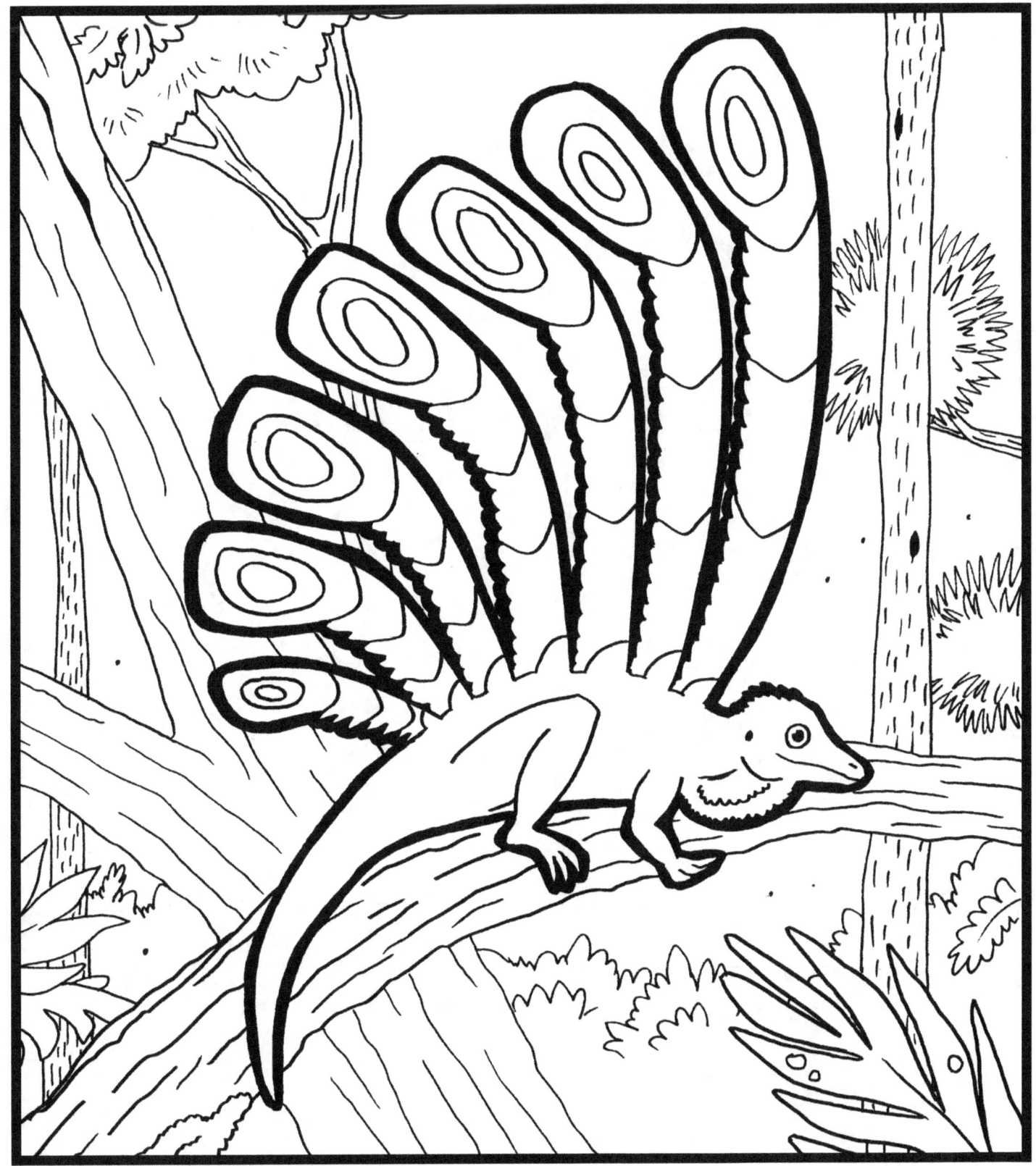

LONGISQUAMA

235 MYA, Mid Triassic, Kyrgyzstan

Since its discovery, the strange structure on this animal's back has been reconstructed as a sail, a pair of gliding wings, and even as a plant that happened to be preserved on top of it. A very prominent part of this little reptile, even its name means "long scales."

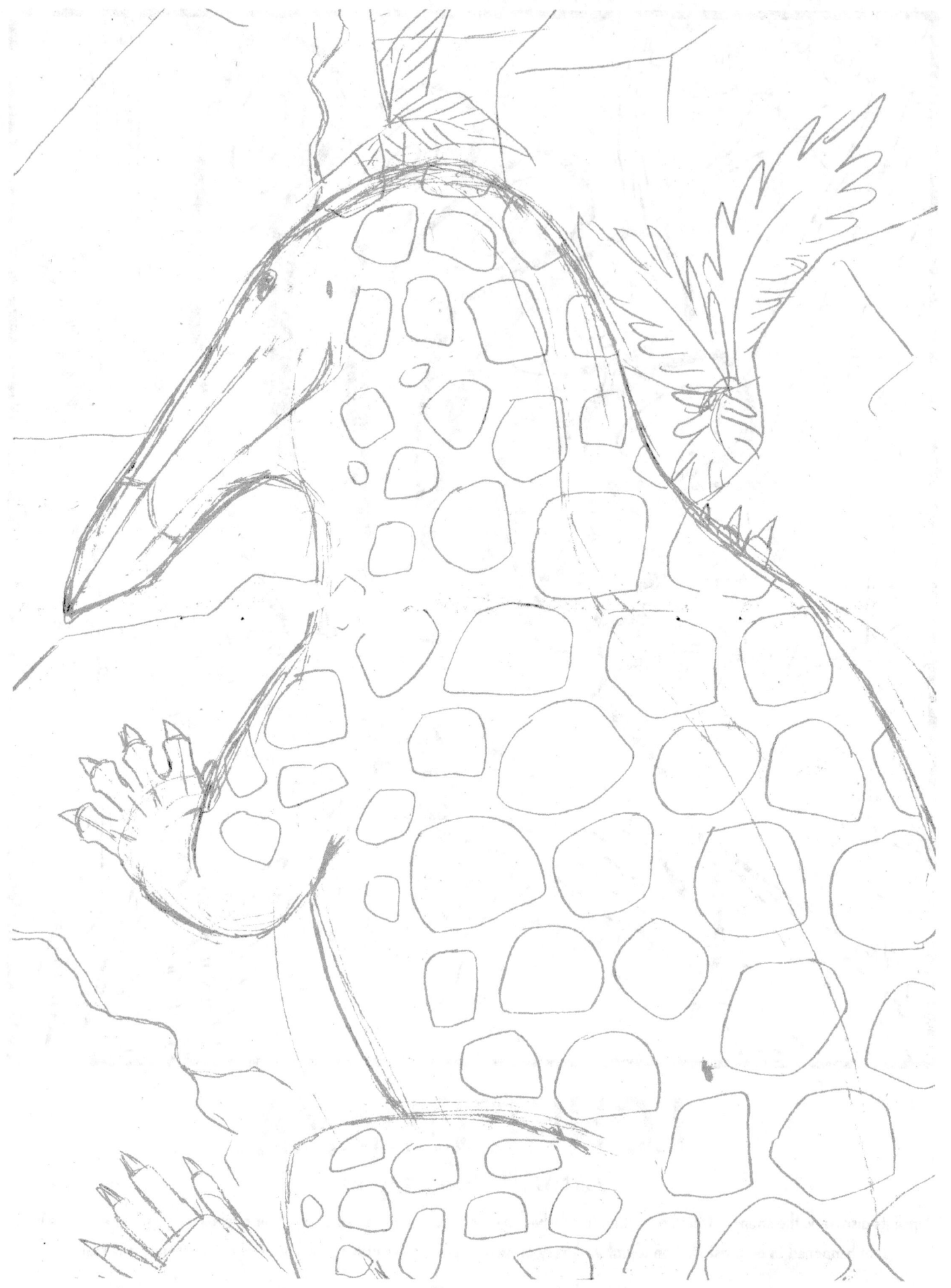

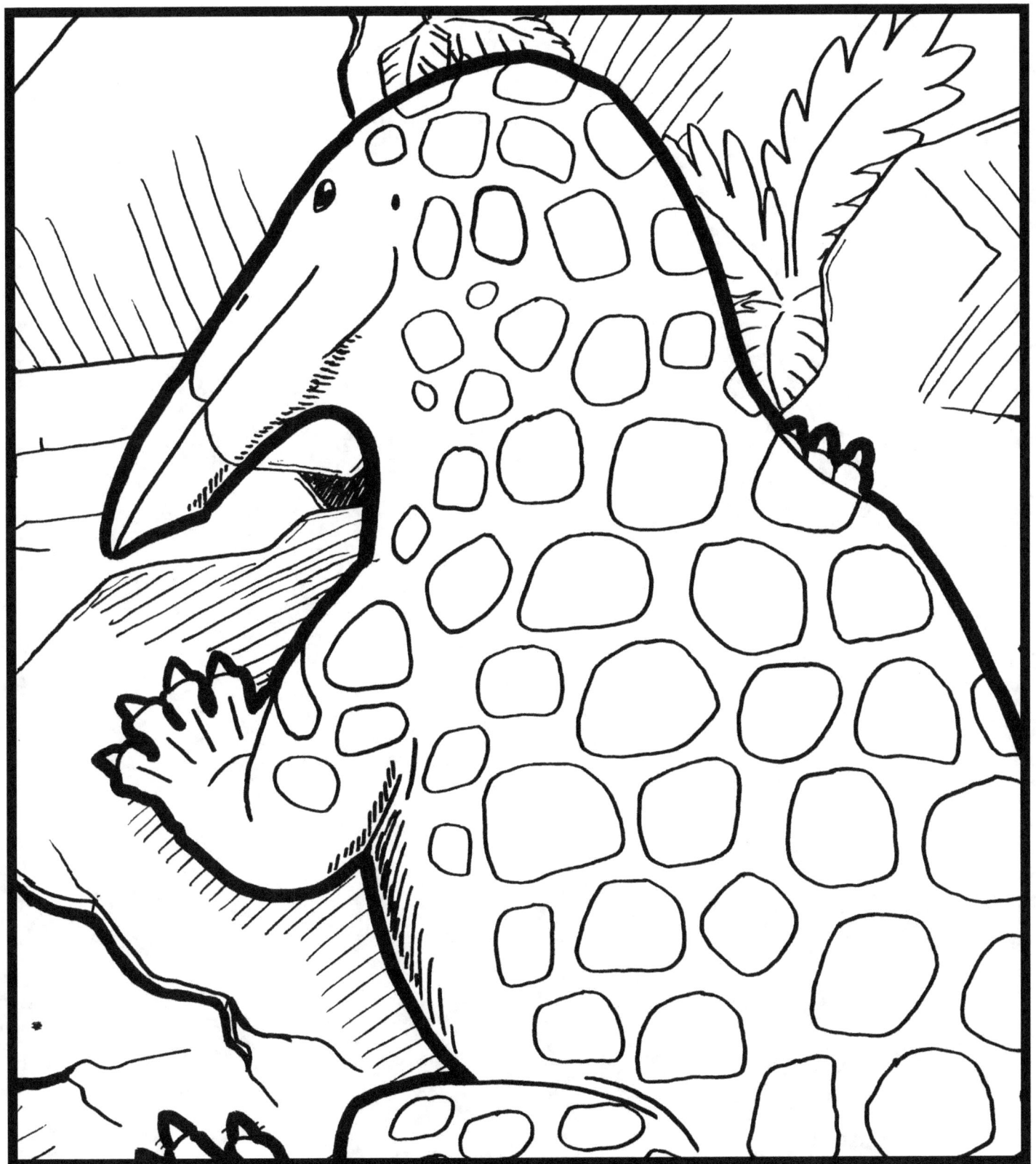

TERATERPETON

235 - 221 MYA, Late Triassic, Canada

Very different-looking from other reptiles of its time, the head was long and beak shaped, and lacked teeth on the ends.
It shared skull features with the marine reptiles, ichthyosaurs and plesiosaurs, but was not closely related to them.
It is not currently known what role this animal played in its ecosystem, so until more is known, it is up to speculation.

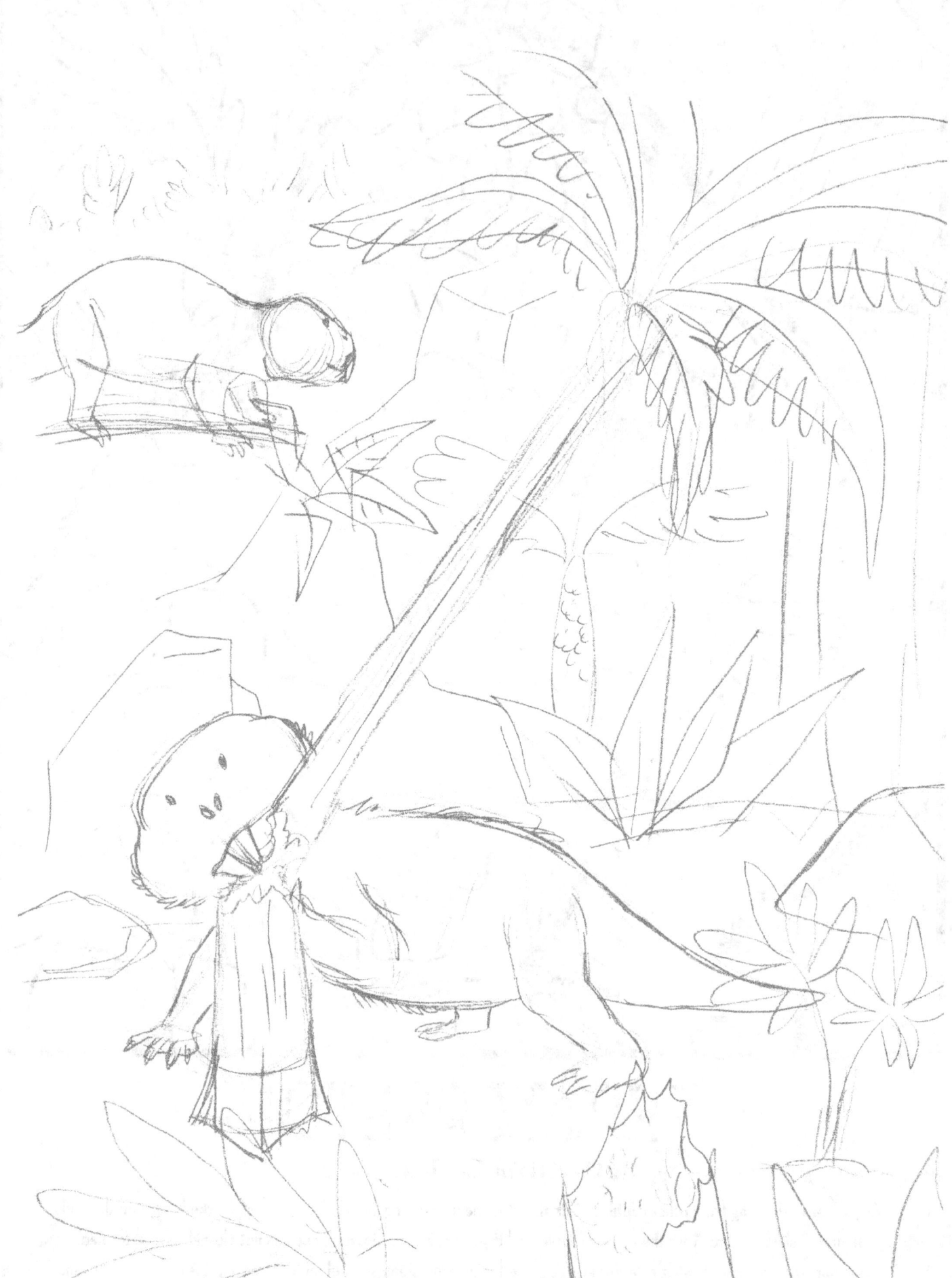

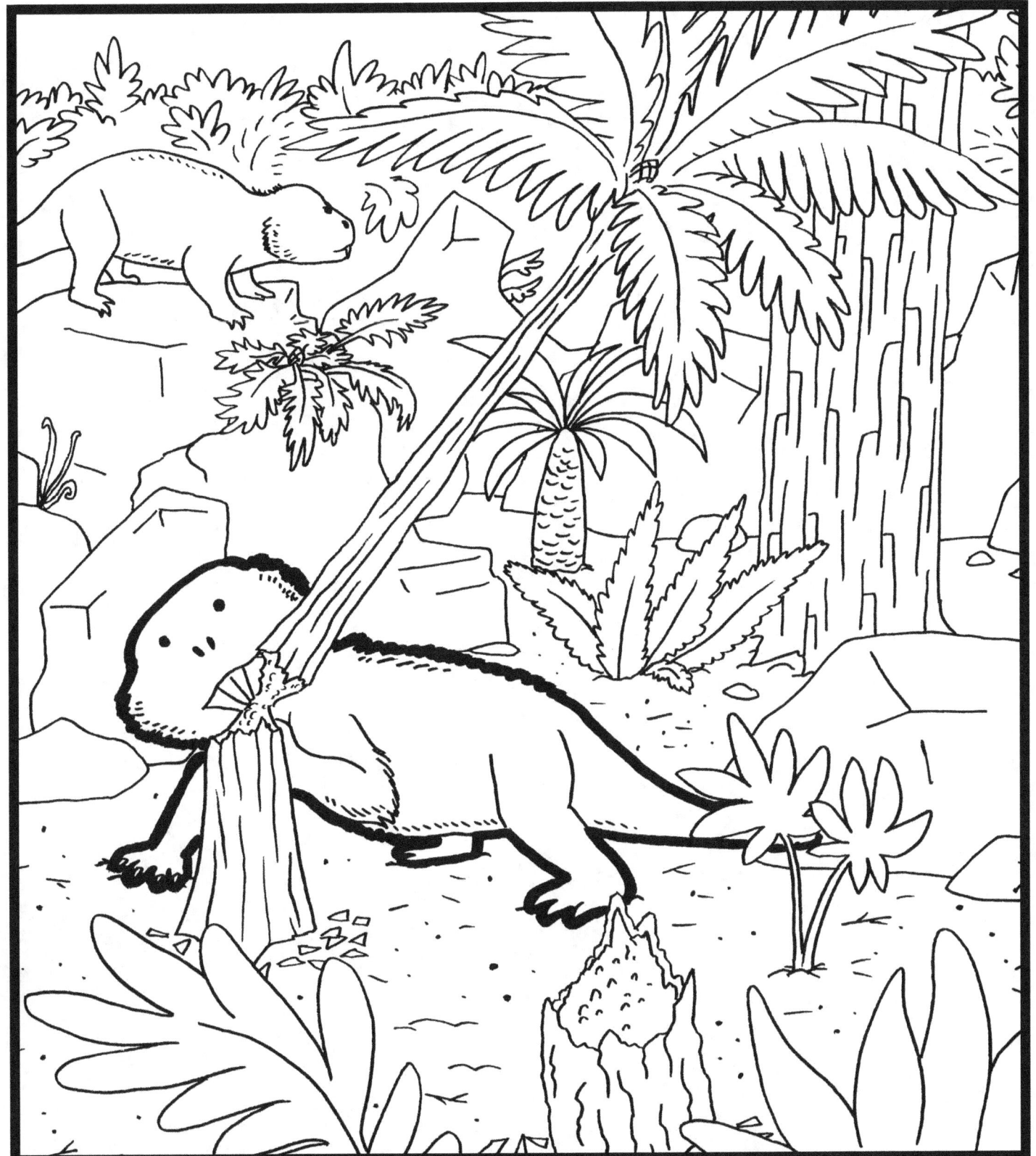

HYPERODAPEDON

231 - 216 MYA, Late Triassic, Argentina, Brazil, India, Scotland

These animals were small herbivores that were widespread throughout the supercontinent of Pangaea.
This had the result of their fossils being found on several of today's continents.
Its diet probably consisted of seed ferns, and it most likely died out when these plants became scarce.

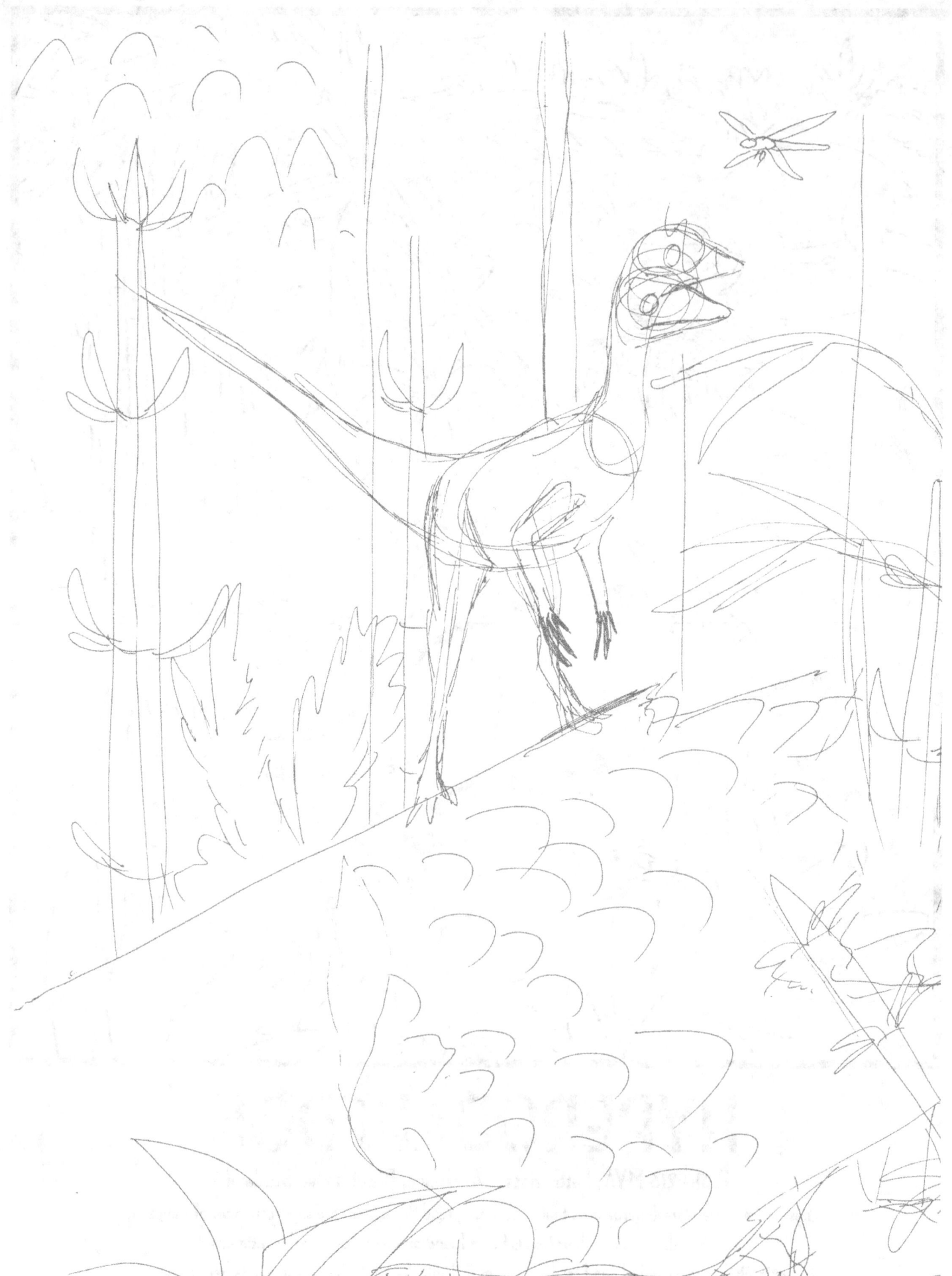

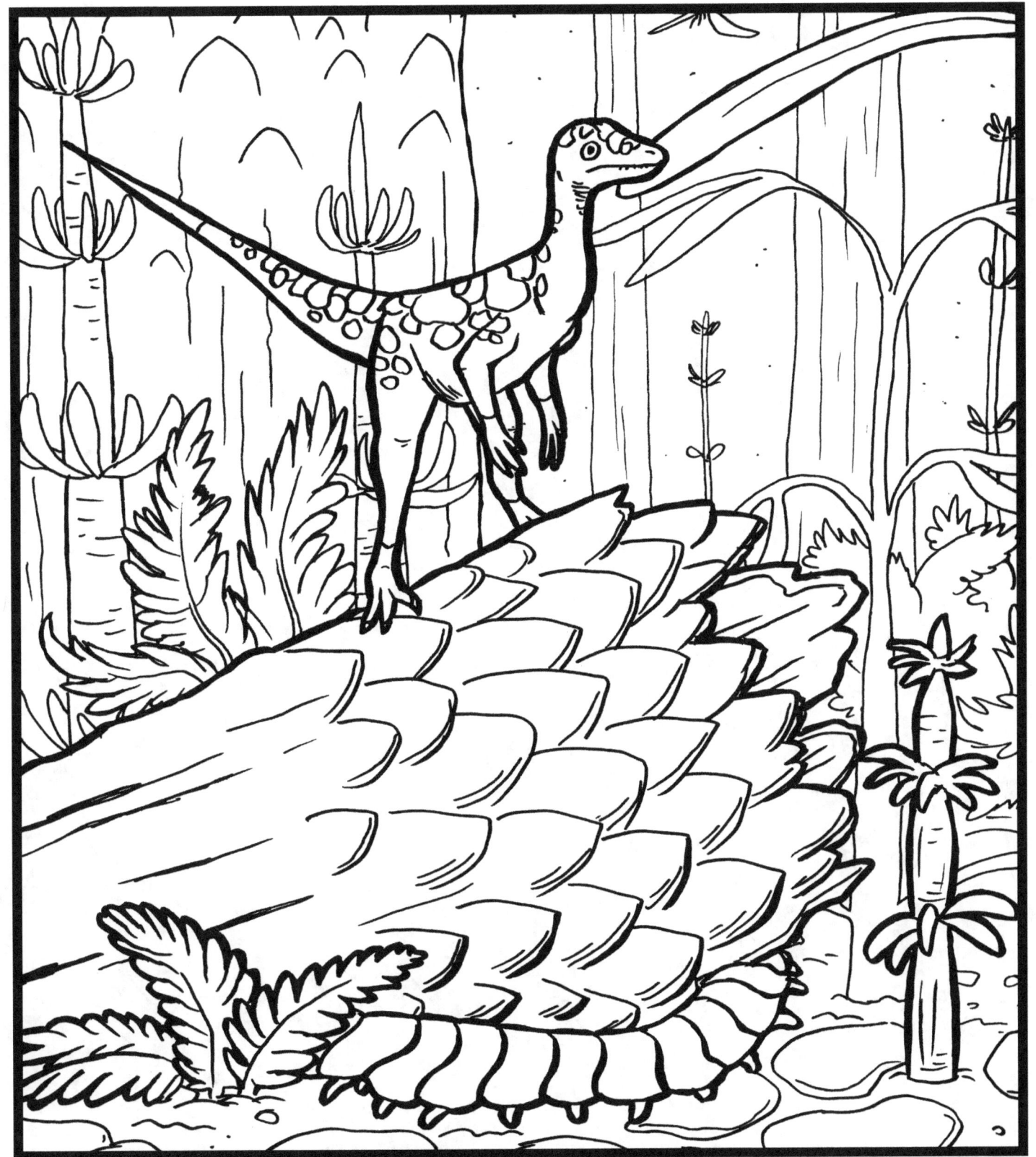

EORAPTOR

231 MYA, Late Triassic, Argentina

One of the first dinosaurs, eoraptor was a small, swift omnivore that stood on two legs. Its name means "dawn plunderer."
During its time, dinosaurs were not yet the dominant land vertebrates they would eventually become.

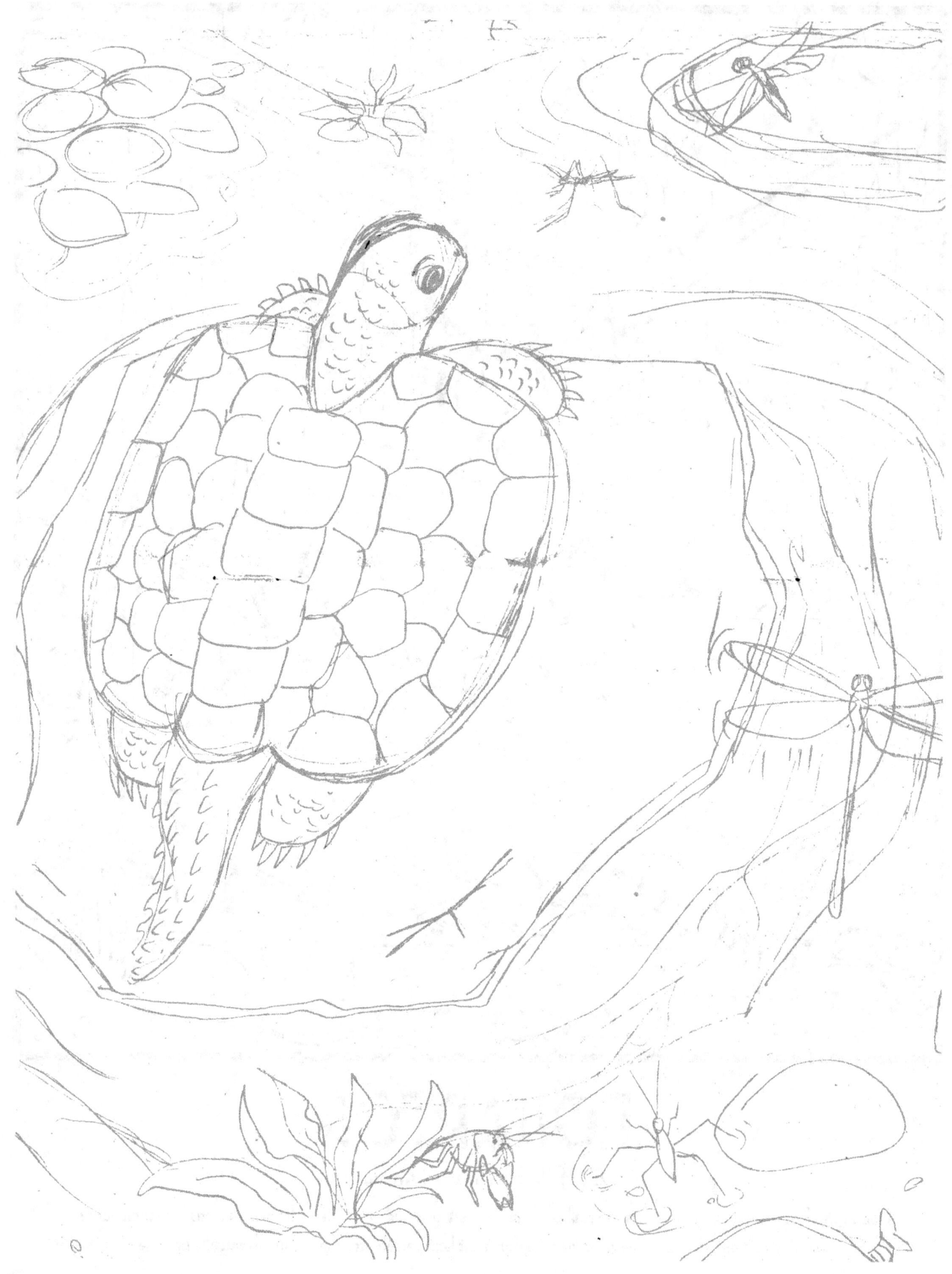

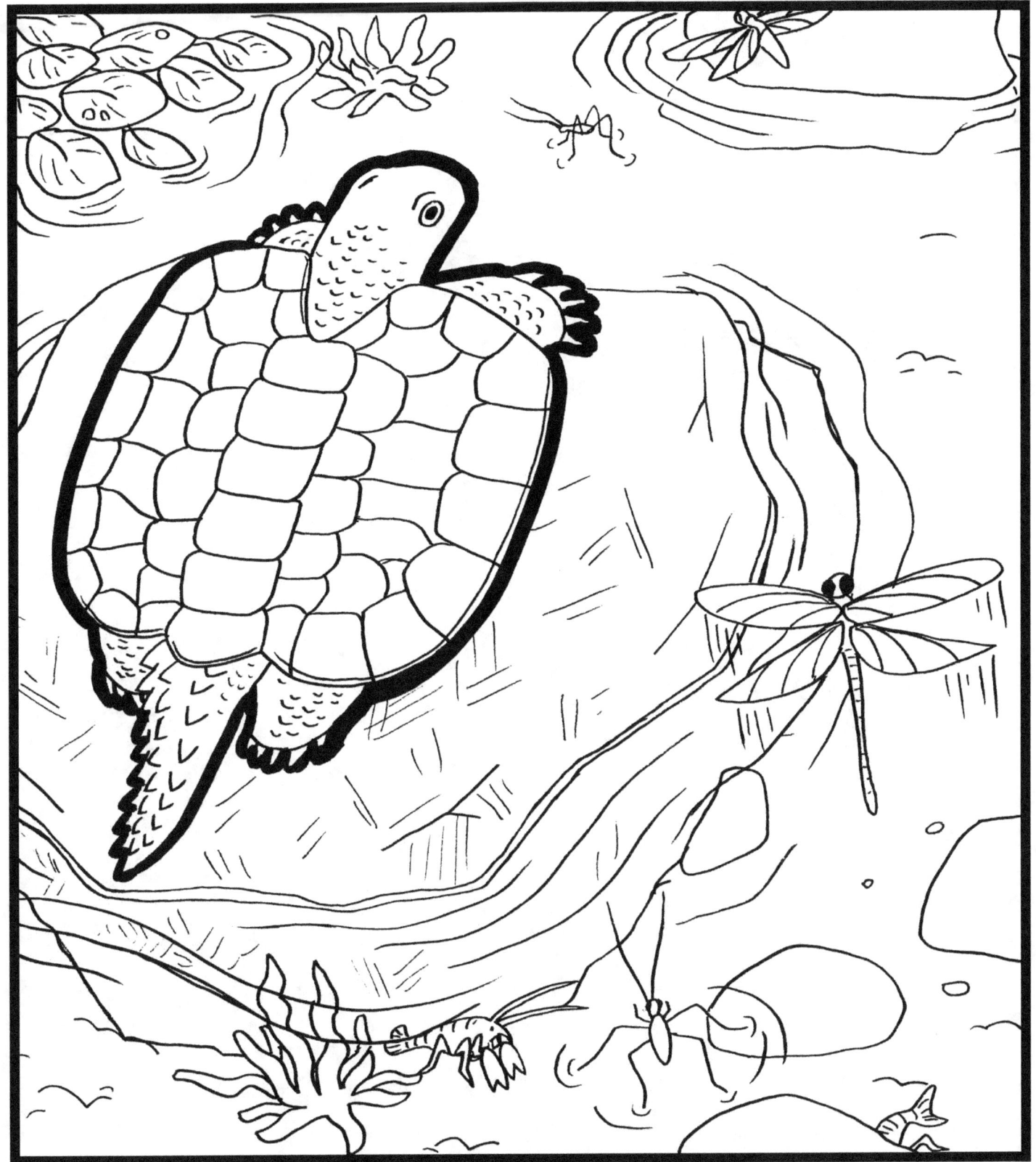

HENODUS

228 - 220 MYA, Late Triassic, Germany

This meter-long placodont looked very similar to a turtle, yet it bore no relation to them.
Its limbs were weak and made venturing onto land difficult, and its hammer-shaped jaws were suitable for filter-feeding and scraping plant
life off the bottom of the ponds and lakes it lived in.

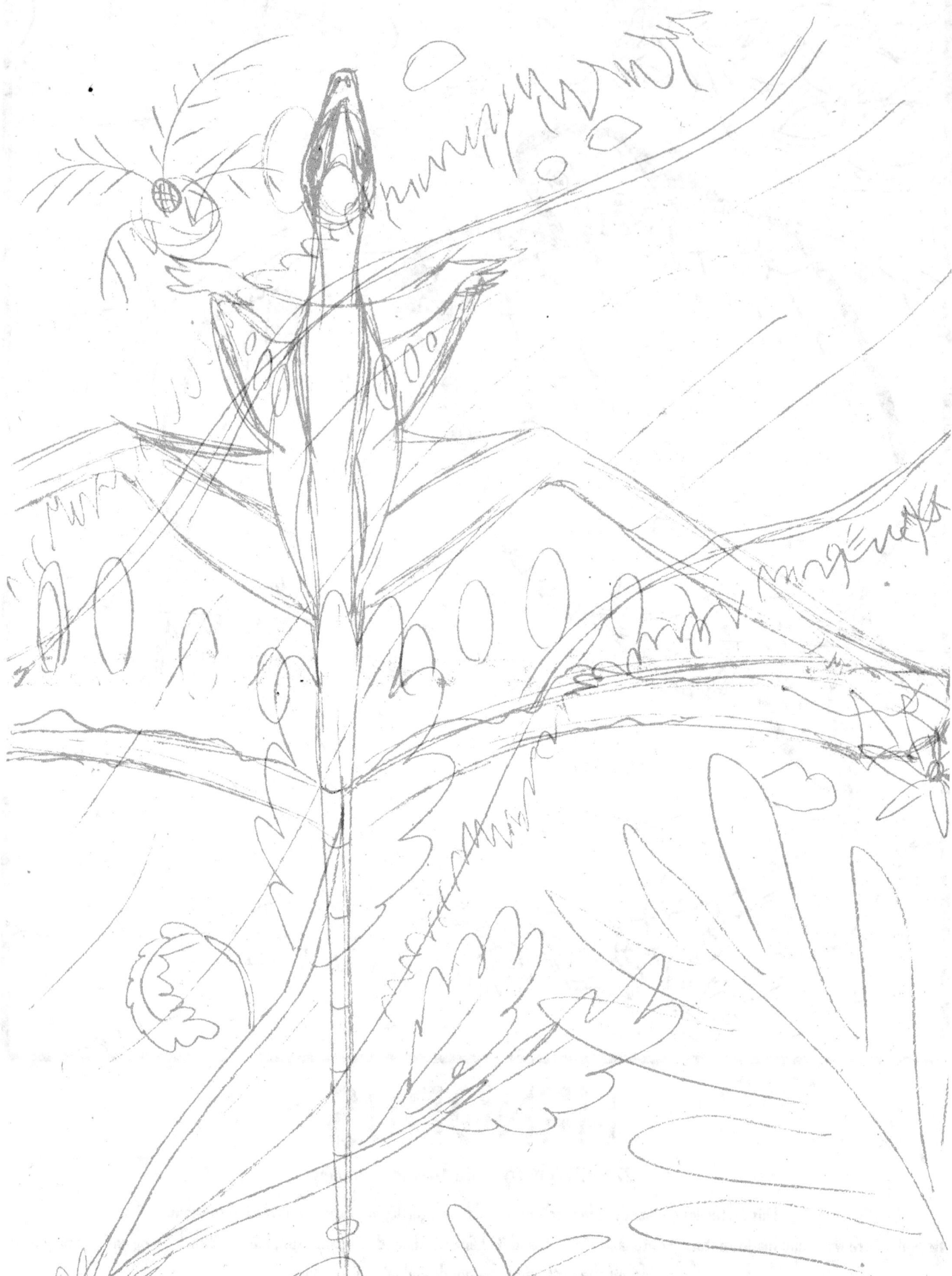

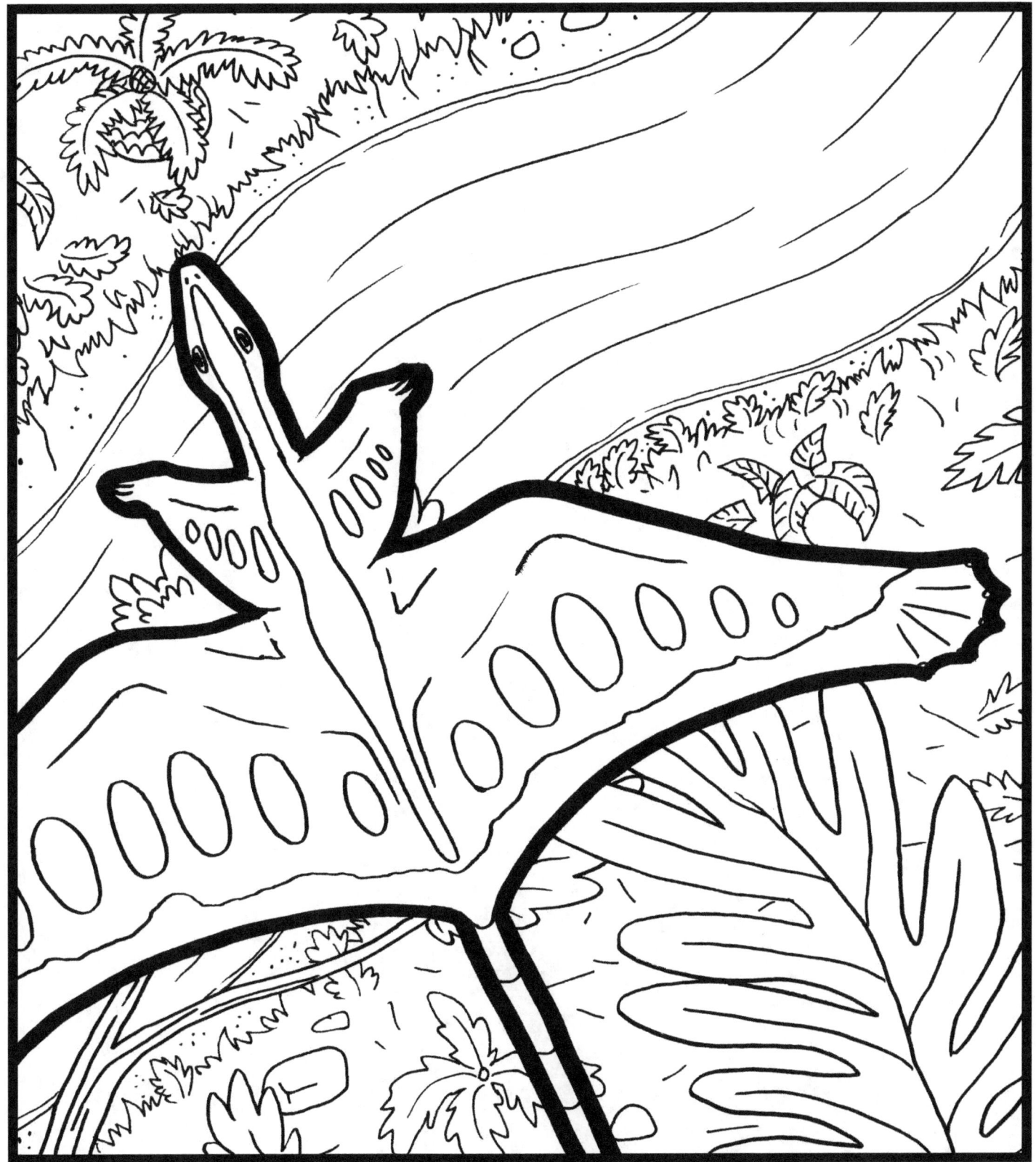

SHAROVIPTERYX

225 MYA, Middle to Late Triassic, Kyrgyzstan

This unusual early reptile was evolution's way of coming up with a novel way of travel by air. Where birds, bats, flying fish, and pterosaurs all use their front limbs or fins in an airfoil, this protorosaur used its hind limbs for that function. It would glide from tree to tree, using the membrane around its forelimbs to help steer its path.

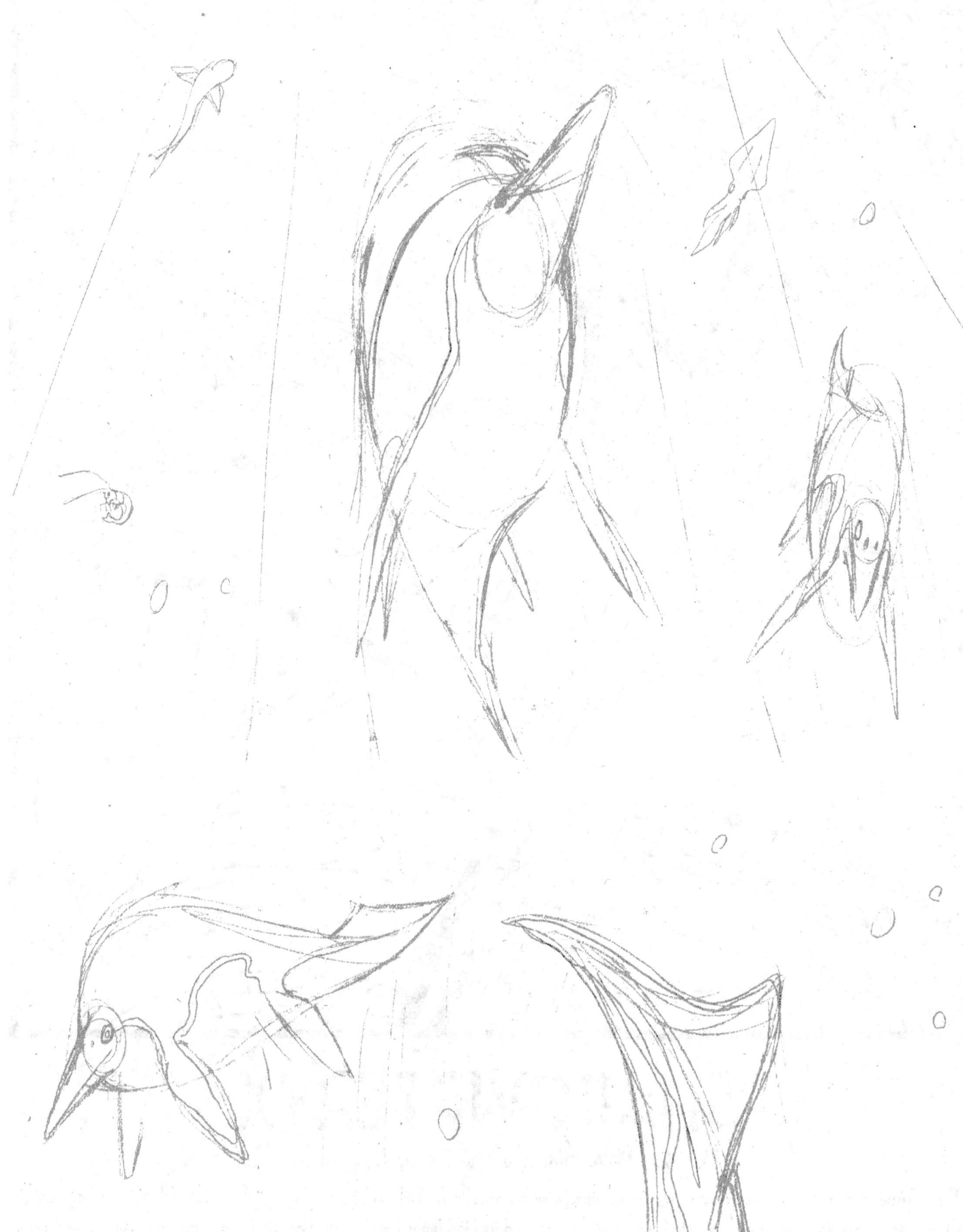

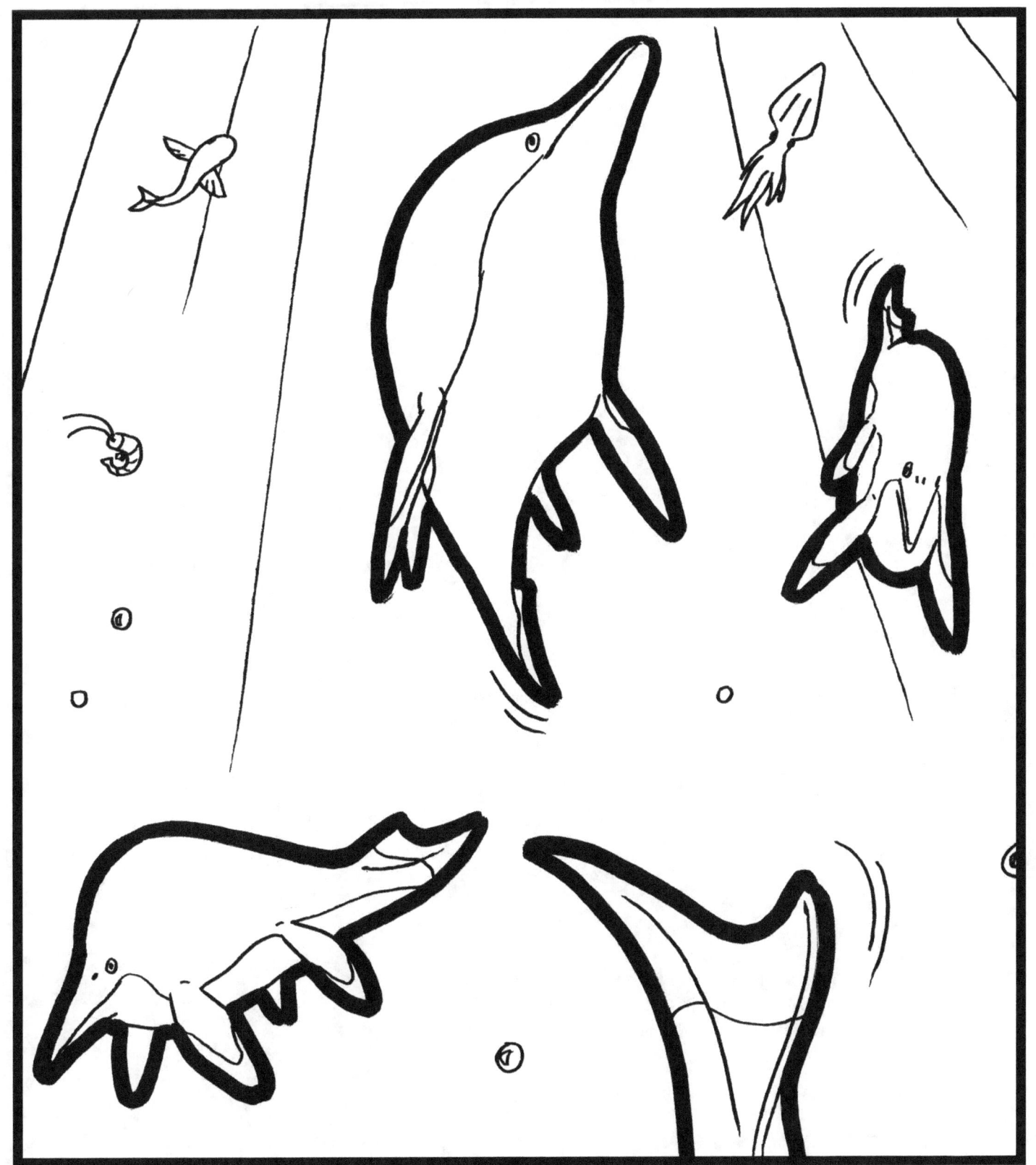

SHONISAURUS

221.5 - 212 MYA, Late Triassic, United States

Shonisaurus is one of the better known ichthyosaurs, and it owes this to the abundance of fossils left behind. The body was very deep and round compared to other marine reptiles, and it could reach a length of up to 15 meters. It did not have a dorsal fin, and the fluke of the tail wasn't quite as developed in this animal as it would be for later icthyosaurs.

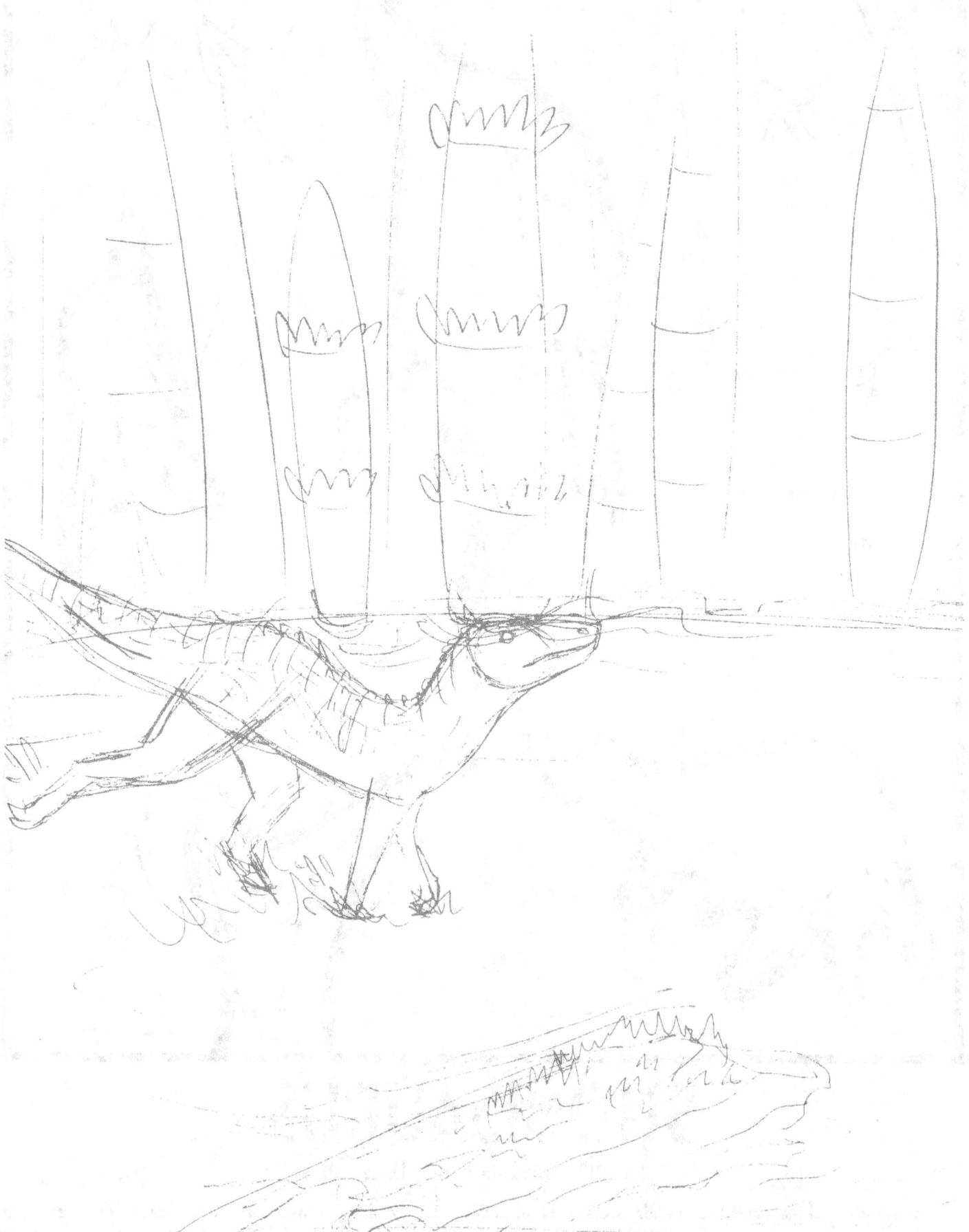

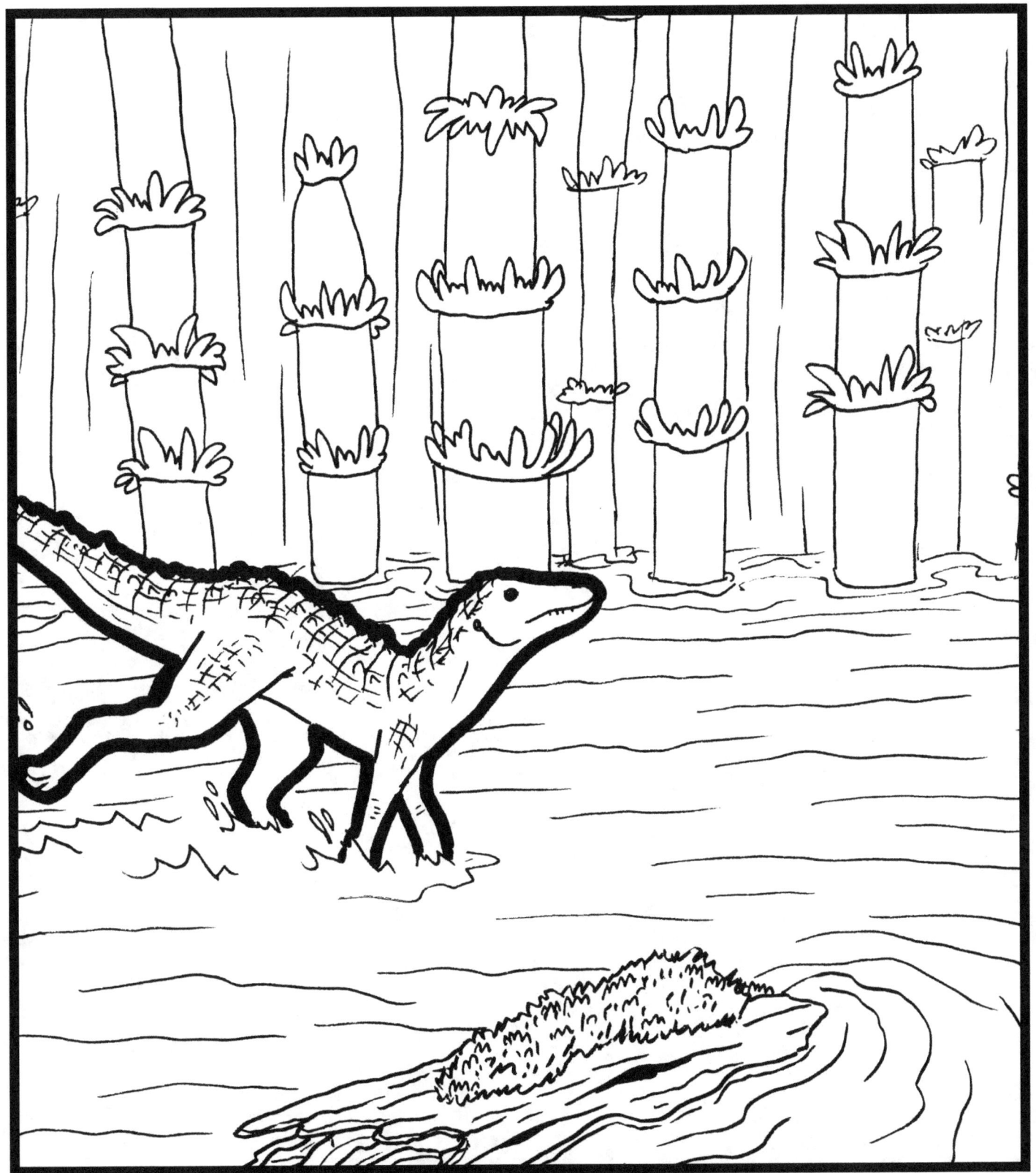

HESPEROSUCHUS

220 MYA, Late Triassic, North America

About a meter and a half long, this animal was a pseudosuchian, which were more closely related to crocodilians than to dinosaurs. They were capable of quickly running on all fours or with only the back legs, and the hands were useful for grasping, digging, and other activities requiring some level of dexterity.

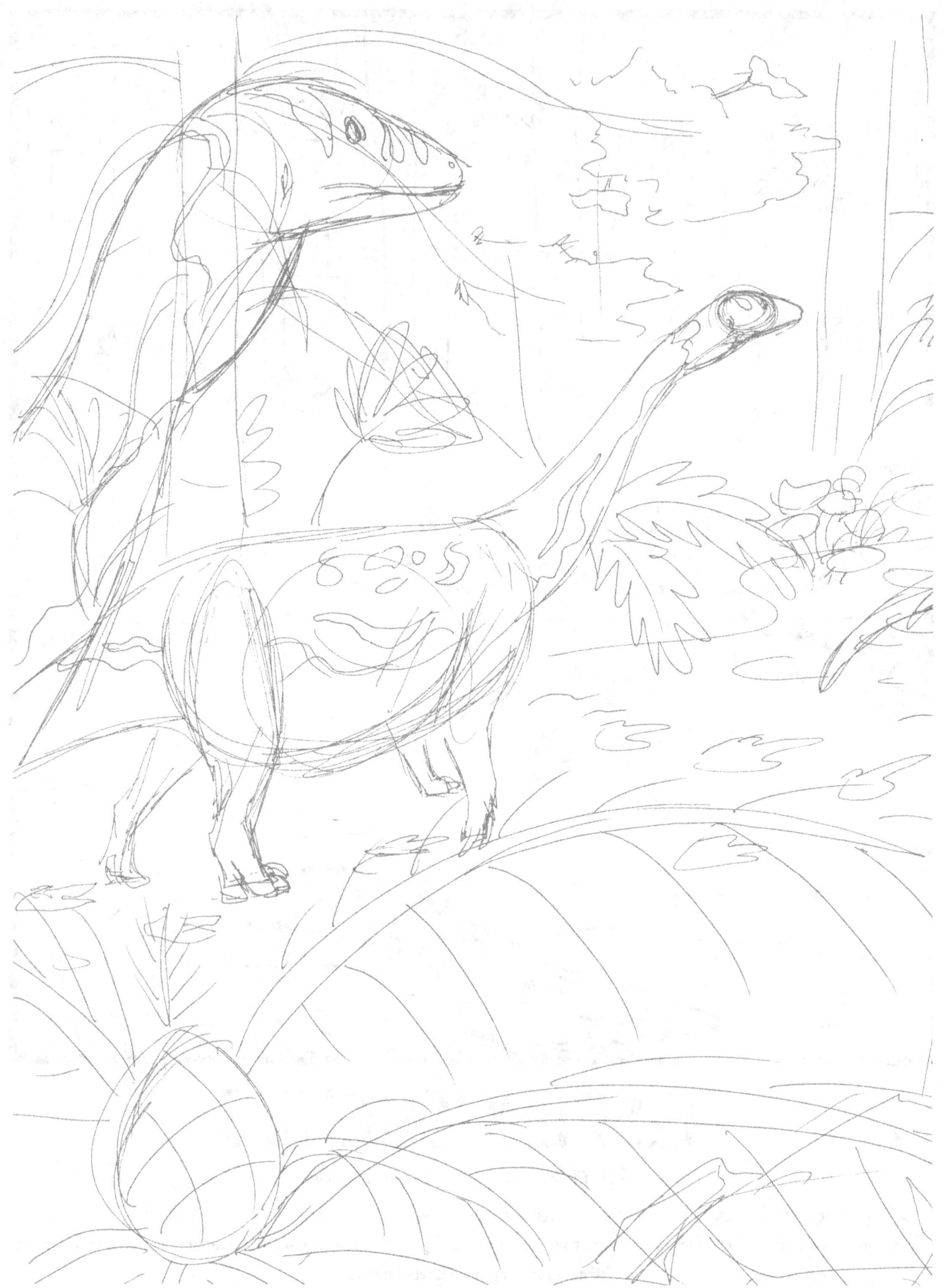

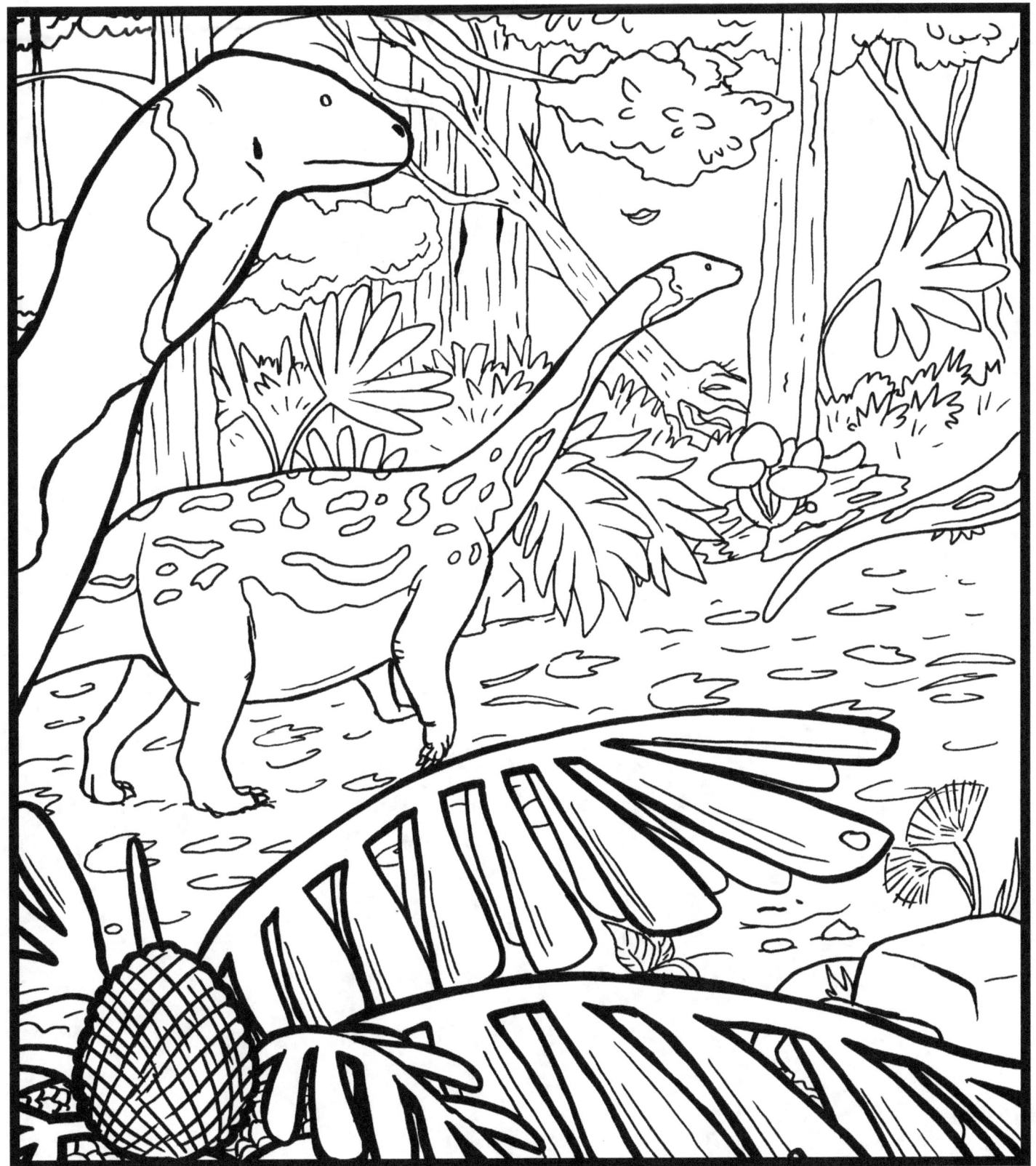

MELANOROSAURUS

216 - 201 MYA, Late Triassic, South Africa

This animal was a basal sauropod, which, at about 8 meters long, was still small compared to the behemoths the future would usher in. Despite this, it had a heavily-built body, robust limbs, and a small head, which would all be traits shared by many sauropods to come.

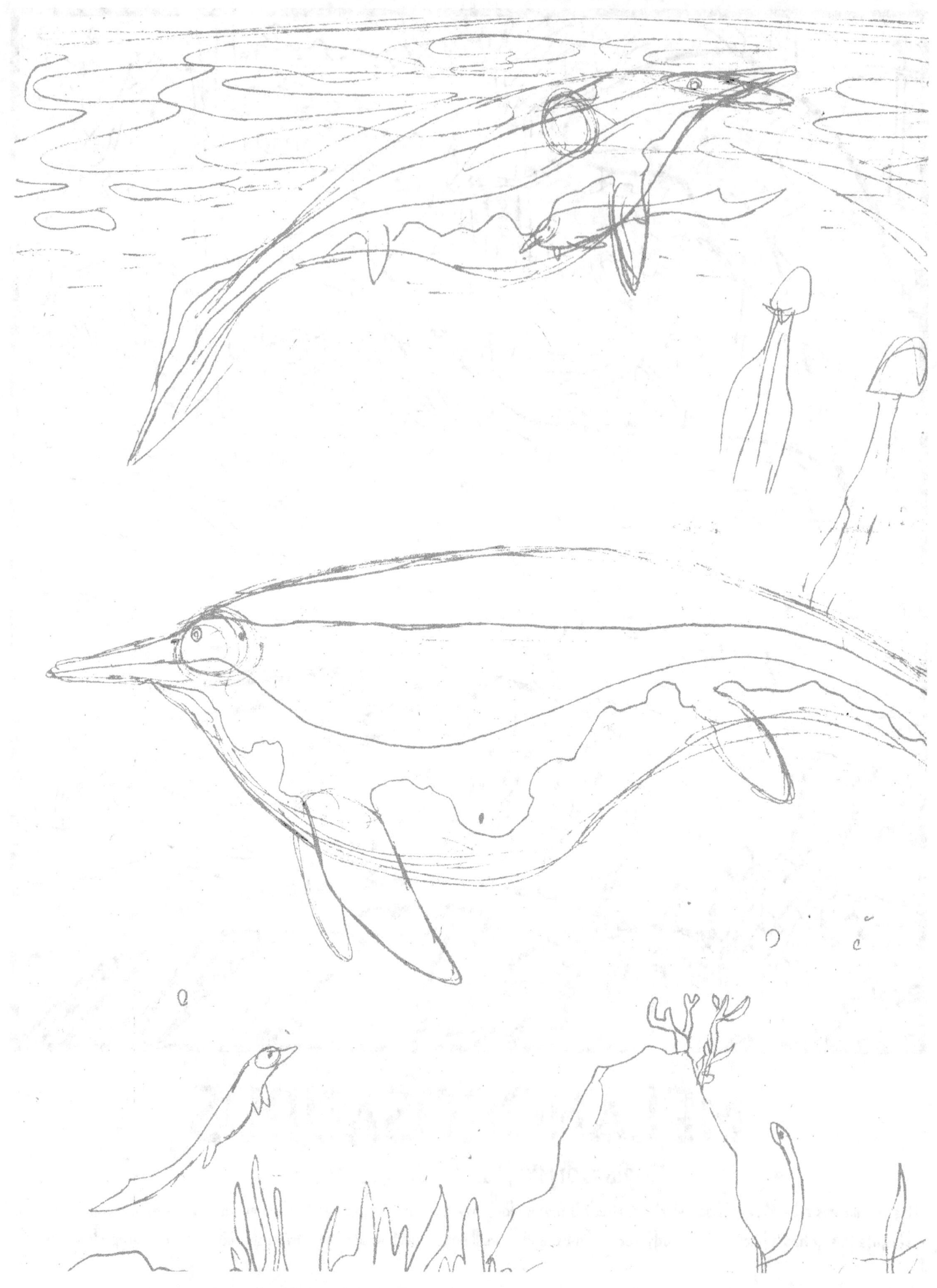

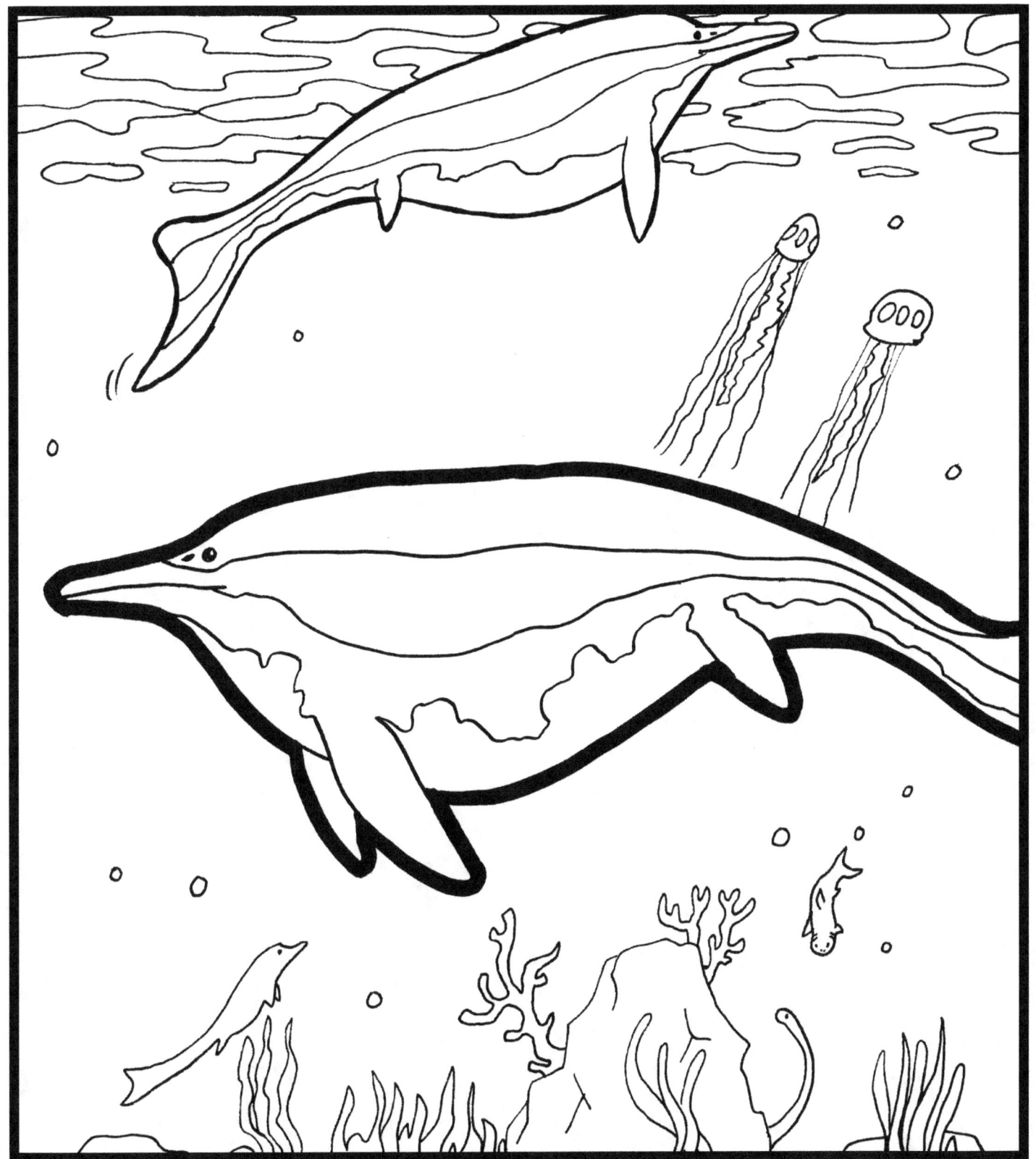

HIMALAYASAURUS

215.5 - 212 MYA, Late Triassic, Tibet

This giant ichthyosaur could reach a length of up to 15 meters, though it is only known from fragmentary remains. It was in the family Shastasauridae, which also included Shonisaurus. Unlike other ichthyosaurs, which had pointed, conical teeth, Himalayasaurus's teeth were large, flat, and suitable for cutting.

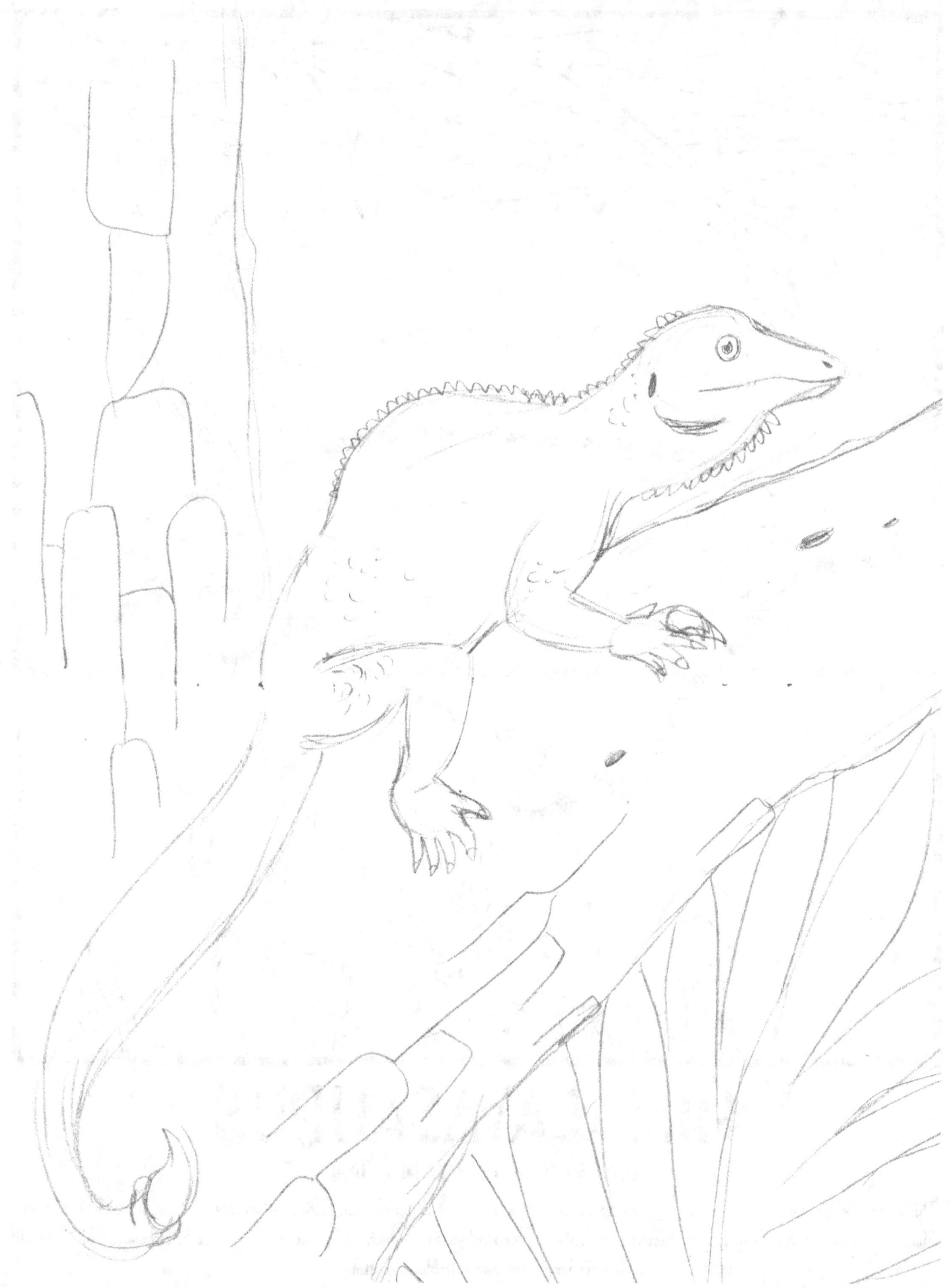

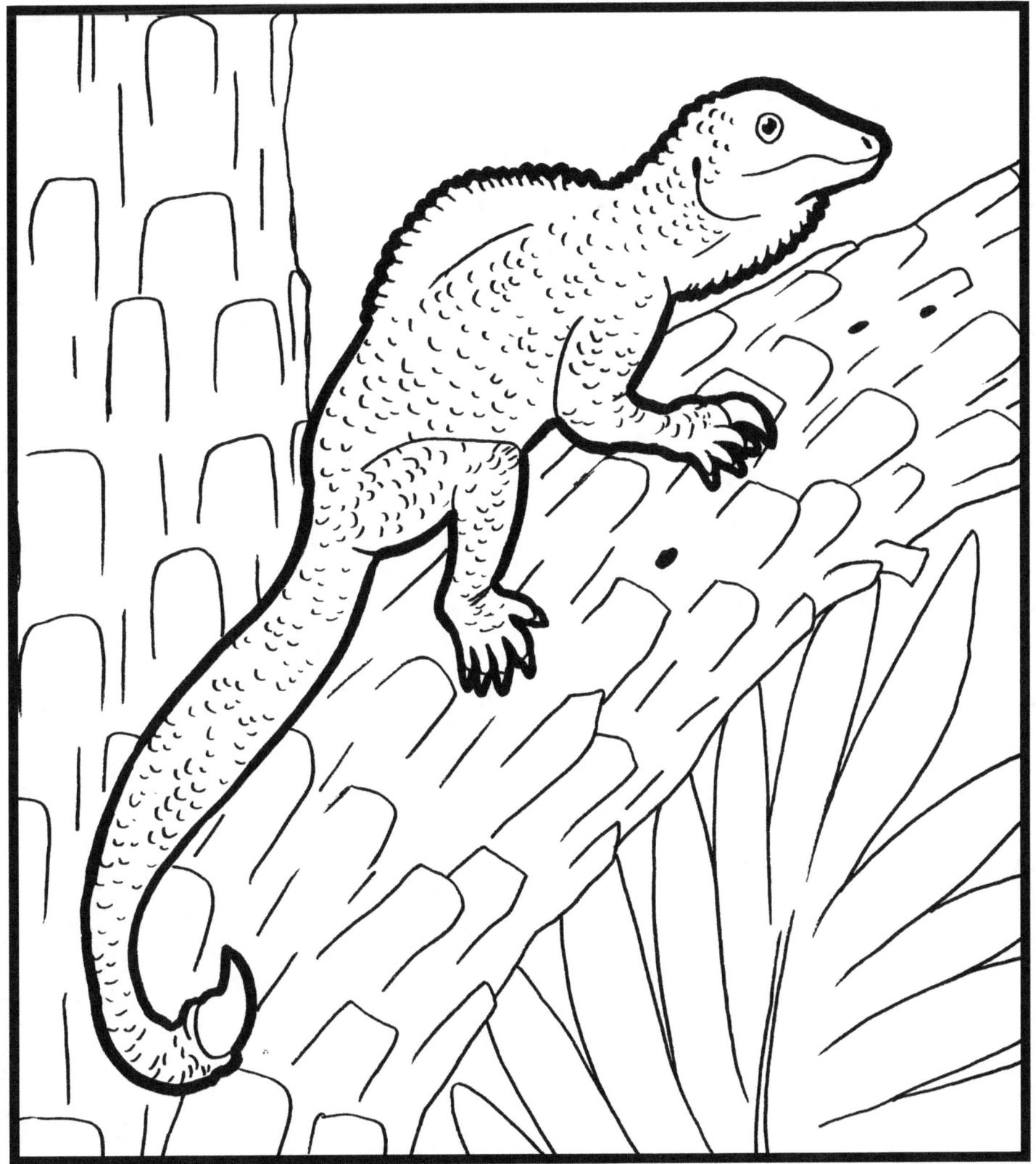

DREPANOSAURUS

212 MYA, Late Triassic, United States and Italy

These reptiles lived in trees, and even had prehensile tails to help them grip onto branches. This tail had a little claw at the end, assisting the animal with keeping its grasp on the tree. They were half a meter long and had a birdlike, triangular head. The large claw on its index finger likely helped it get to insects hiding within tree bark.

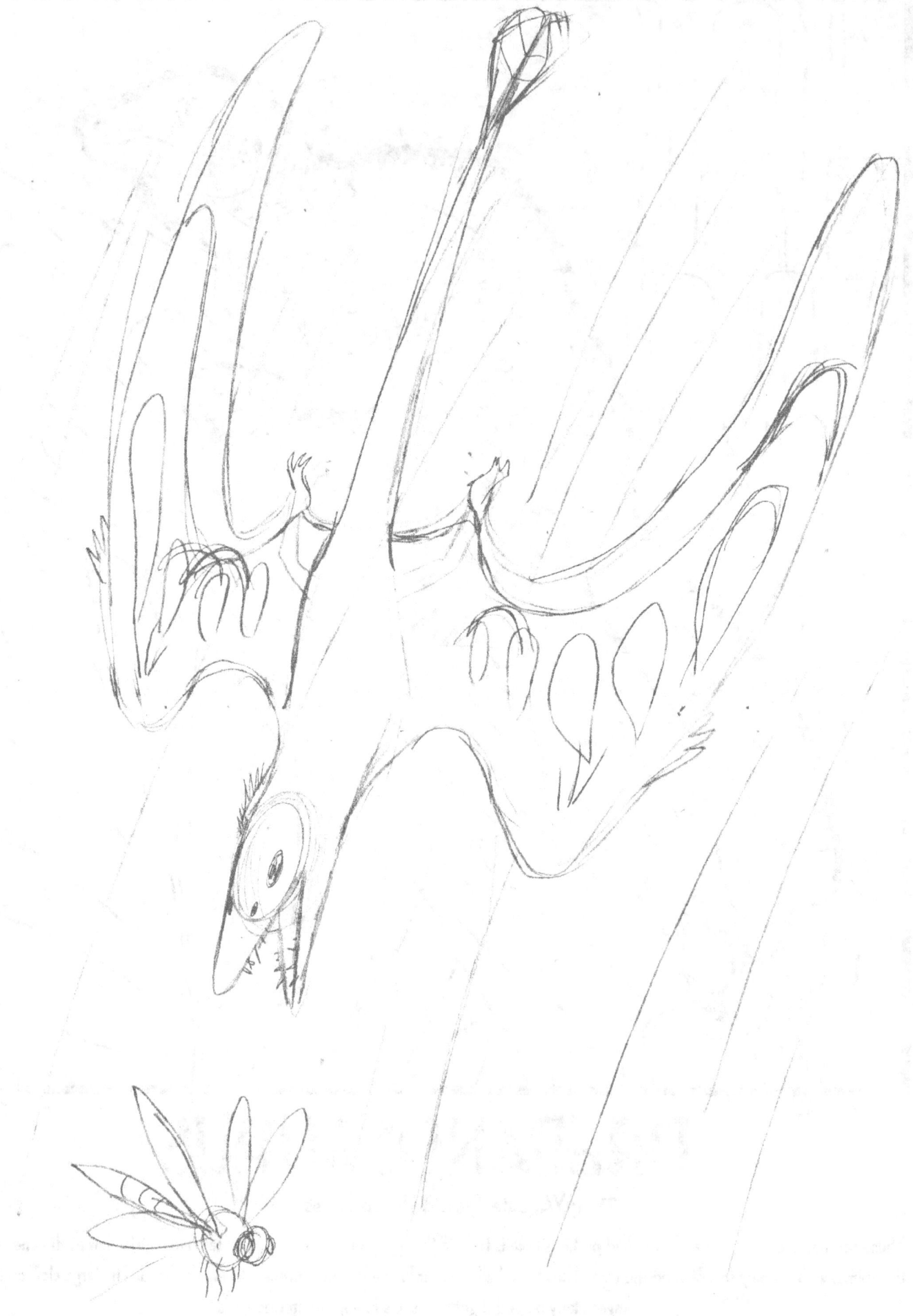

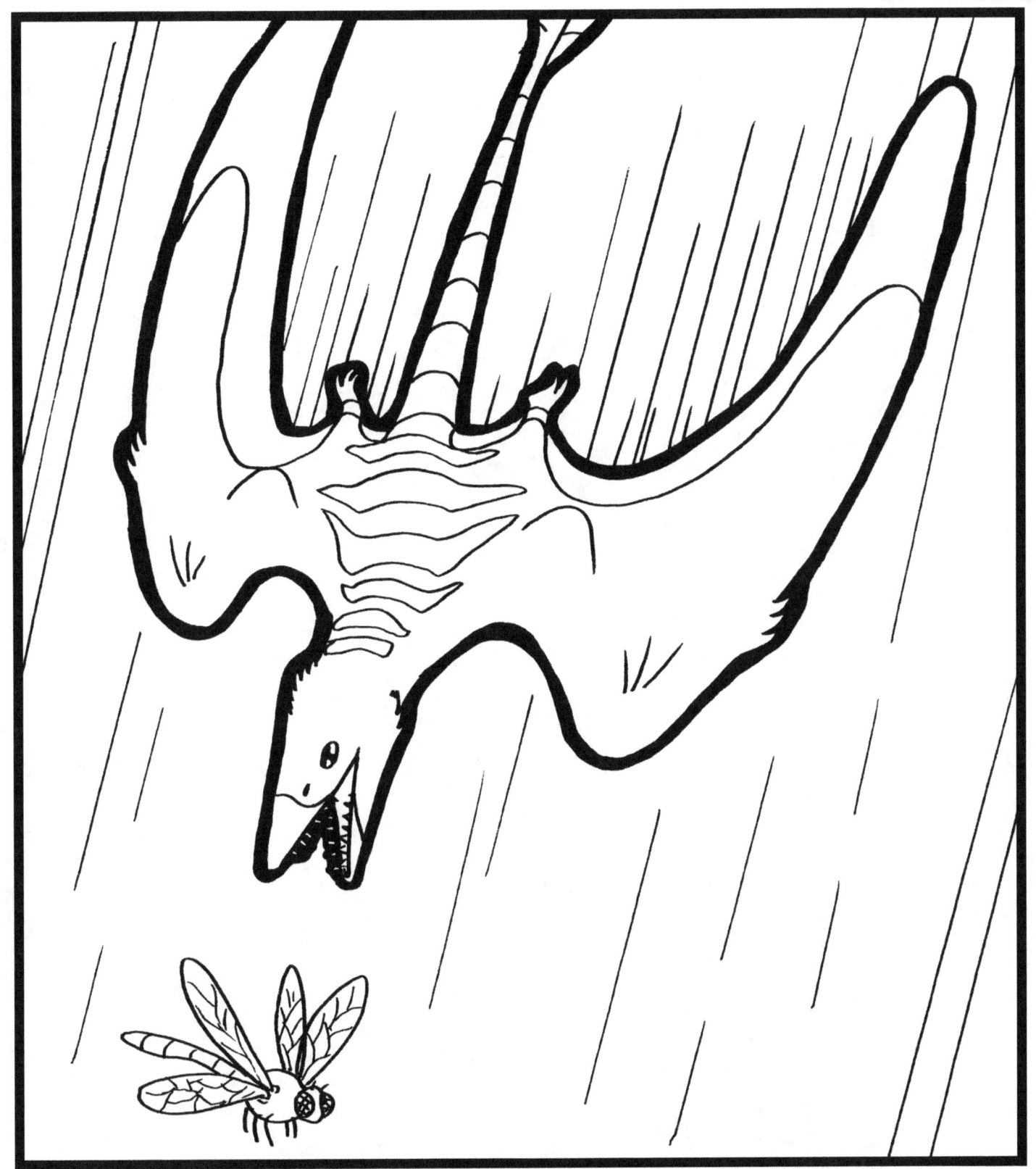

EUDIMORPHODON

210 - 203 MYA, Late Triassic, Italy, France, Luxembourg

A small pterosaur with a wingspan of just over a meter, its teeth had the best fit of all other pterosaurs when its mouth was closed. It is one of the most abundant fossils in Italy. It was also one of the earliest pterosaurs, and like many early members of the group, was a piscivore.

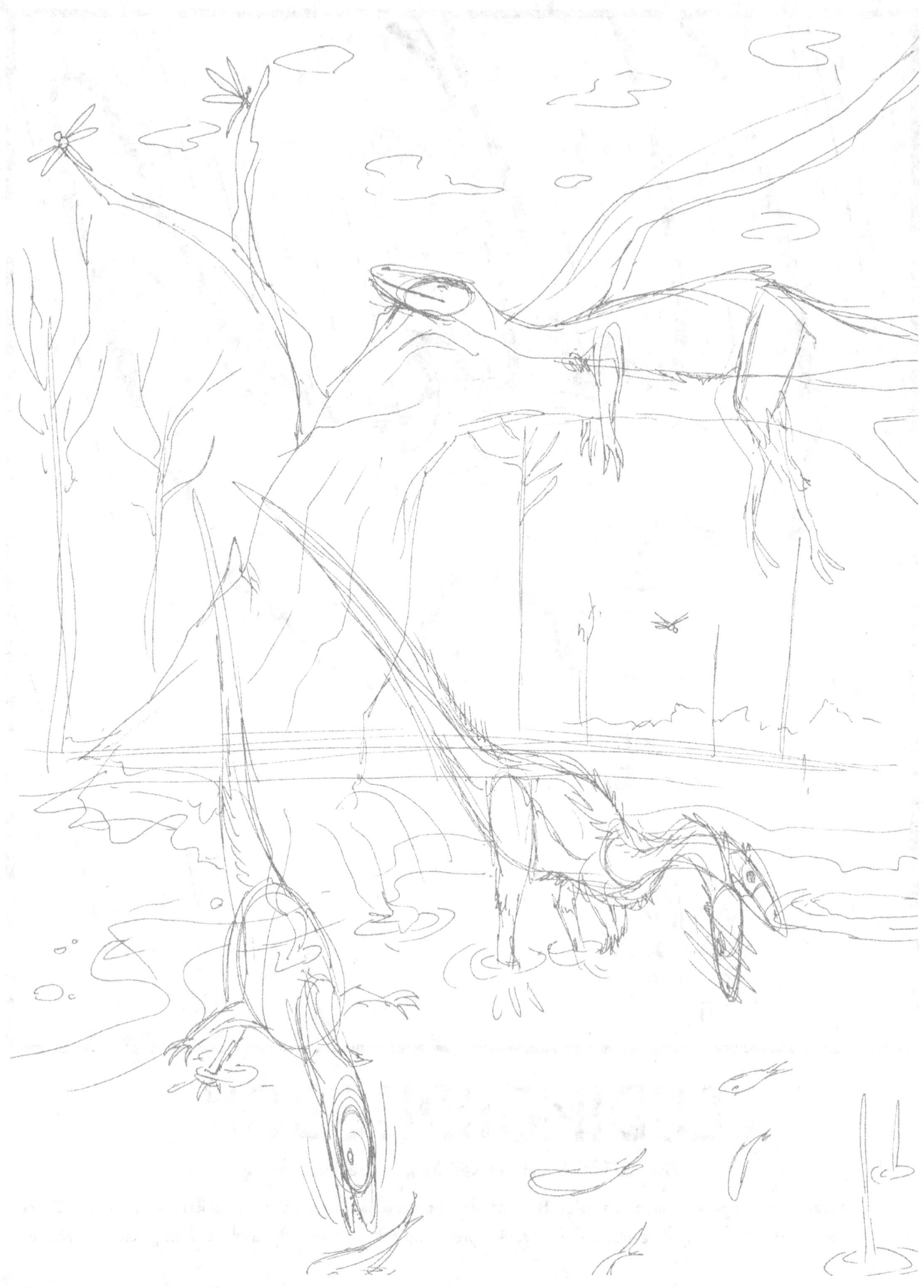

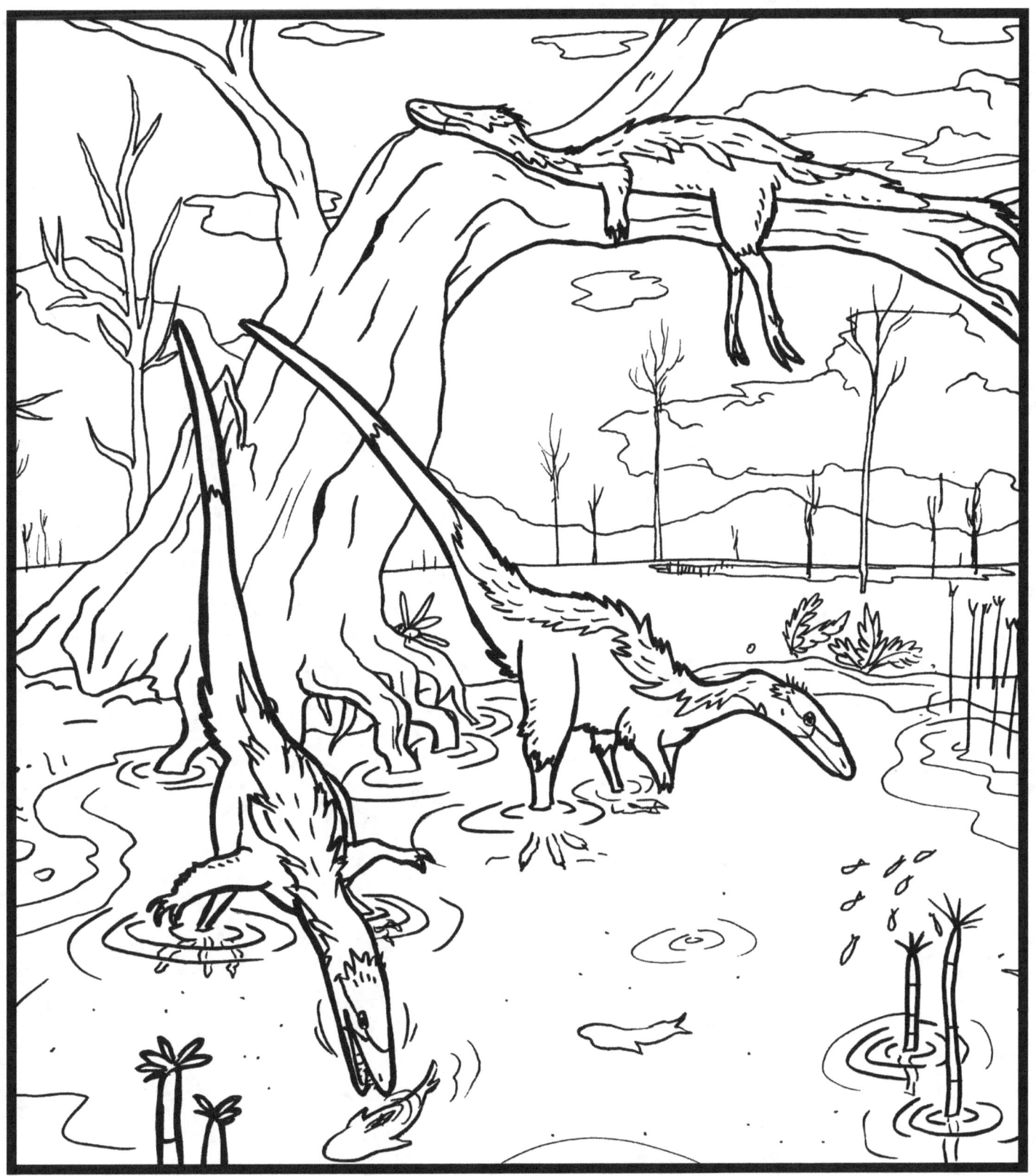

COELOPHYSIS

203 - 196 MYA, Late Triassic to Early Jurassic, United States

This 3 meter-long speedy carnivore had eyesight akin to the hawks and eagles of today, with excellent color vision and poor low-light vision. They hunted small, swift prey, and may have ventured into shallow water to catch fish. Their environment consisted of floodplains with distinct wet and dry seasons.

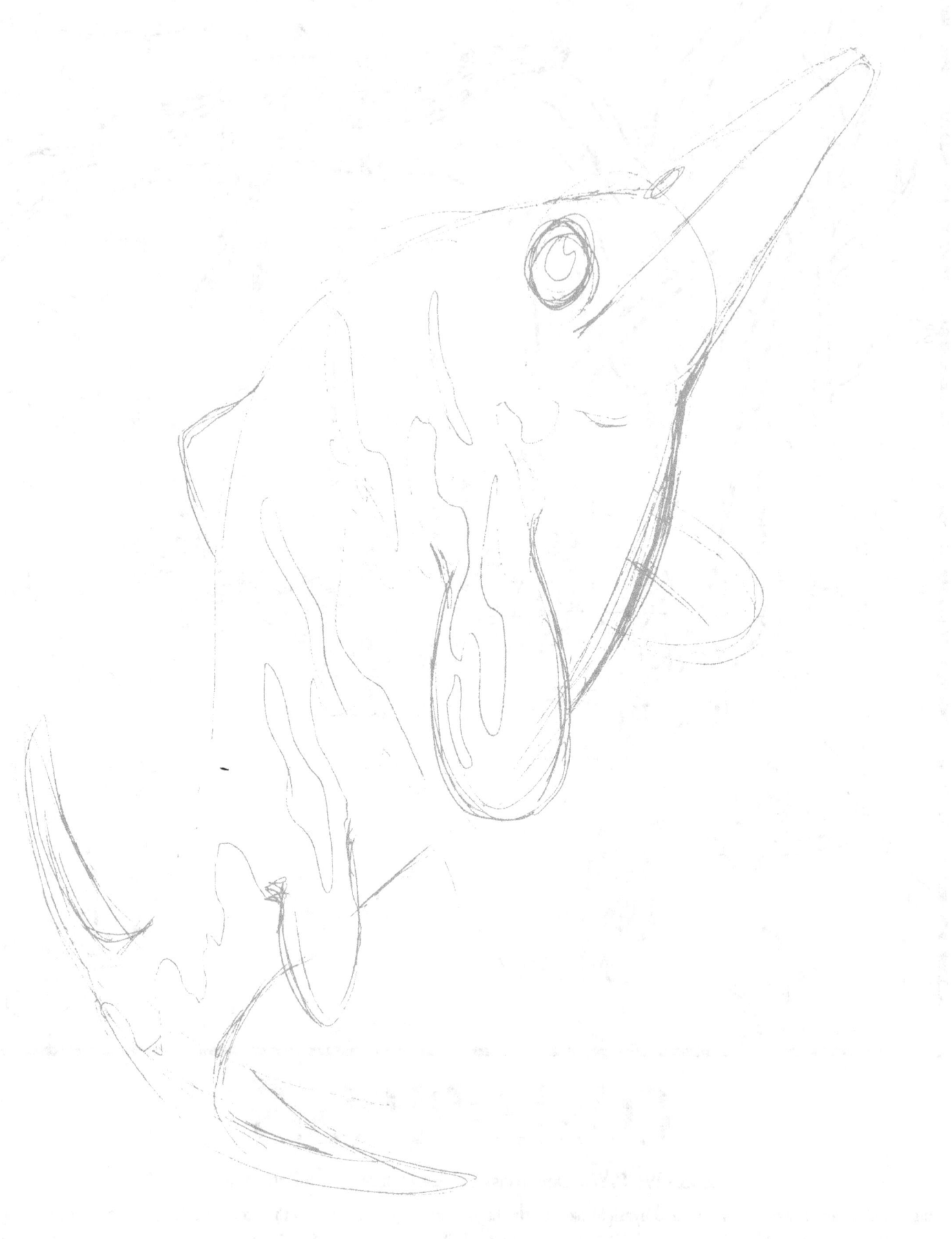

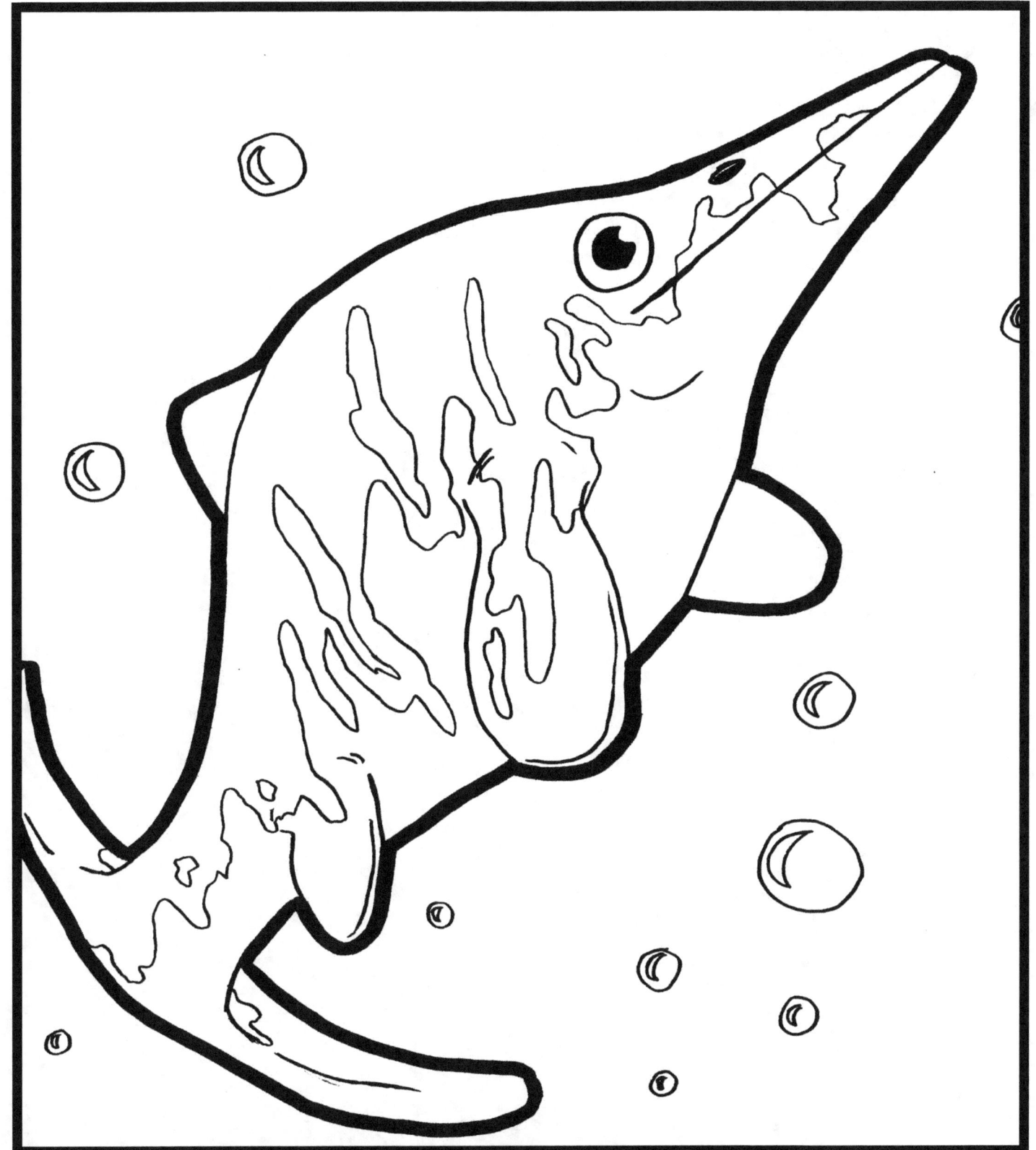

OPTHALMOSAURUS

201.5 - 145.5 MYA, Jurassic, England, France, Greenland, Mexico, United States

At 6 meters long, this common, dolphin-shaped squid-eater lived through almost the entire Jurassic period. Its eyes had a diameter of 23 centimeters, and took up nearly all of the space inside the skull, with even the name of this "eye-lizard" referencing their size. These huge eyes allowed them to hunt in deep waters or at night, when squid tend to be active.

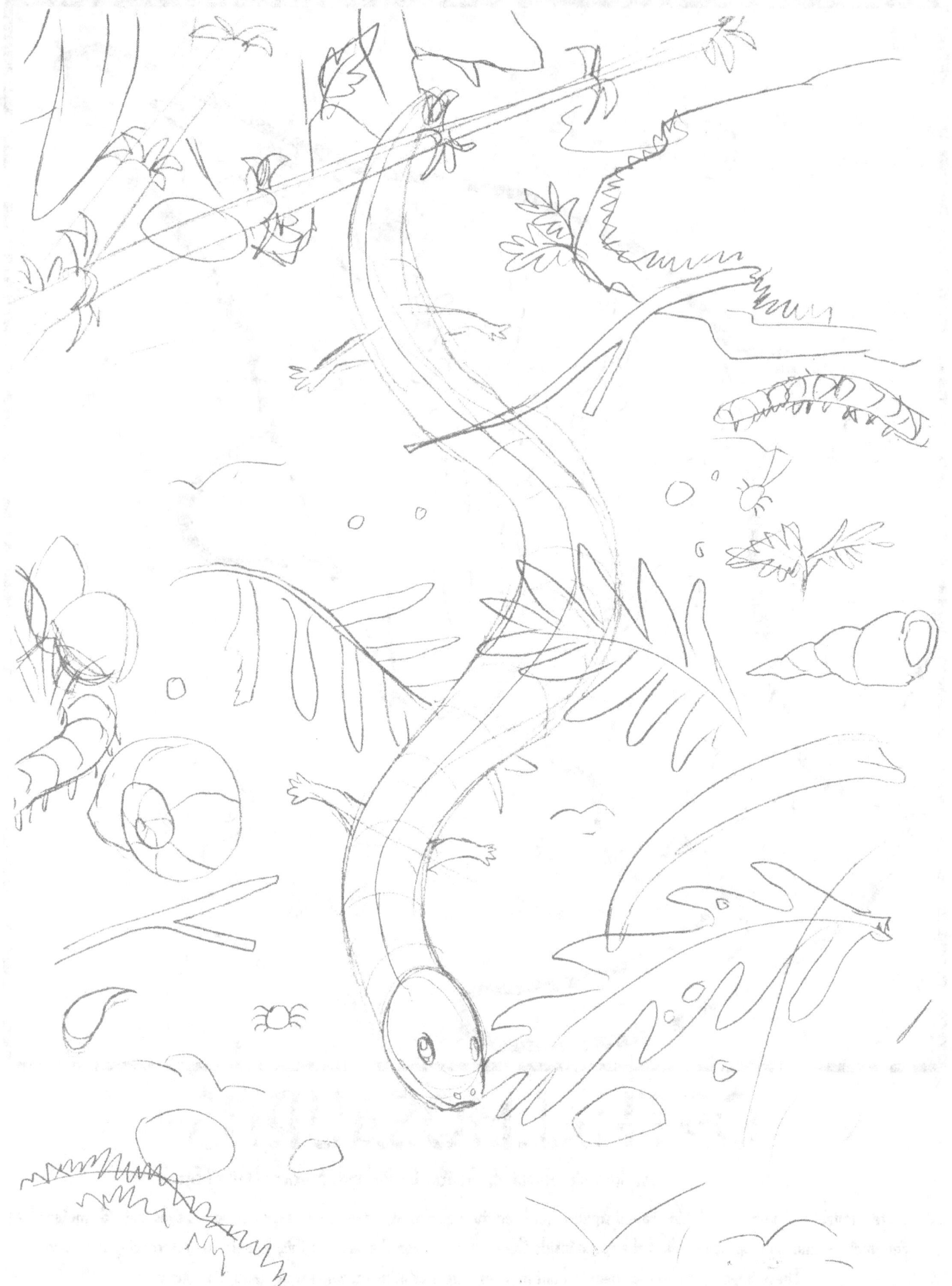

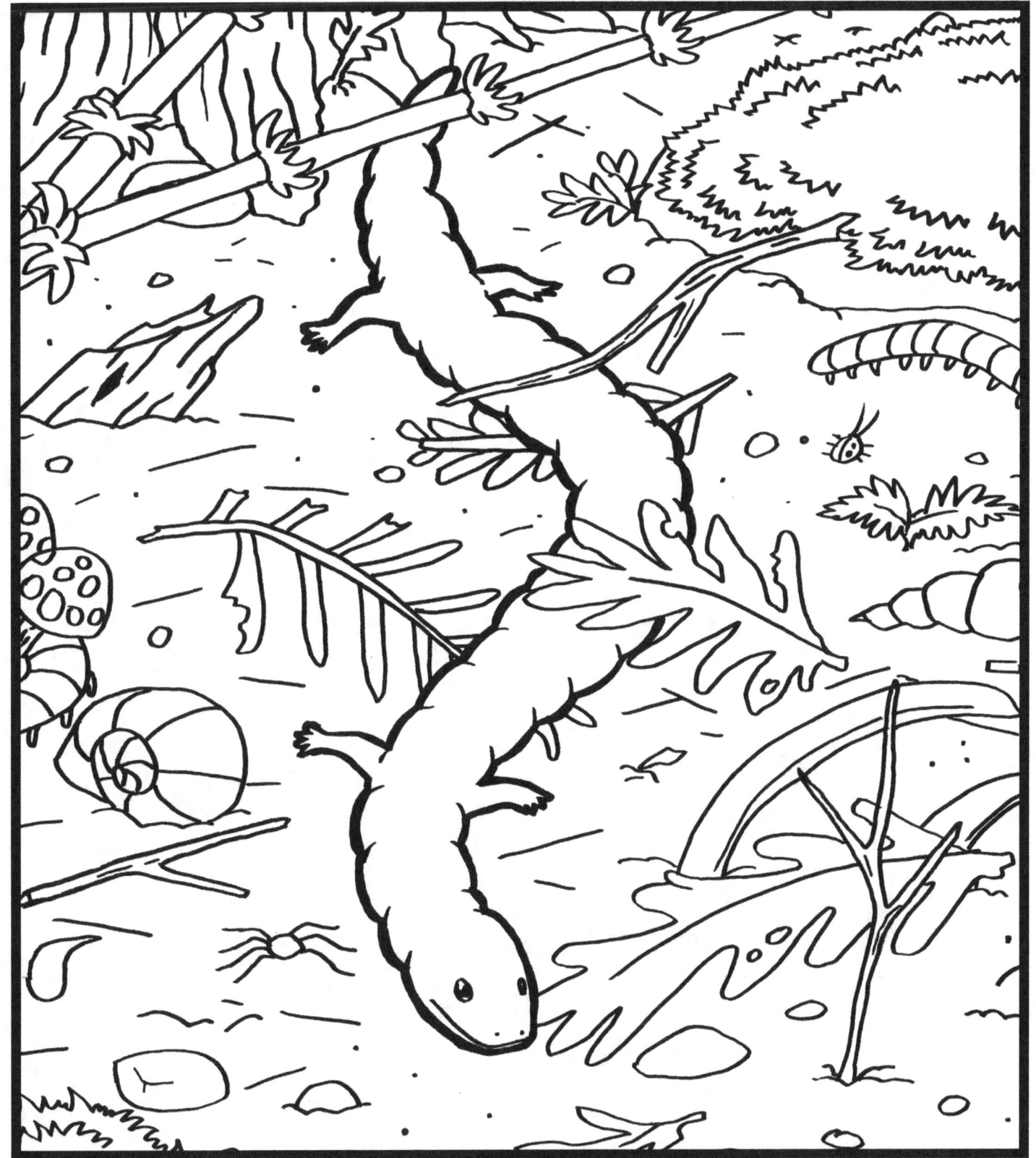

EOCAECILIA

199.6 - 175.6 MYA, Early Jurassic, United States

Although it appeared to look like some sort of snake ancestor, this was an amphibian that likely lived in leaf litter, searching for invertebrates and staying out of sight of larger animals. It reached a length of up to 15 centimeters, and the fossil specimen of this animal was remarkably well-preserved for something as ancient as it.

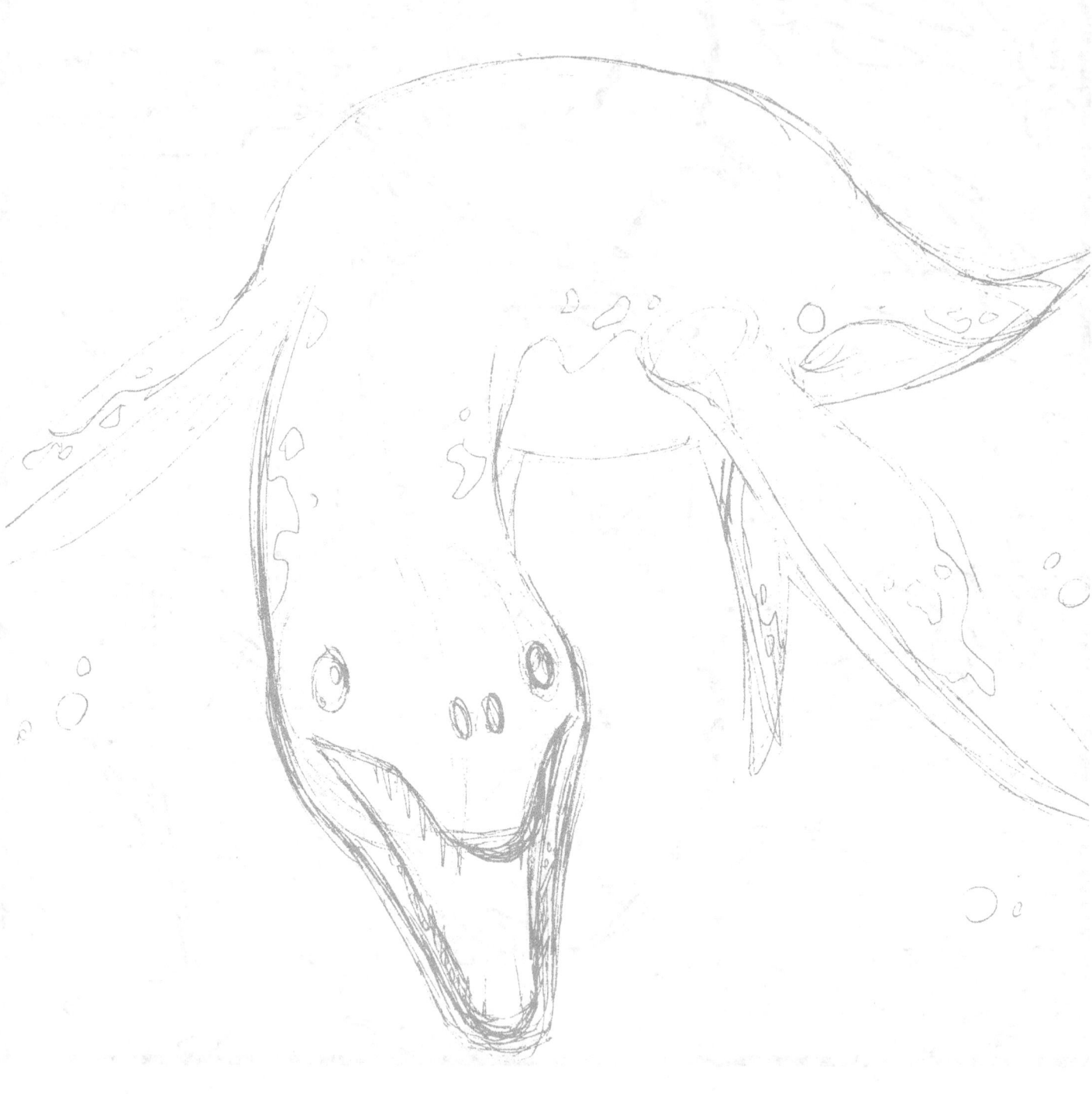

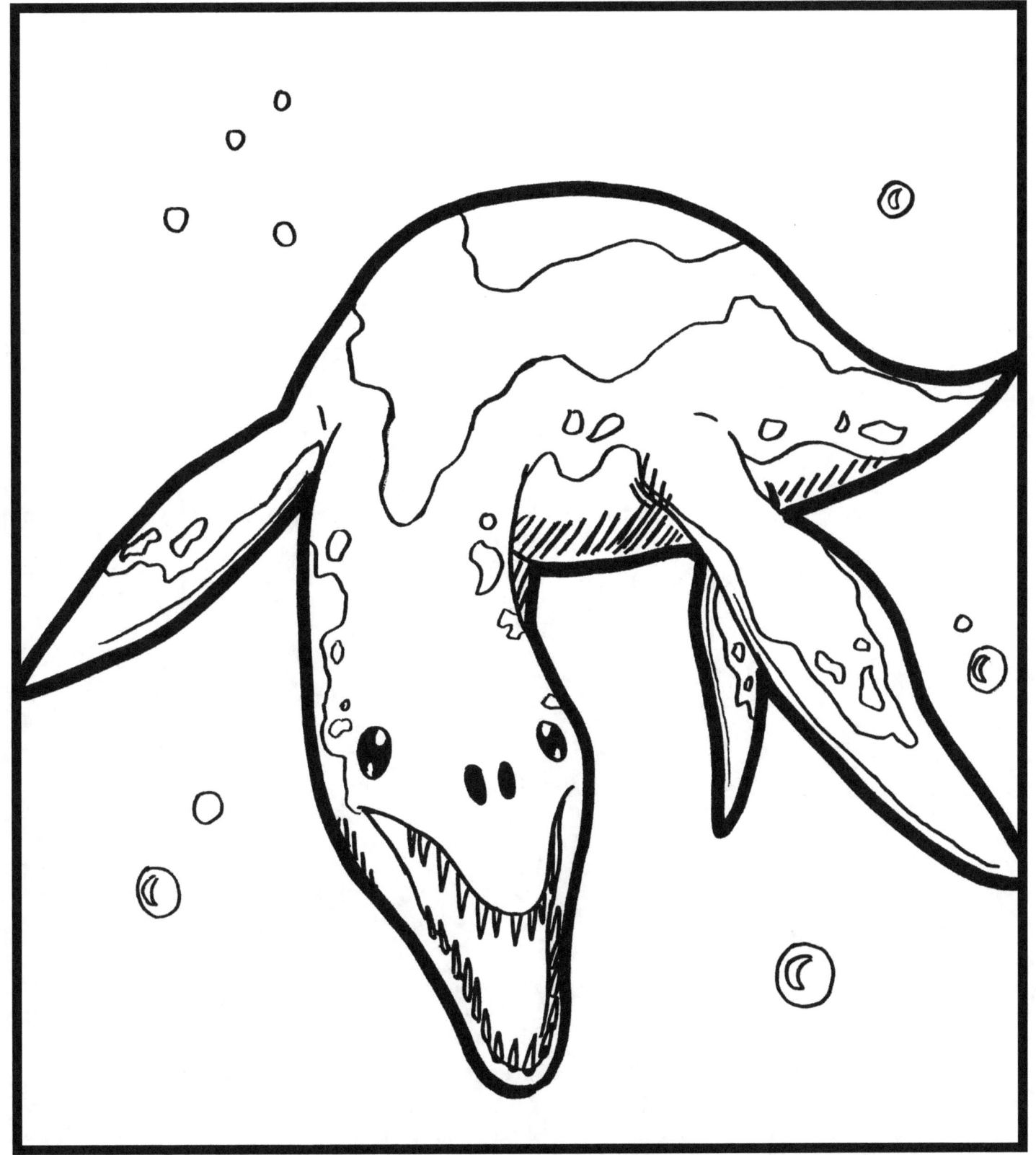

MACROPLATA

199 - 195 MYA, Early Jurassic, England

Compared to other pliosaurs at the time, this marine reptile had a long neck that was twice as long as its skull. It lived off a diet of fish, and used its powerful swimming muscles to lunge at individuals that broke away from synchronized shoals.

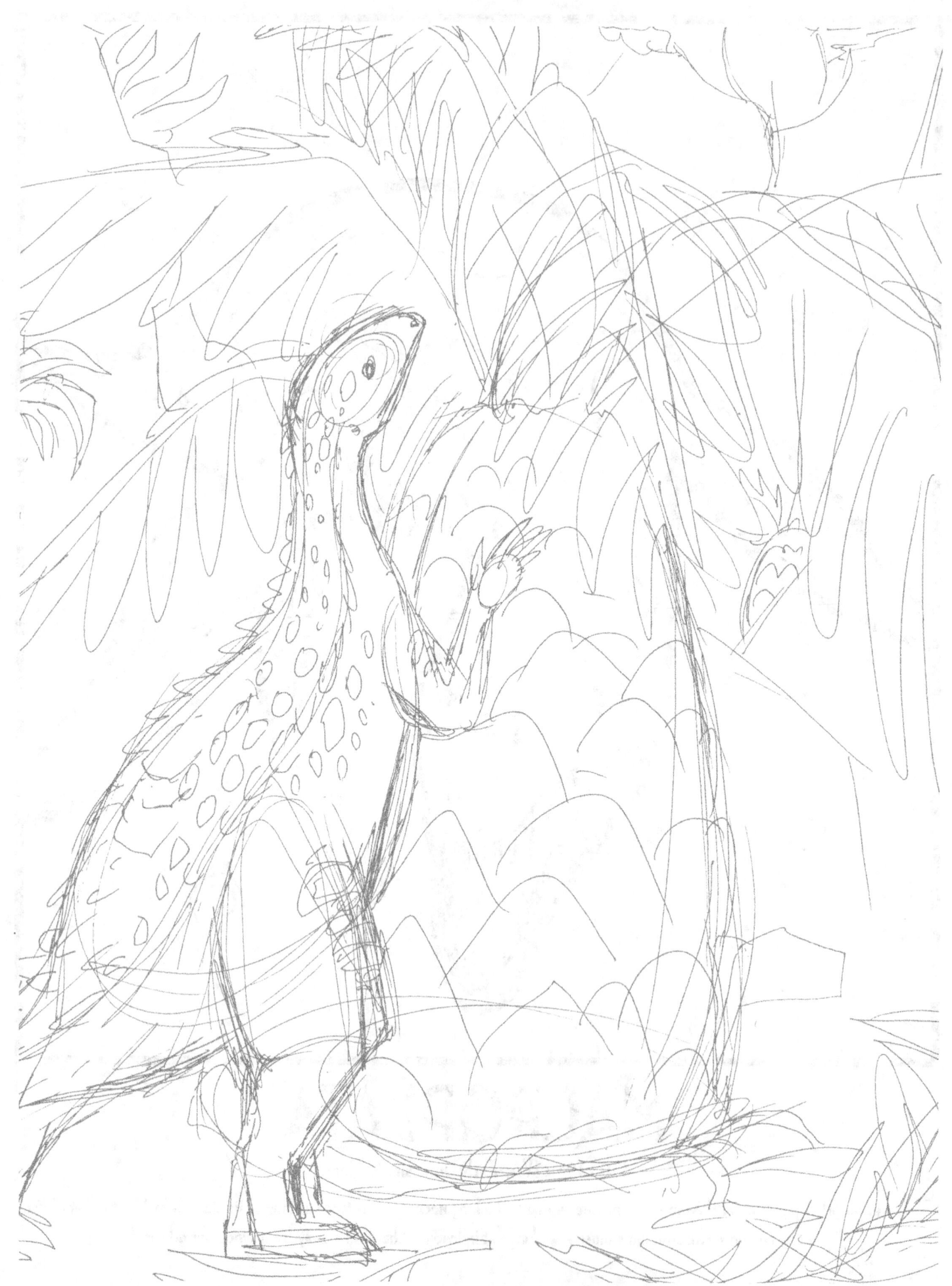

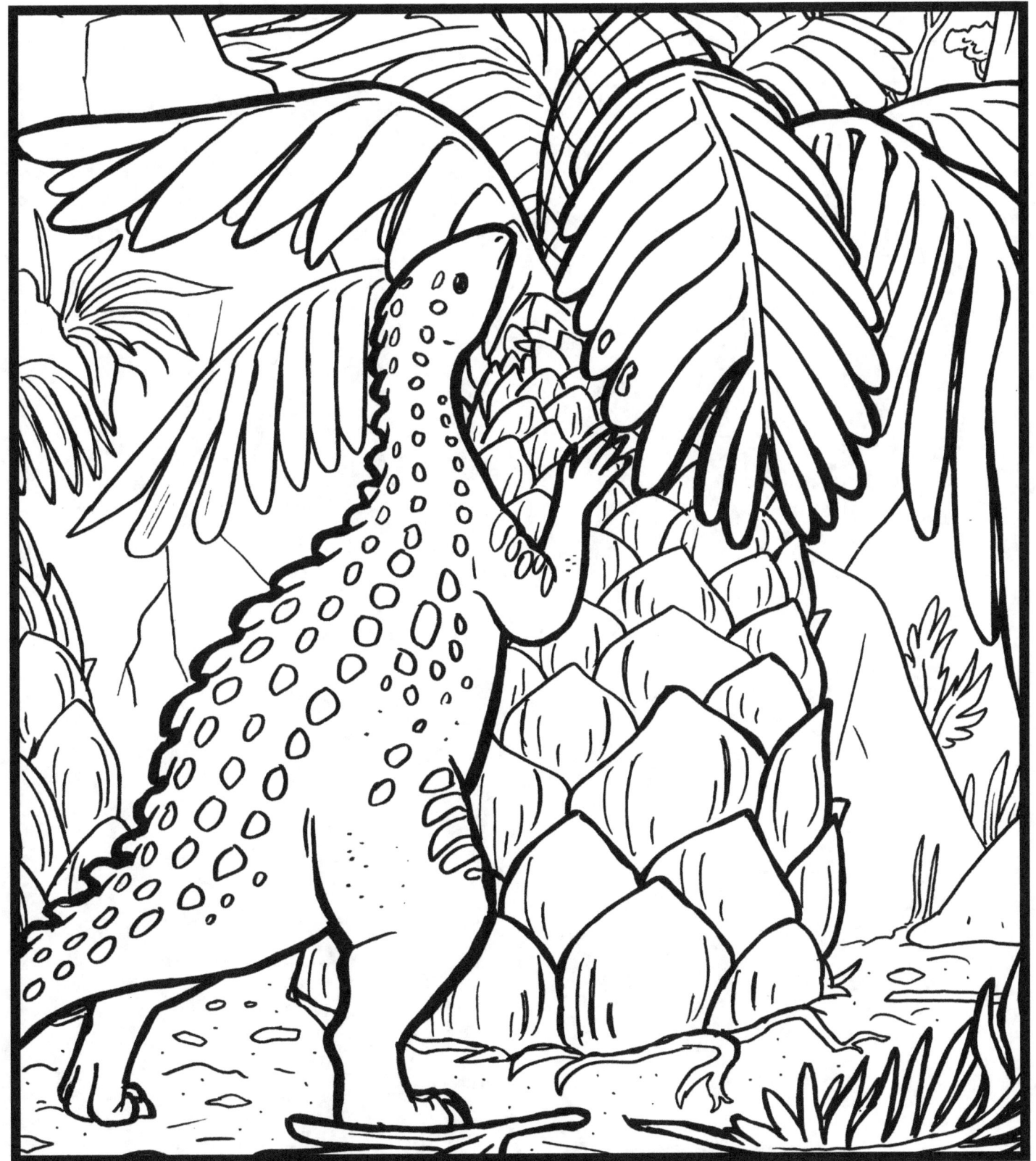

SCUTELLOSAURUS

196 MYA, Early Jurassic, United States

This early "little shielded lizard" had several hundred osteoderms running down the length of its back and tail.
Despite this, it was a small, lightly-built bipedal herbivore, at just over a meter long.

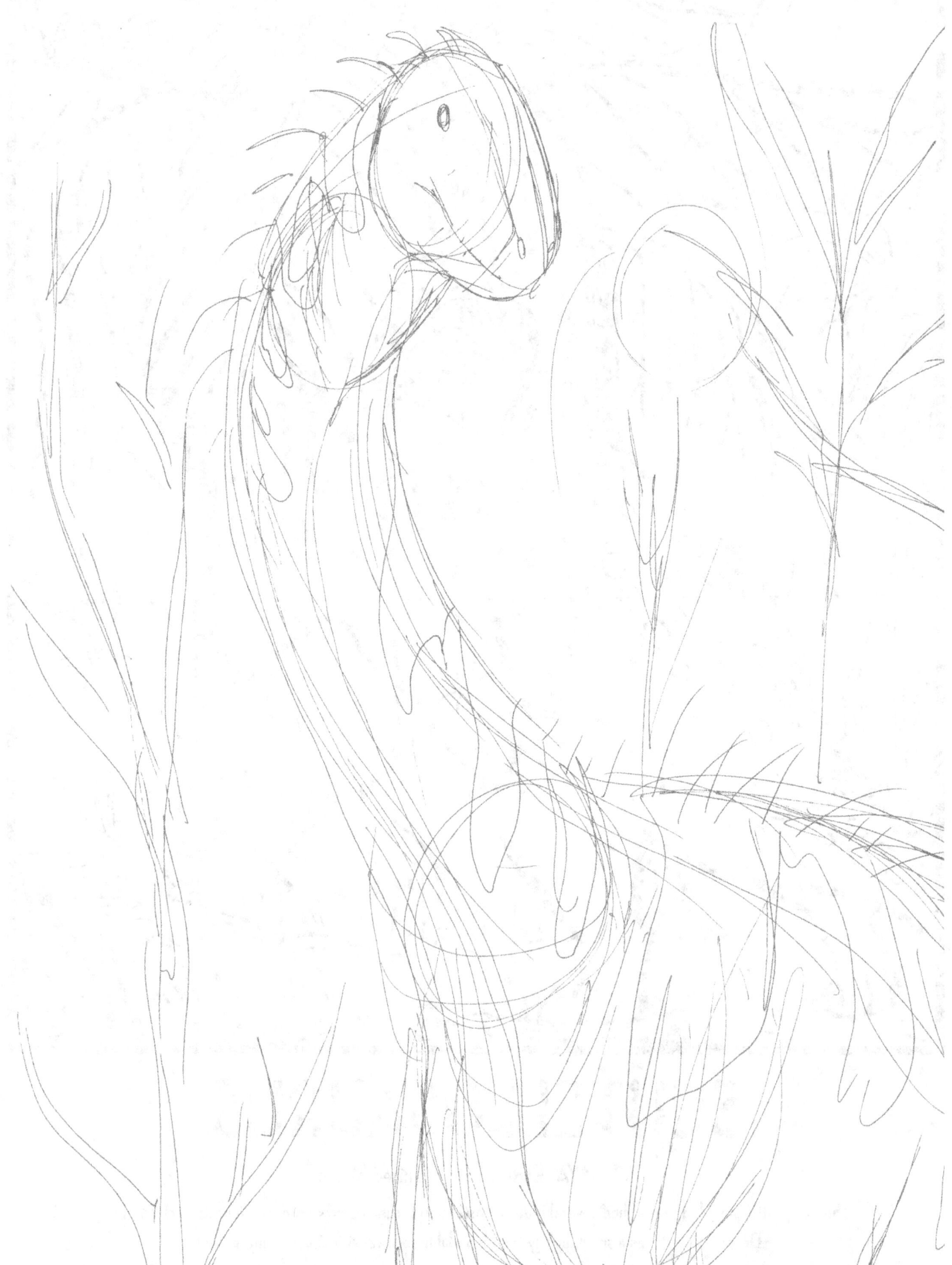

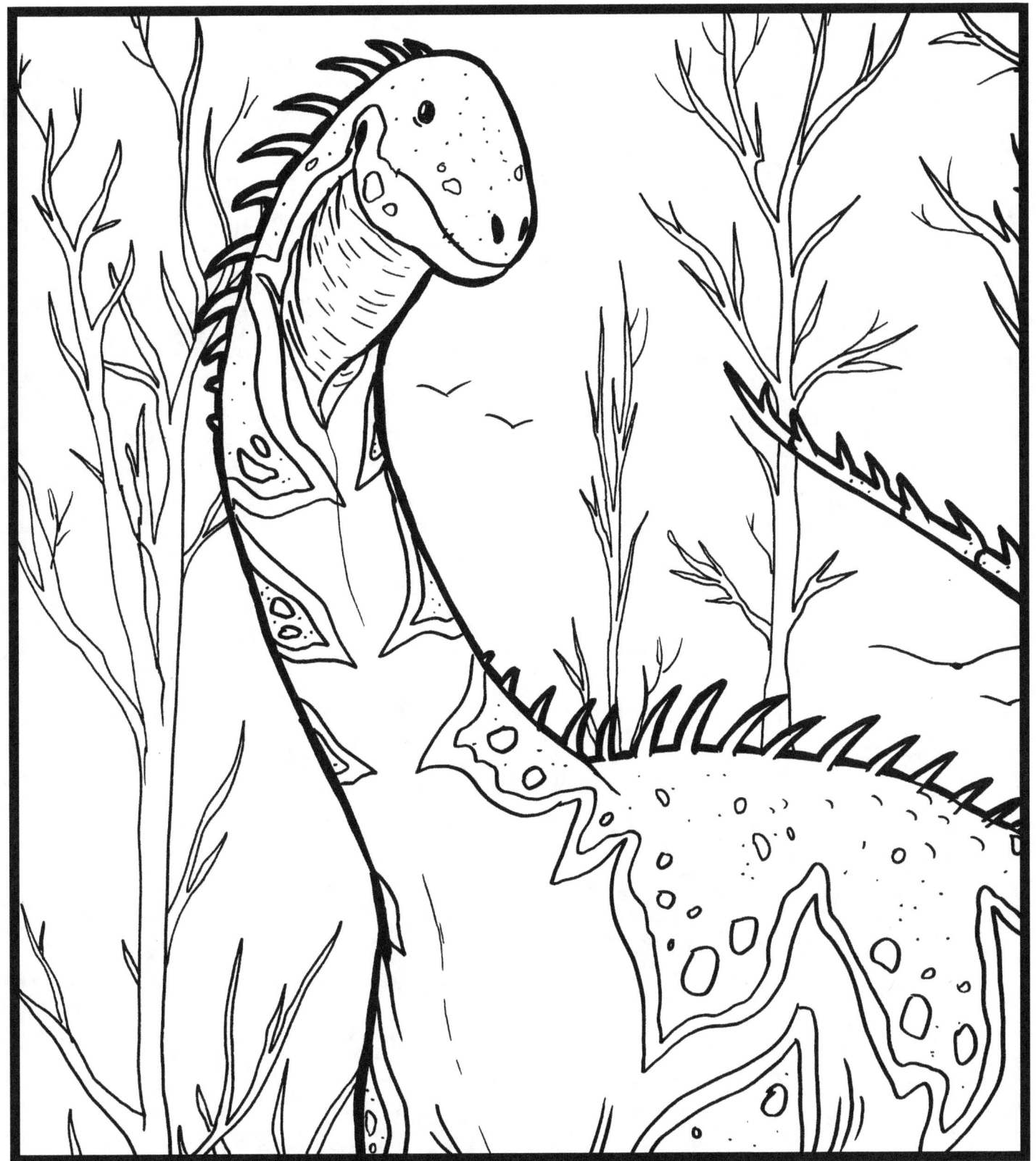

BARAPASAURUS

196 - 183 MYA, Early Jurassic, India

A sauropod with one of the most completely known skeleton of its group in the early Jurassic, the size of Barapasaurus was comparable to later members, at around 14 meters in length. Even this early in sauropod evolution, the skeleton showed hints of developing ways of overcoming the stresses of its body's sheer weight, such as the hollowing of the vertebrae.

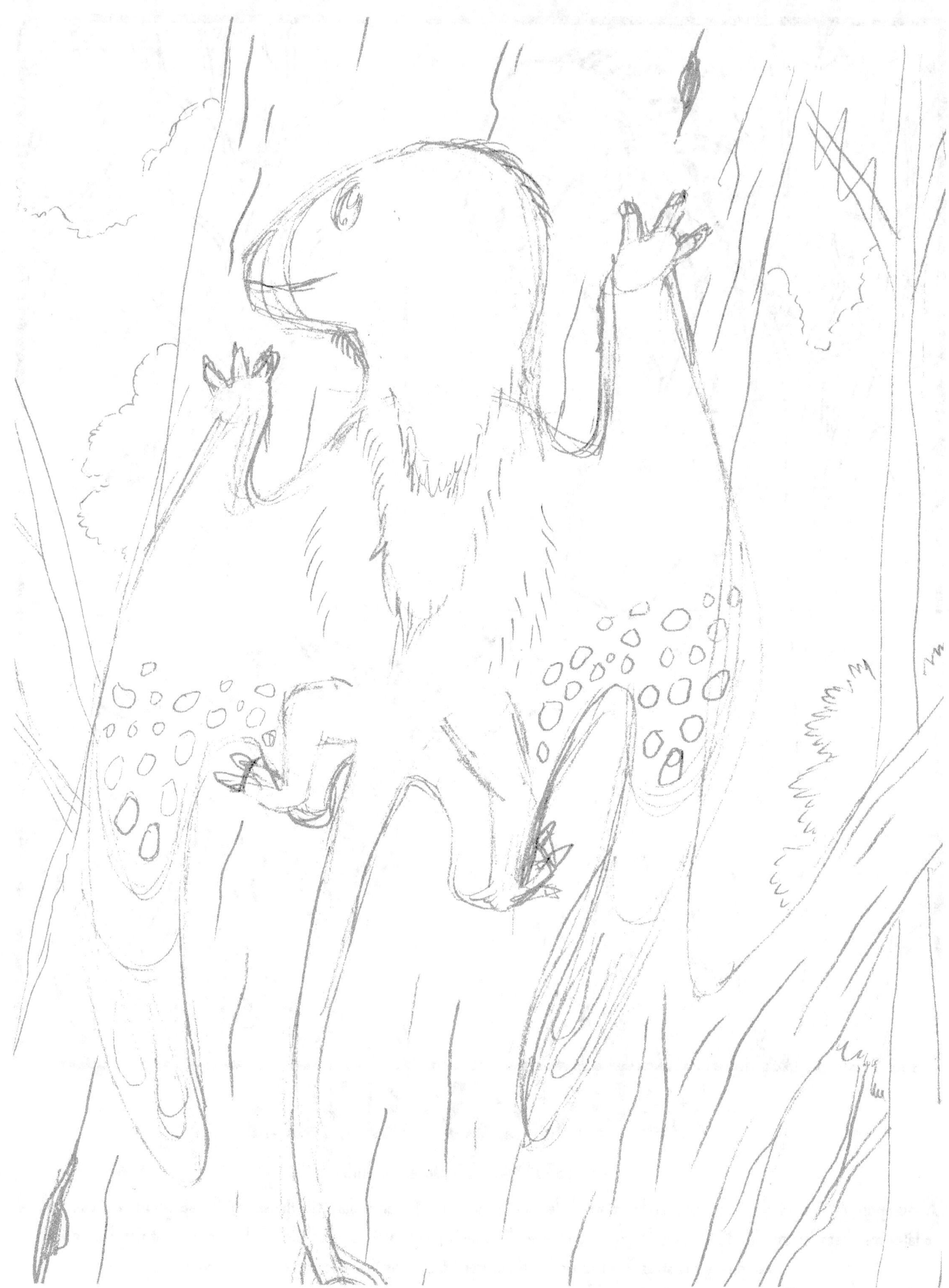

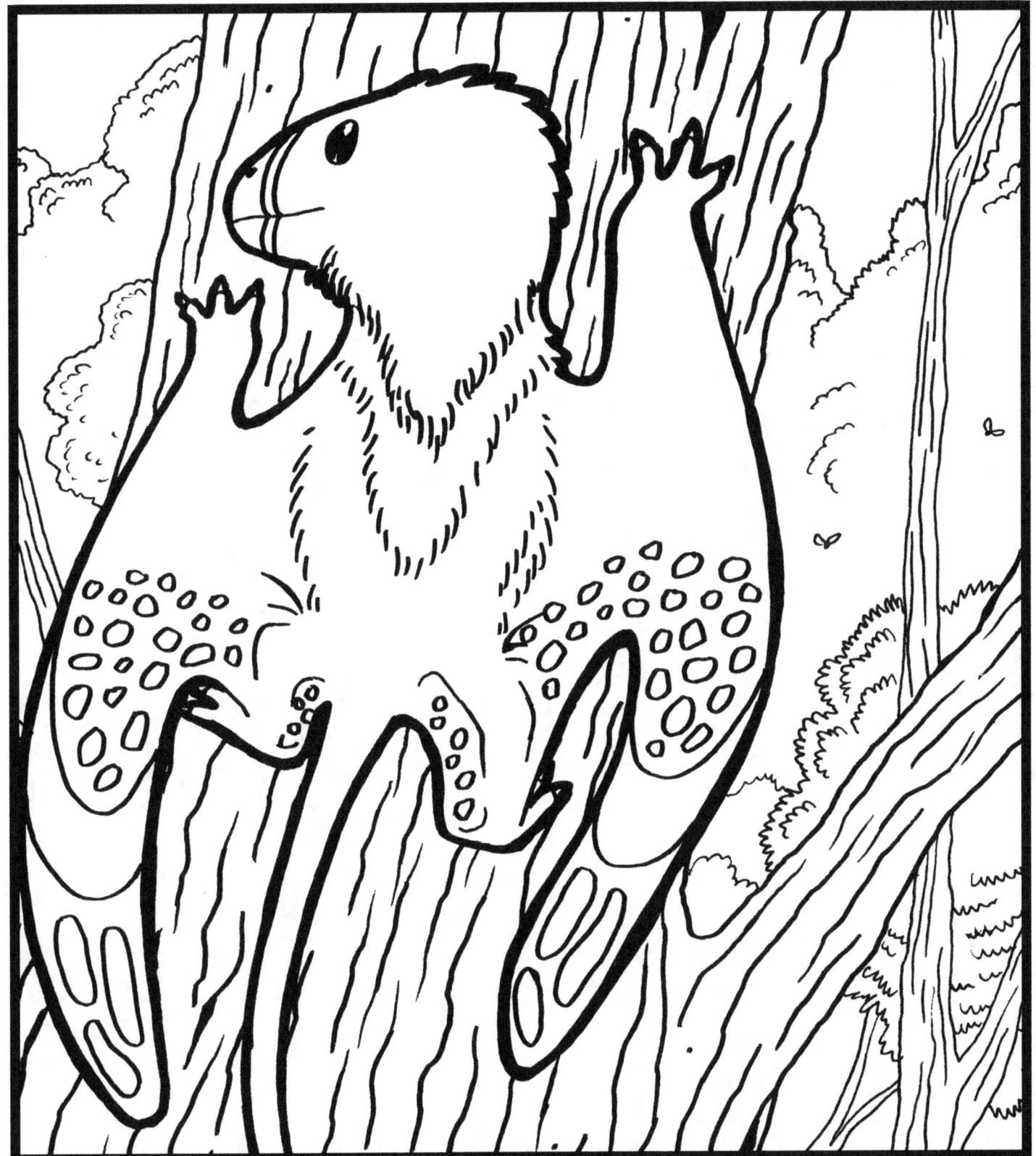

DIMORPHODON

195 - 190 MYA, Early Jurassic, Mexico, England

This pterosaur was active on the ground, in the air, and was even adept at climbing trees. Before its was discovered, it was assumed that pterosaurs, along with many other prehistoric animals, were sluggish, unintelligent beasts that were fated to extinction. Its well-developed limbs showed that it could get around on the ground or in the trees just fine.

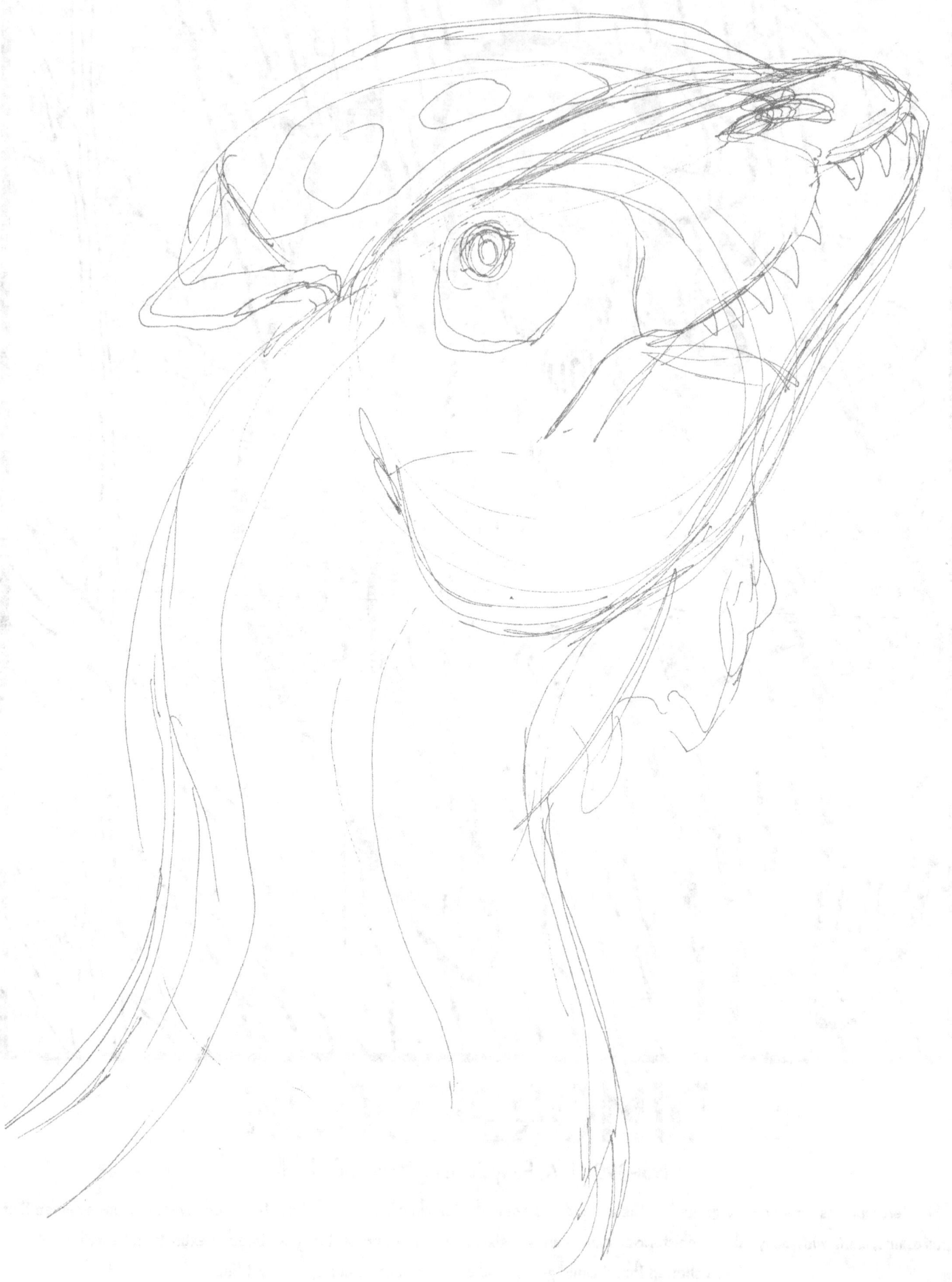

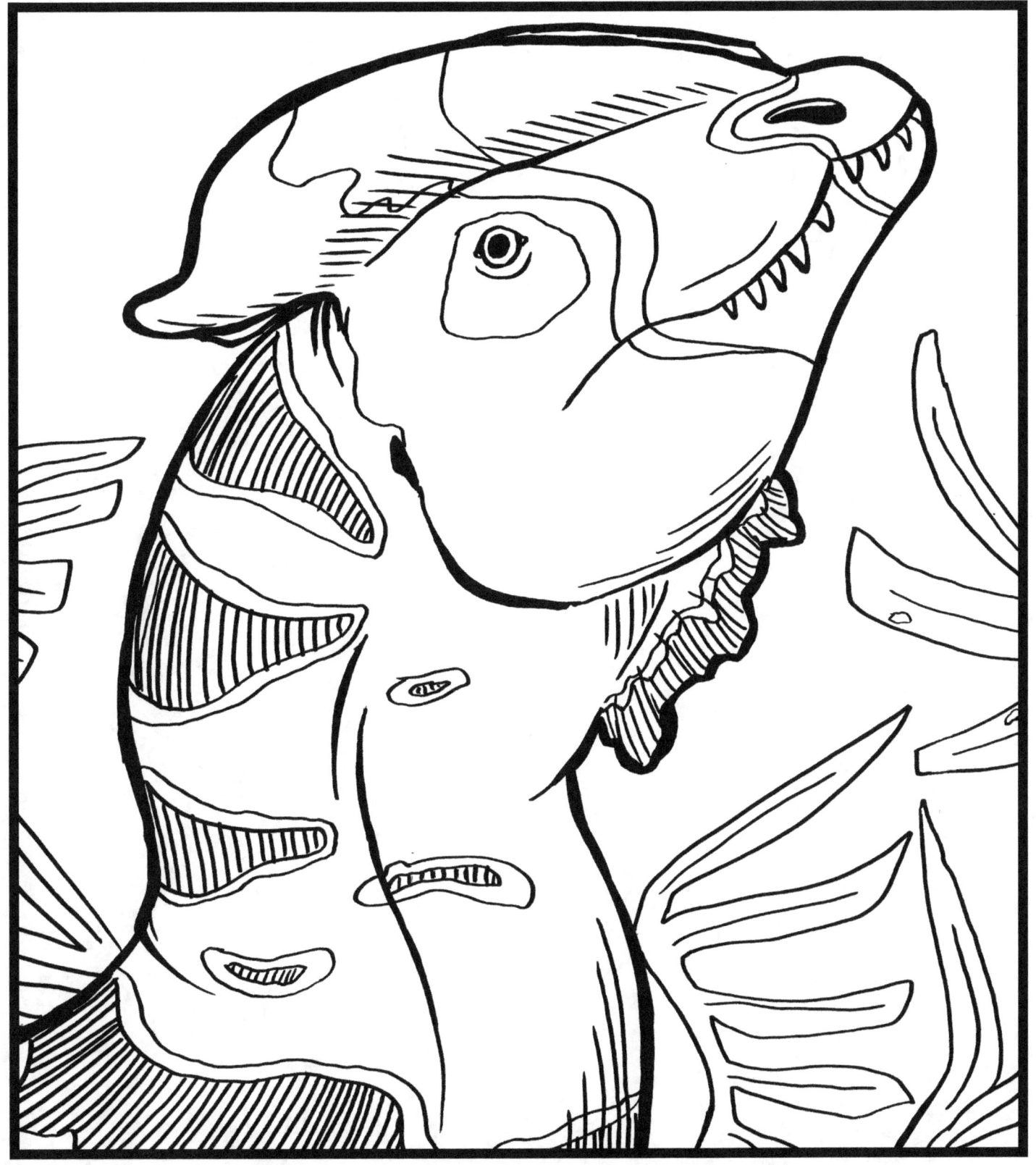

DILOPHOSAURUS

193 MYA, Early Jurassic, United States

This predator was one of the largest carnivorous dinosaurs of its time, at 7 meters in length. No significant differences have been found between fossils that would suggest dimorphism between male and female skeletal systems. Because of similarities it seems to share with spinosaurs and its proximity to water in life, it has been suggested that it was a piscivore.

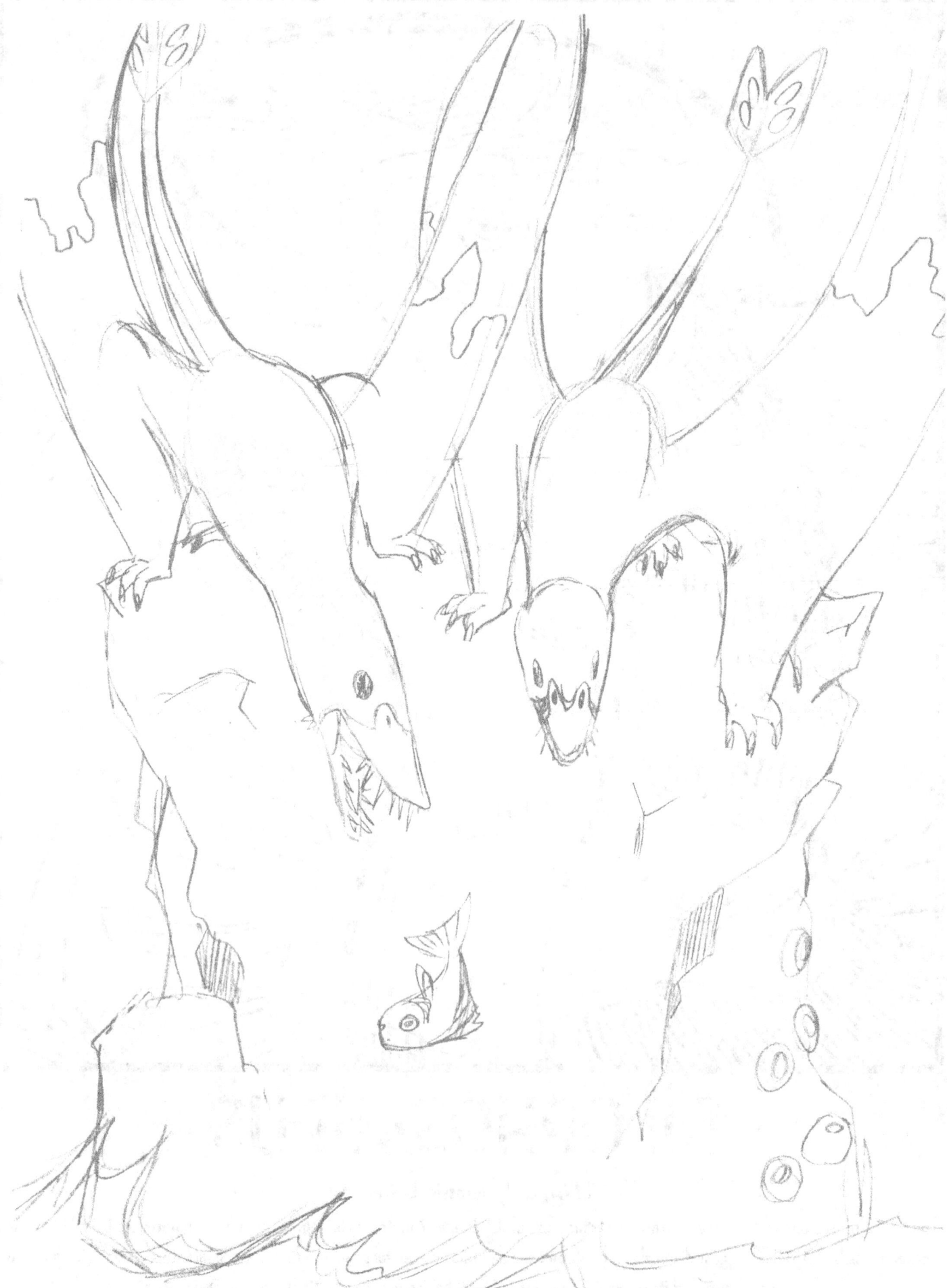

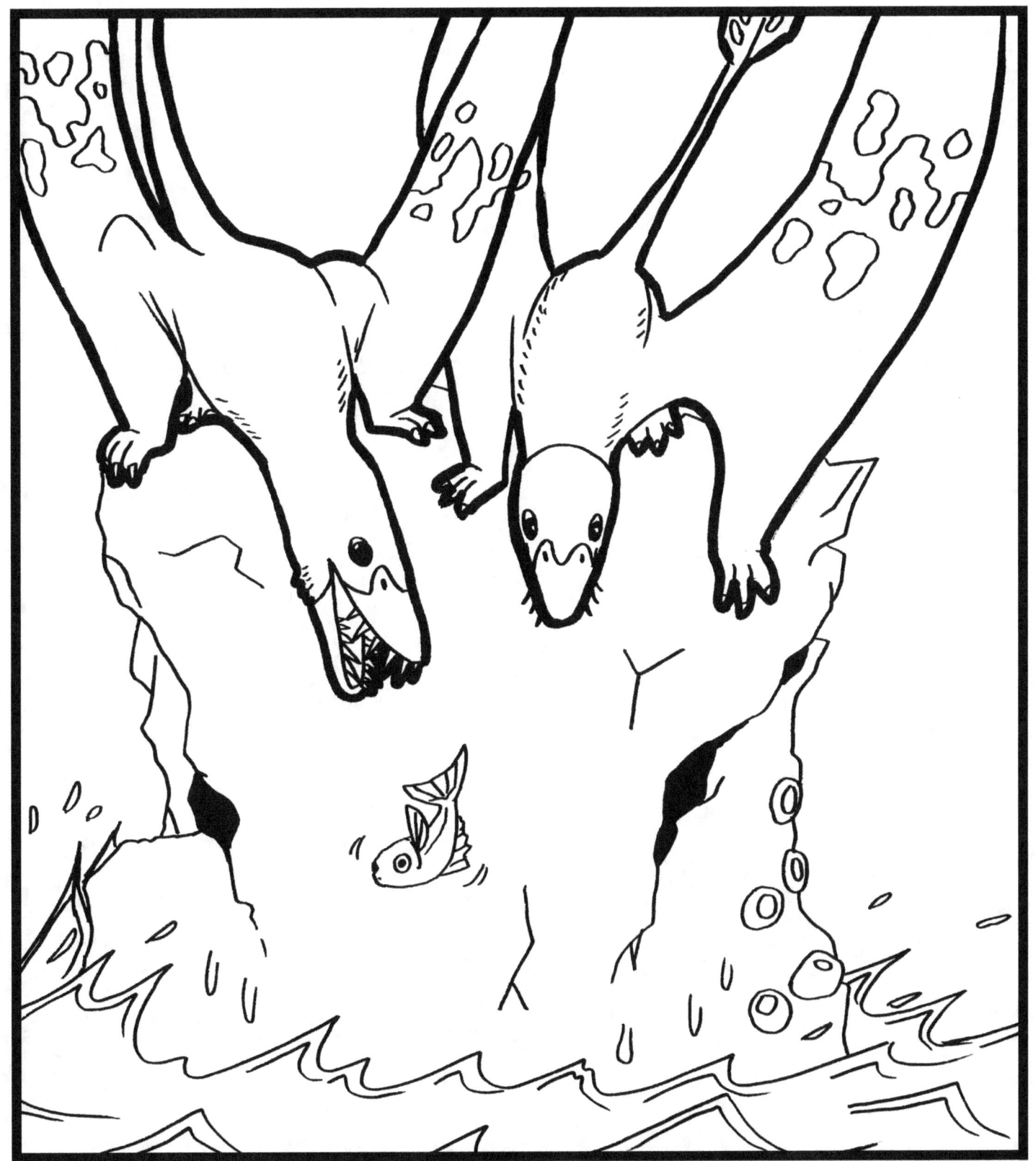

DORYGNATHUS

180 MYA, Early Jurassic, Germany and France

This pterosaur's most notable trait were its long, conical teeth, which pointed away from the mouth, which it used to hold onto slippery fish, squid, and other sea creatures. It was also a member of the Rhamphorhynchinae, a family of long-tailed, short necked pterosaurs that also generally lacked a bony crest on the head.

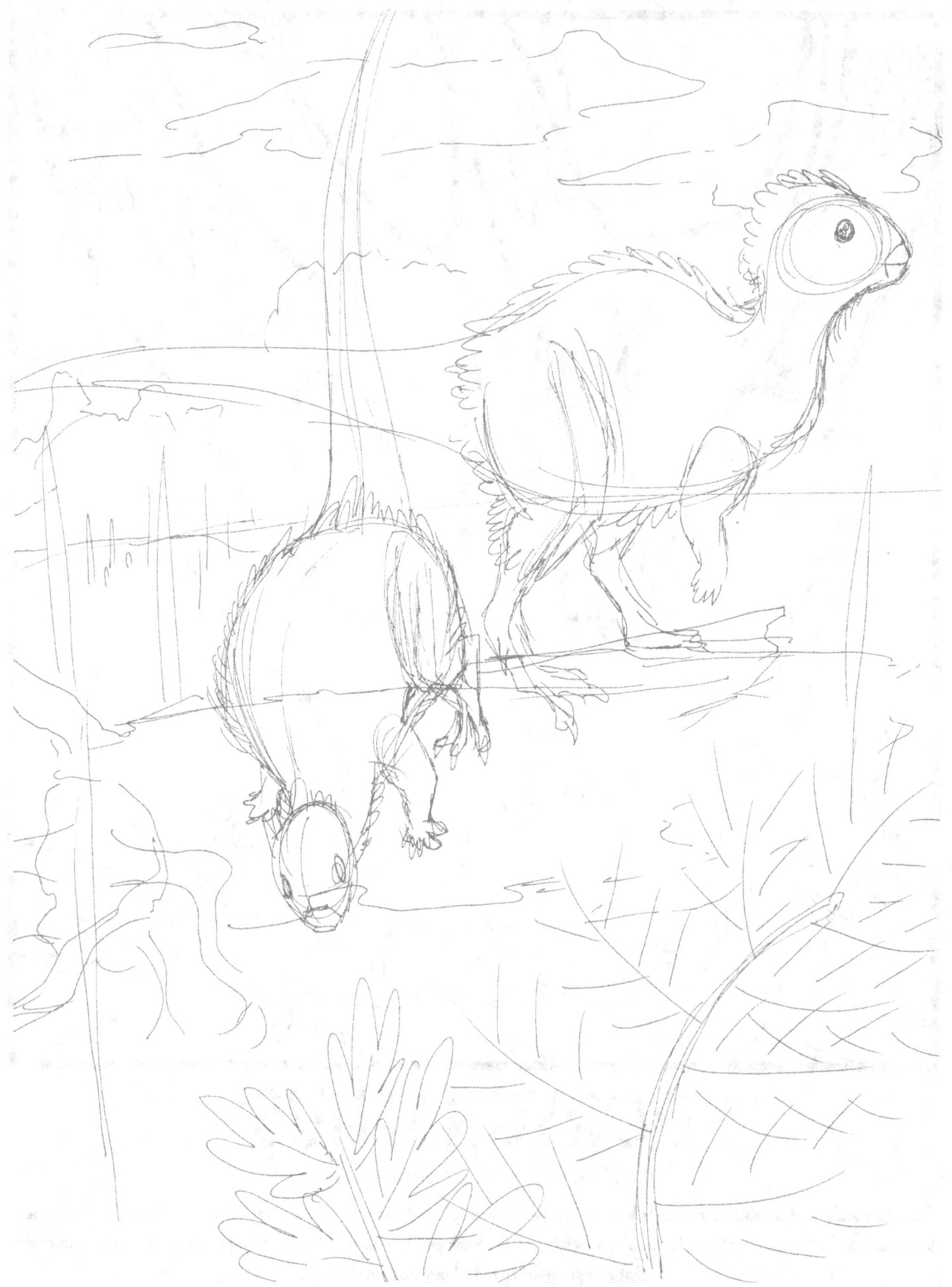

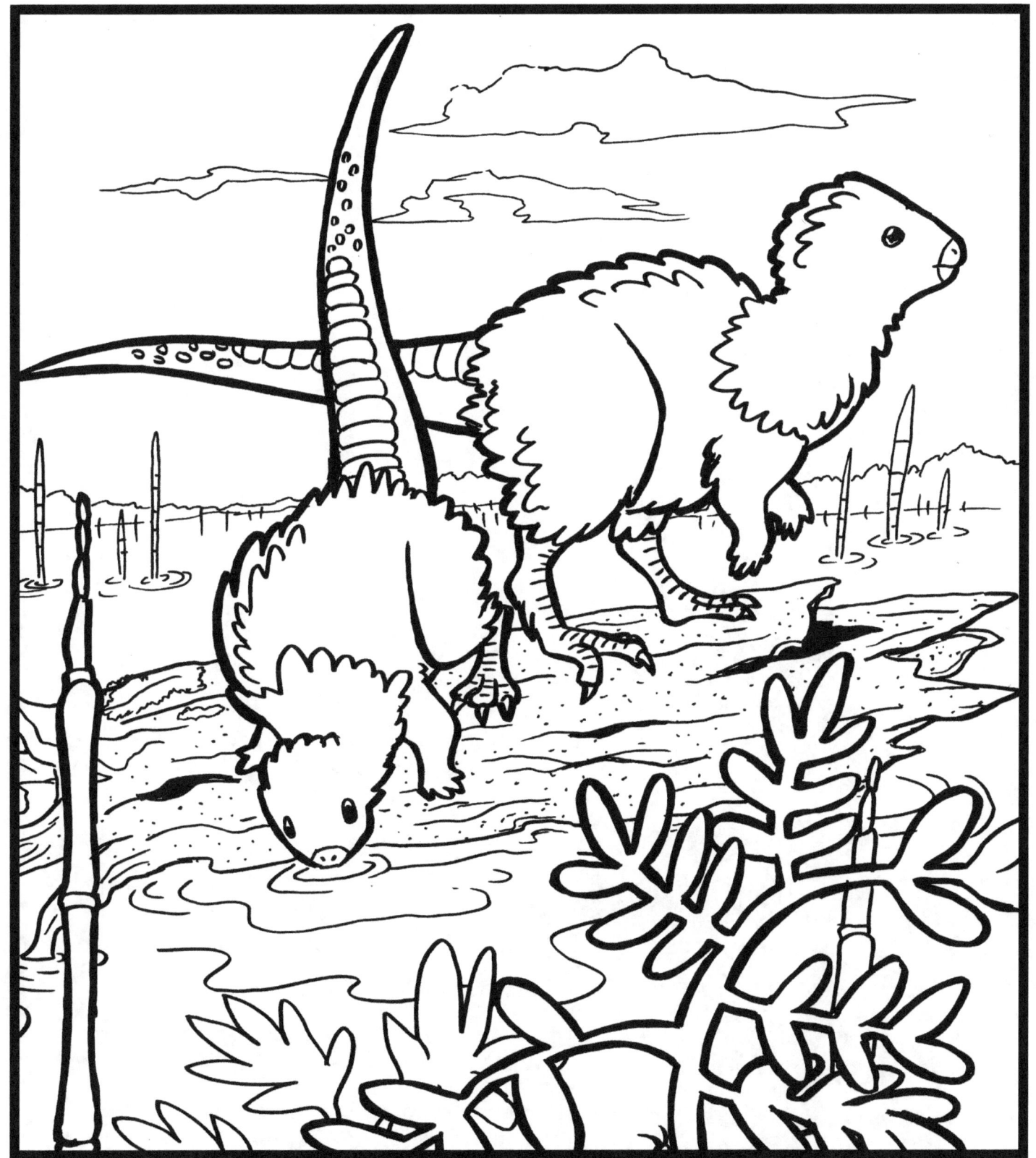

KULINDADROMEUS

169 - 144 MYA, Middle to Late Jurassic, Russia

This little herbivore was 1.5 meters long, and had a body covered by a layer of dinofuzz, while its tail had a scaly covering.
This discovery of an ornithischian dinosaur having a feathery coat points at this trait originating in a shared common ancestor of both
bird-hipped and lizard-hipped groups of dinosaur, rather than being exclusive to the theropods.

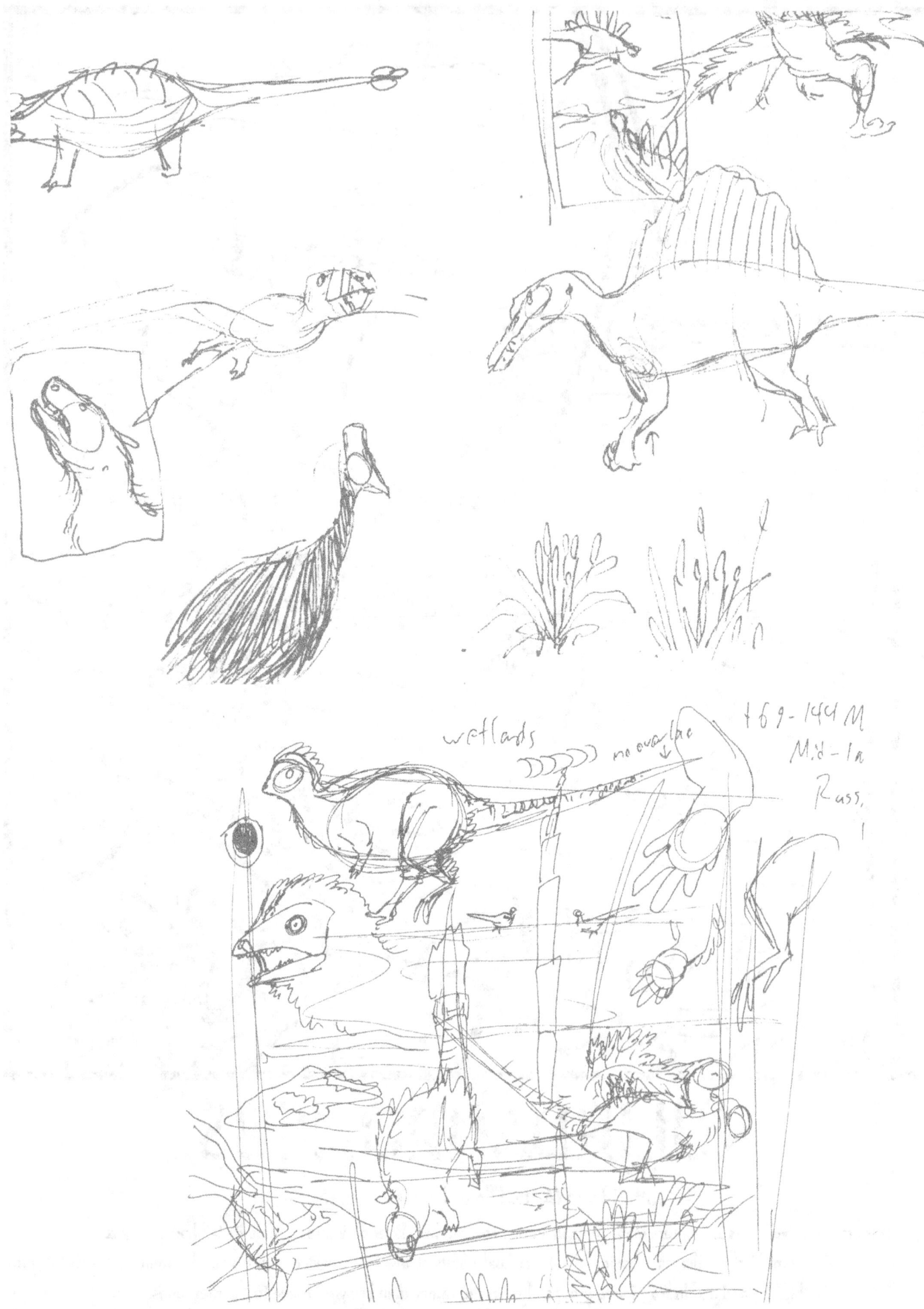

wetlands　　no overlap　†69-144M
Mid-la
Russ.

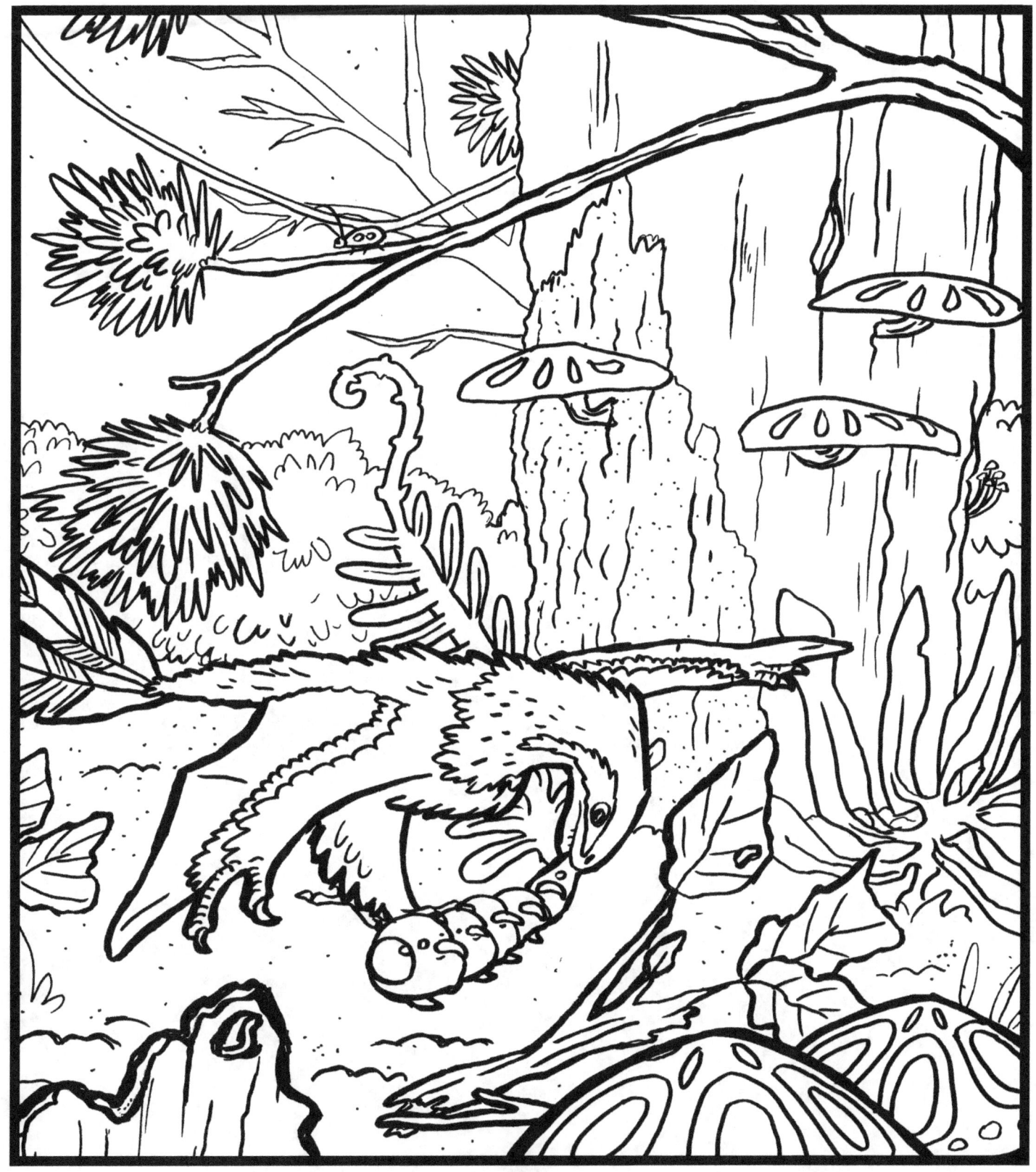

YI

160 MYA, Middle Jurassic, China

Yi is not only notable for having the shortest genus name of any dinosaur, but for the way its wings are presented. Yi had a feathery coating, and bones at its wrist supported a membrane that may have been used in gliding, with some flapping to control its descent. This was a very small dinosaur, weighing less than 400 grams.

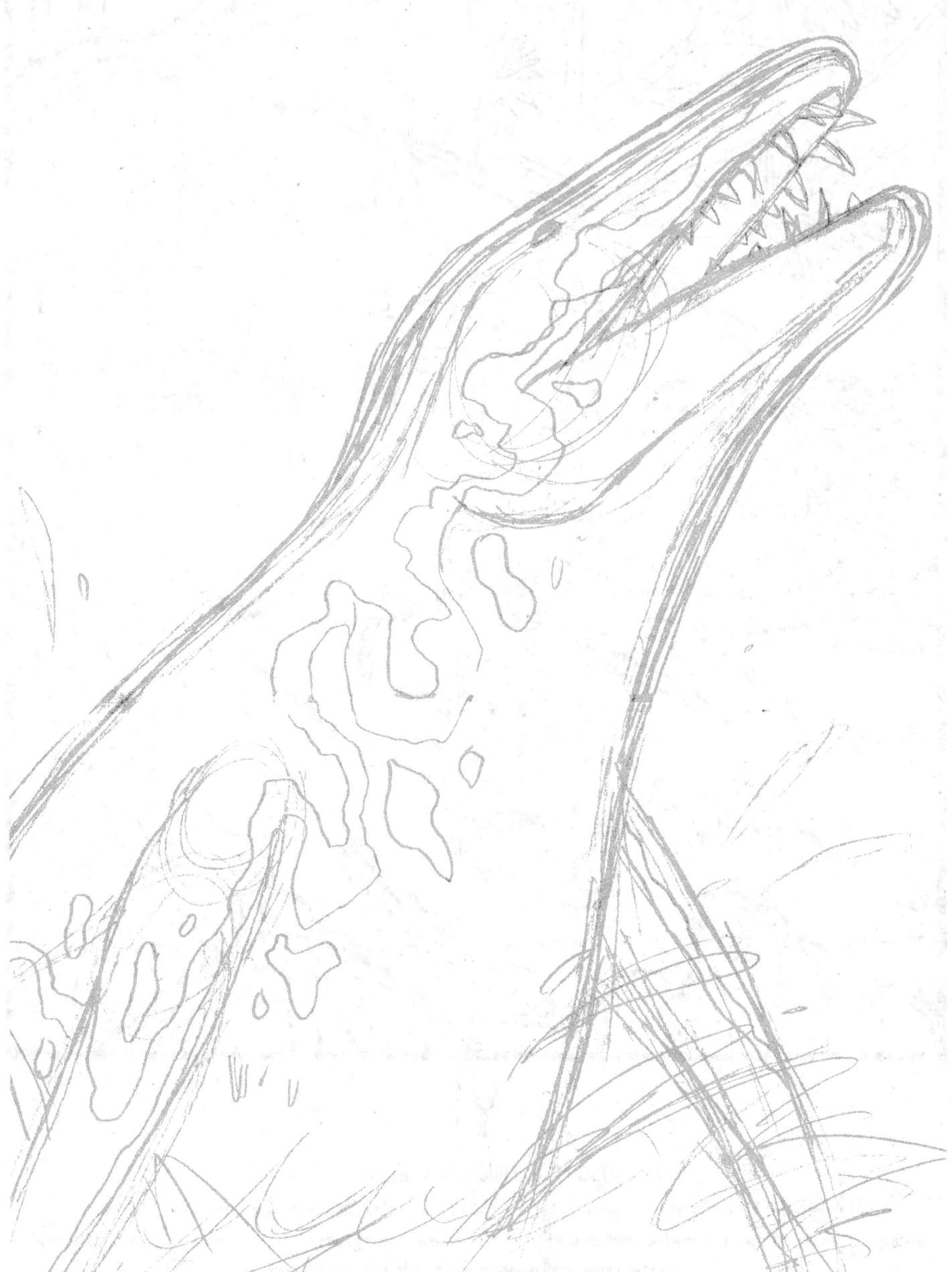

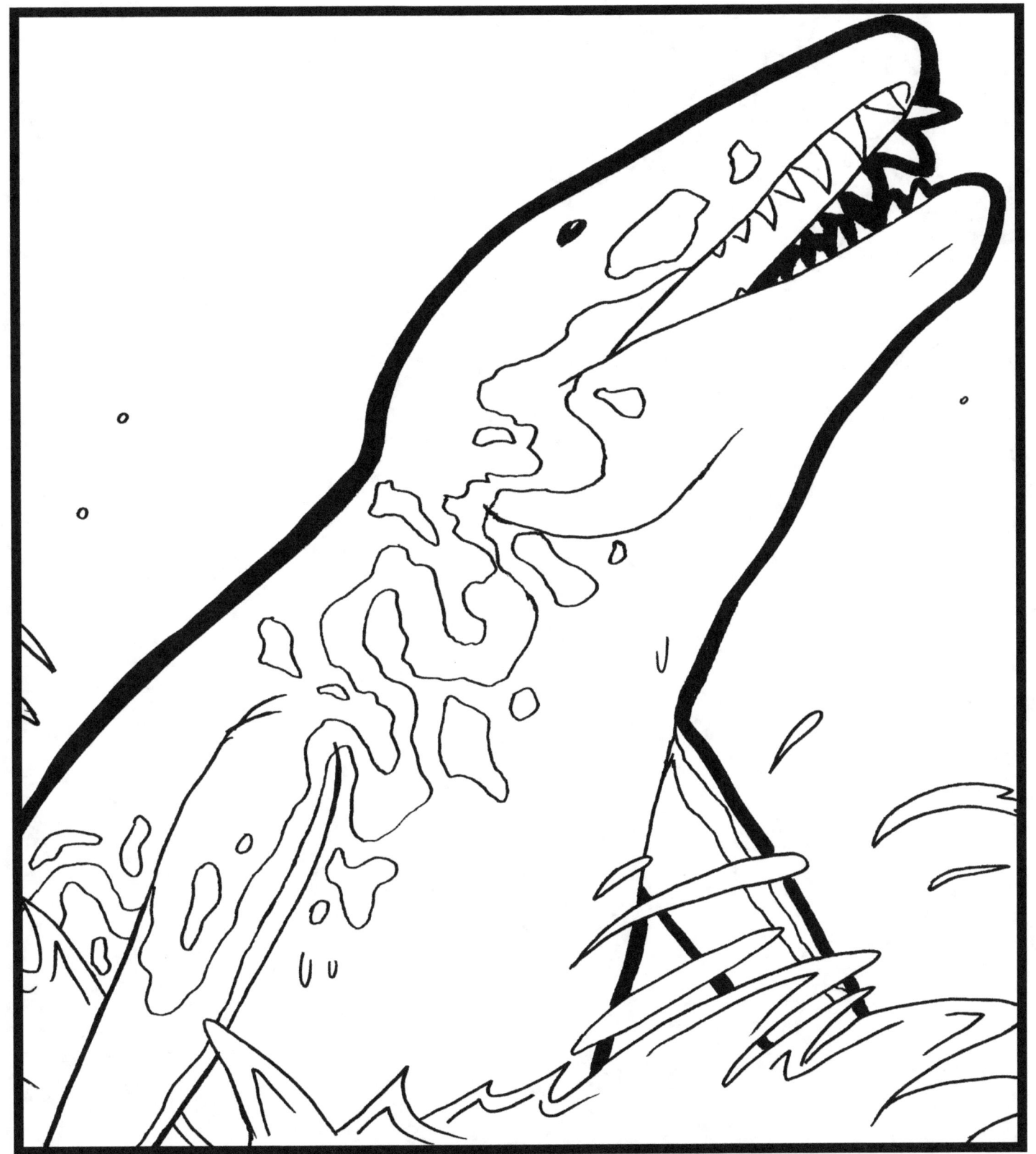

LIOPLEURODON

160 - 155 MYA, Middle to Late Jurassic, England, France, Germany, Argentina, Mexico

Many specimens of this Jurassic apex predator have been discovered. Its head was a fifth of the total size of its body, and was filled with long and strong conical teeth. It used its sense of smell to find its prey, which it would ambush in a quick burst of speed.

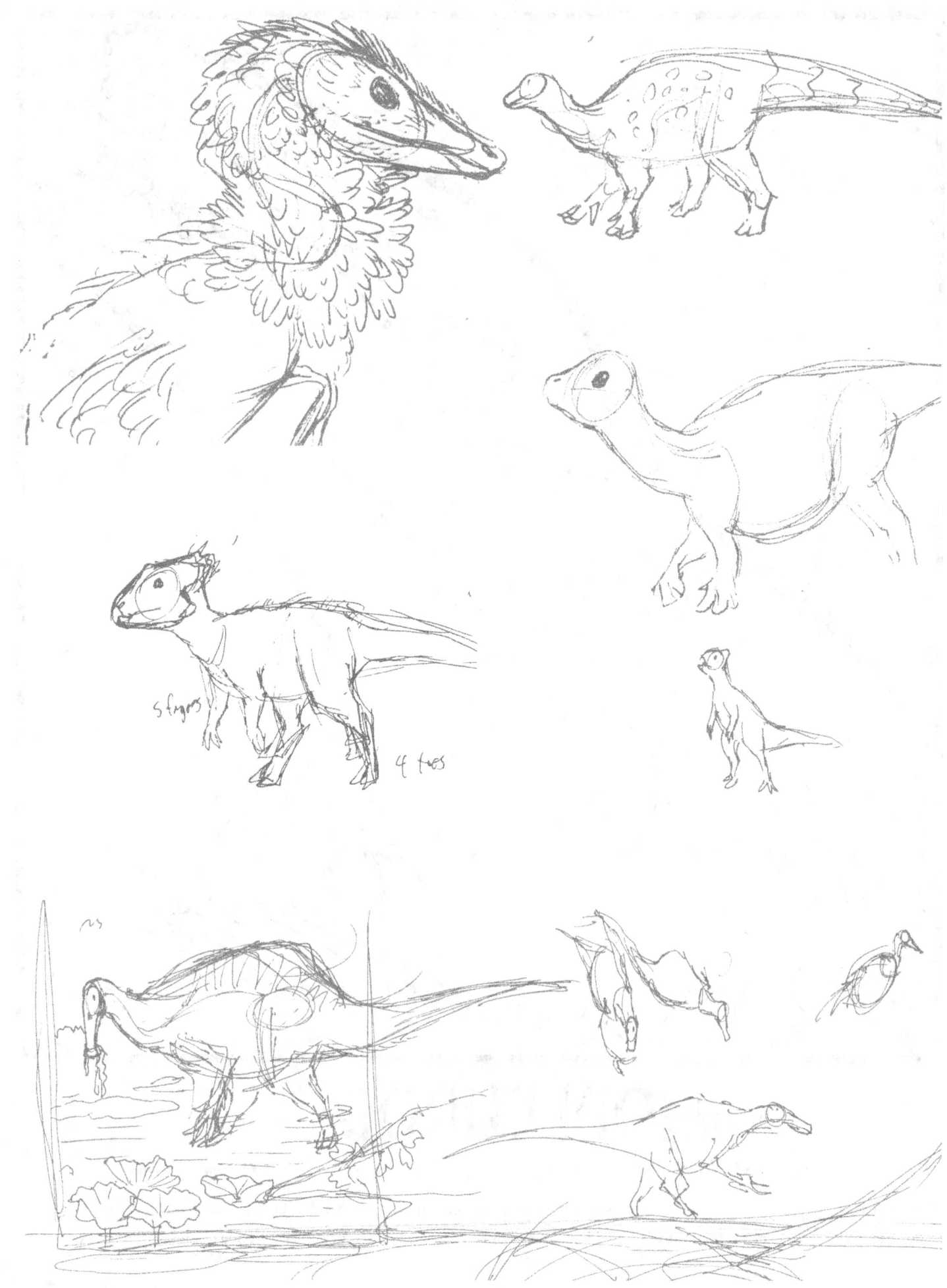

5 fingers

4 toes

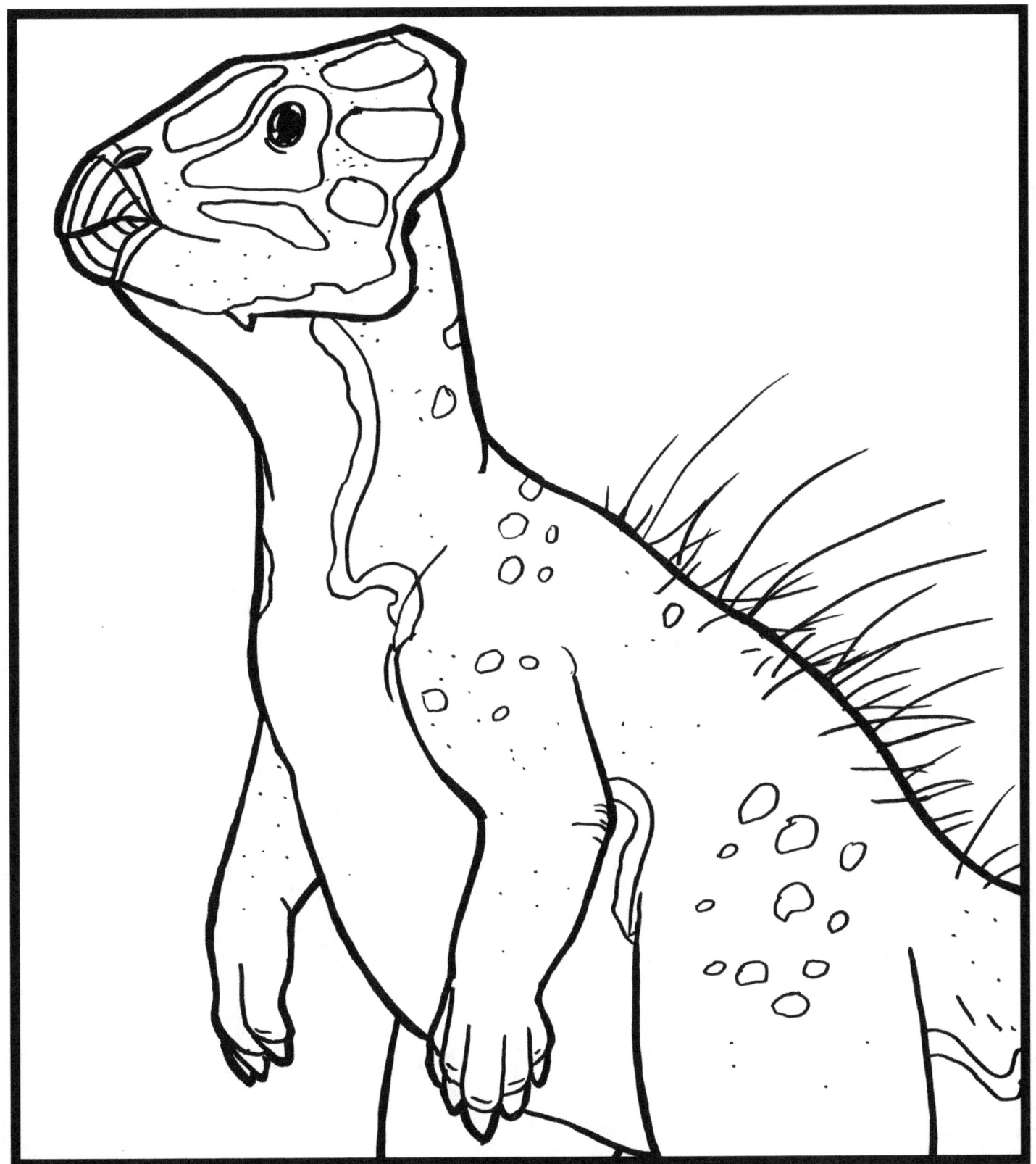

YINLONG

158 MYA, Late Jurassic, China

At just over a meter long, this ancestor to the ceratopsians was much smaller than many of its future descendants.
The crest wasn't nearly as pronounced as it would become in time, and this little generalist herbivore was still a bipedal dinosaur.
It might have had a set of quills along its tail and back, as some later species would be discovered with evidence of such.

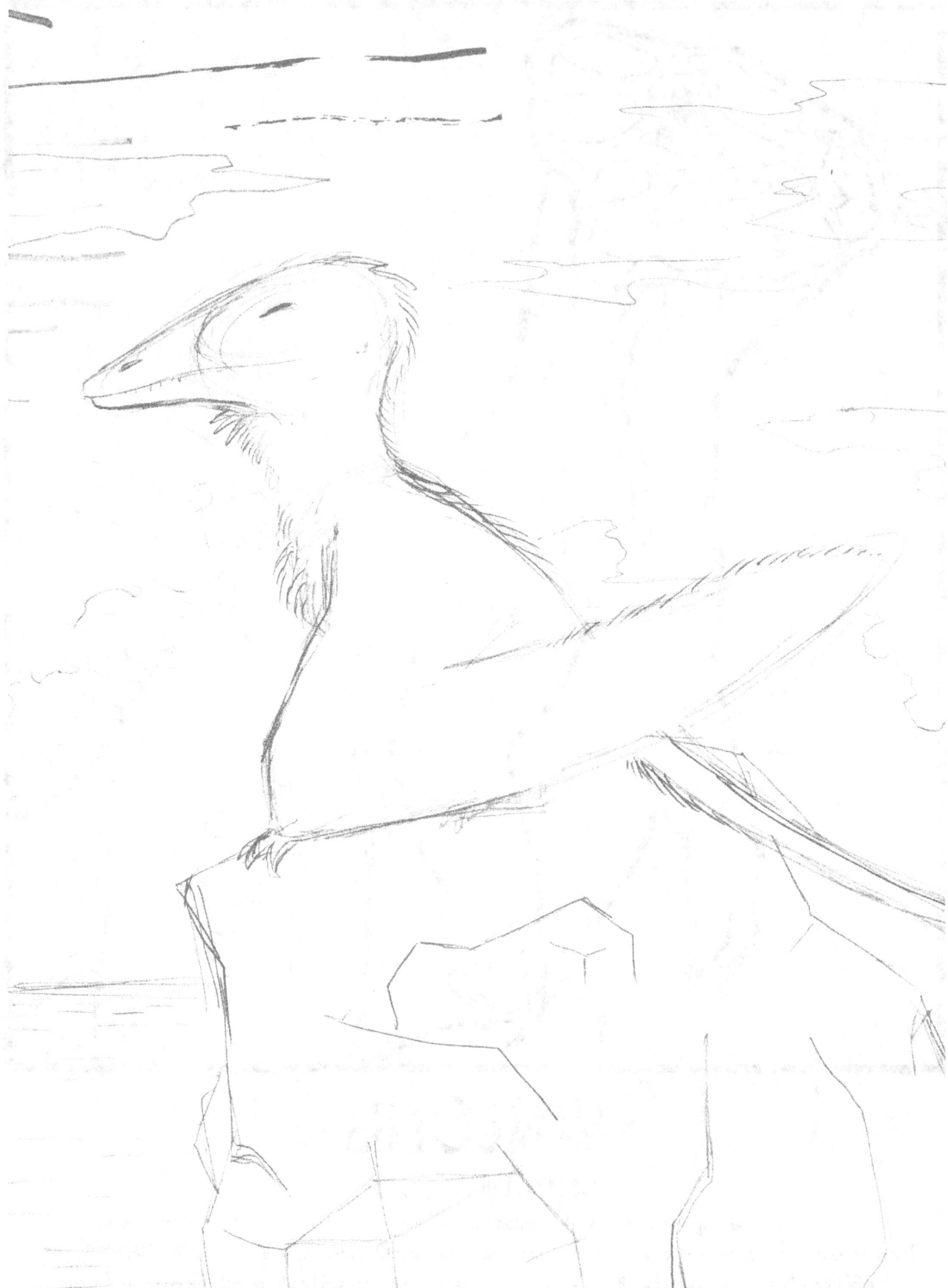

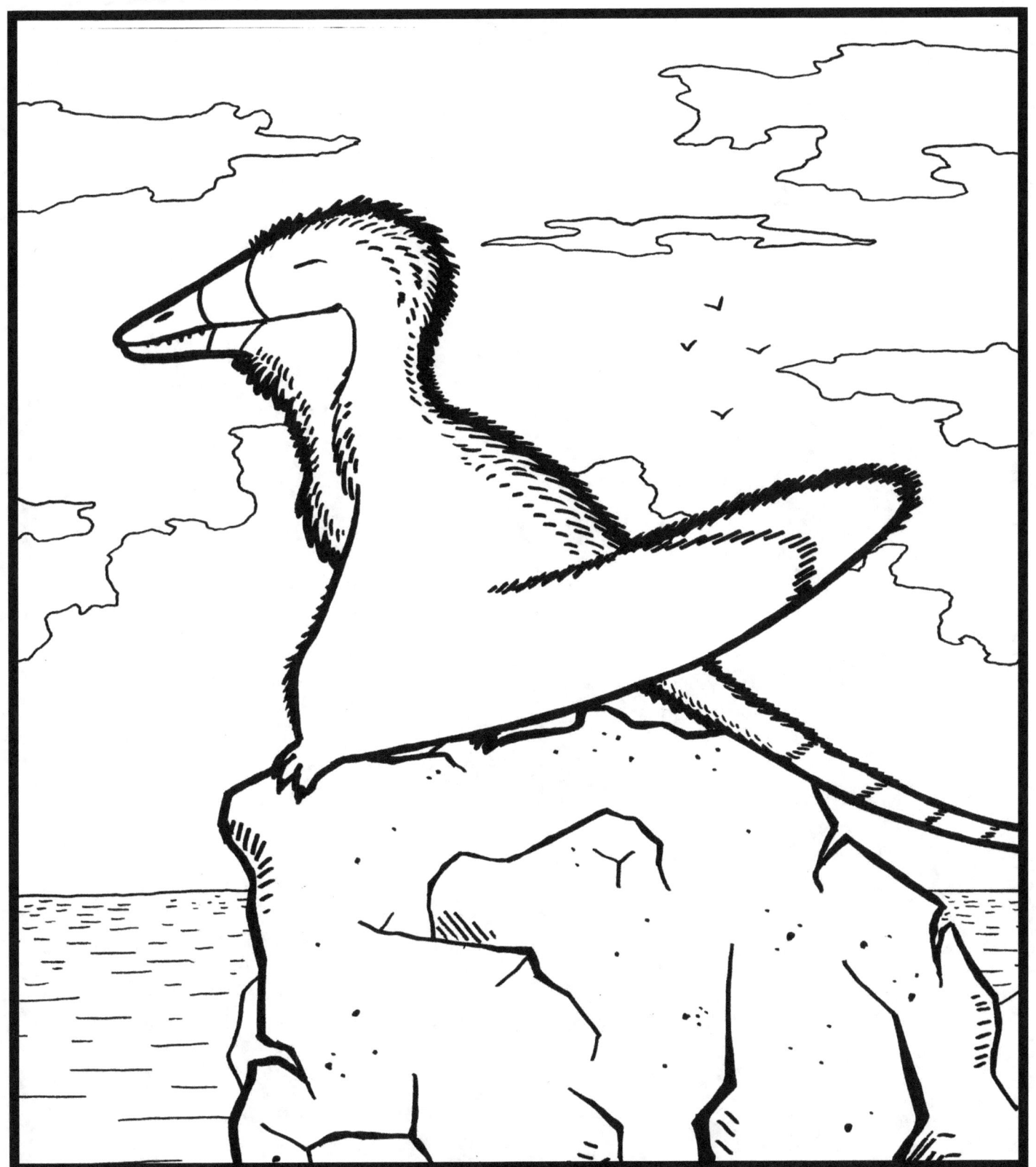

SORDES

155.7 MYA, Late Jurassic, Kazakhstan

This was the first pterosaur fossil to be discovered with direct evidence for a thick covering of fuzz, called pycnofibers. Today we know that all pterosaurs had this type of covering on nearly their entire bodies. This tiny short-necked pterosaur's tail was over half of its total length, and it also lacked a head-crest. It probably lived off of insects and small vertebrates

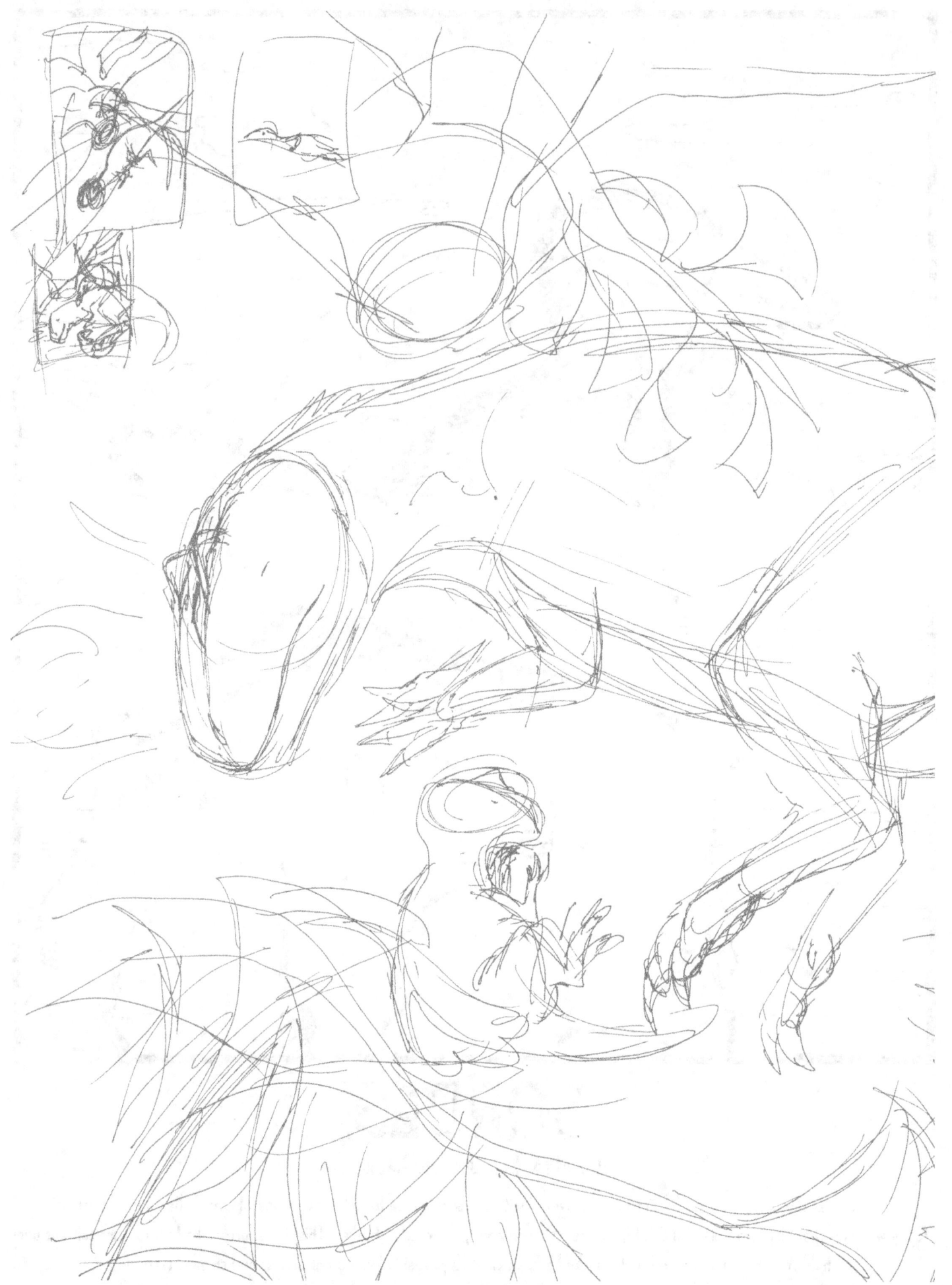

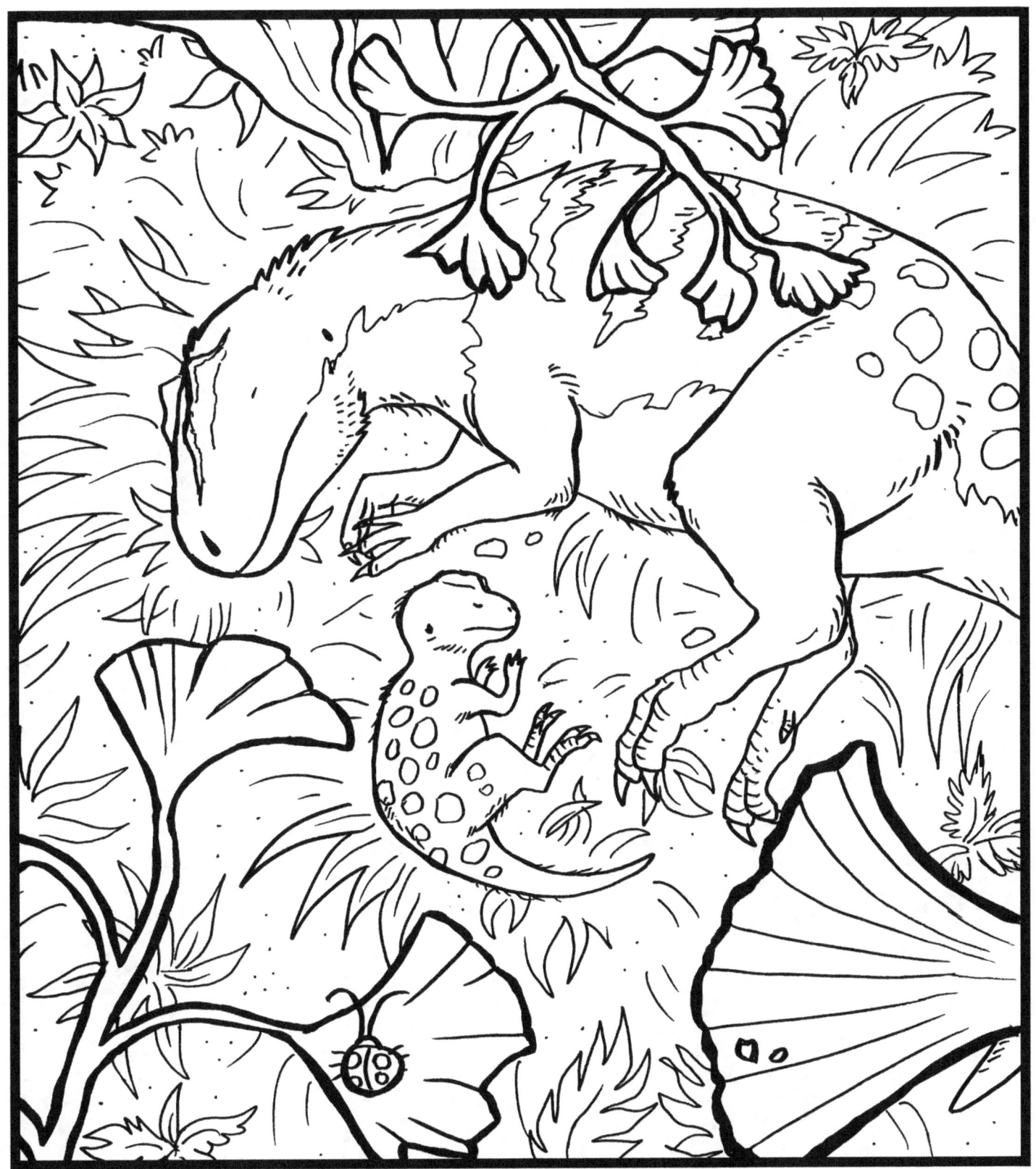

ALLOSAURUS

155 - 150 MYA, Late Jurassic, United States, Portugal

A large bipedal predator averaging 8.5 meters in length, Allosaurus is one of the better-known, and more common, theropods. It wasn't as built for speed as other theropods, but was better suited as an ambush predator in its semiarid, floodplain, environment.

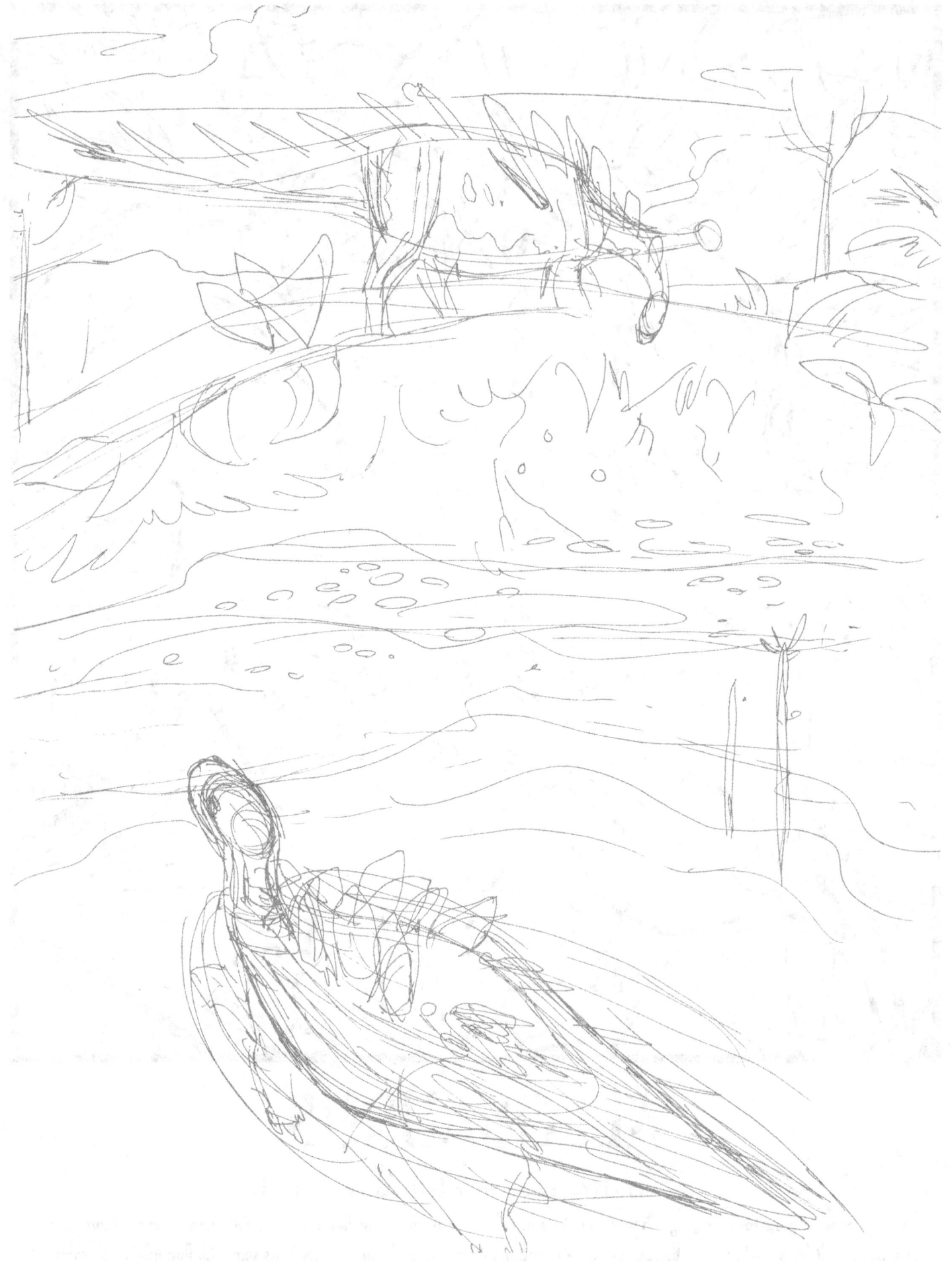

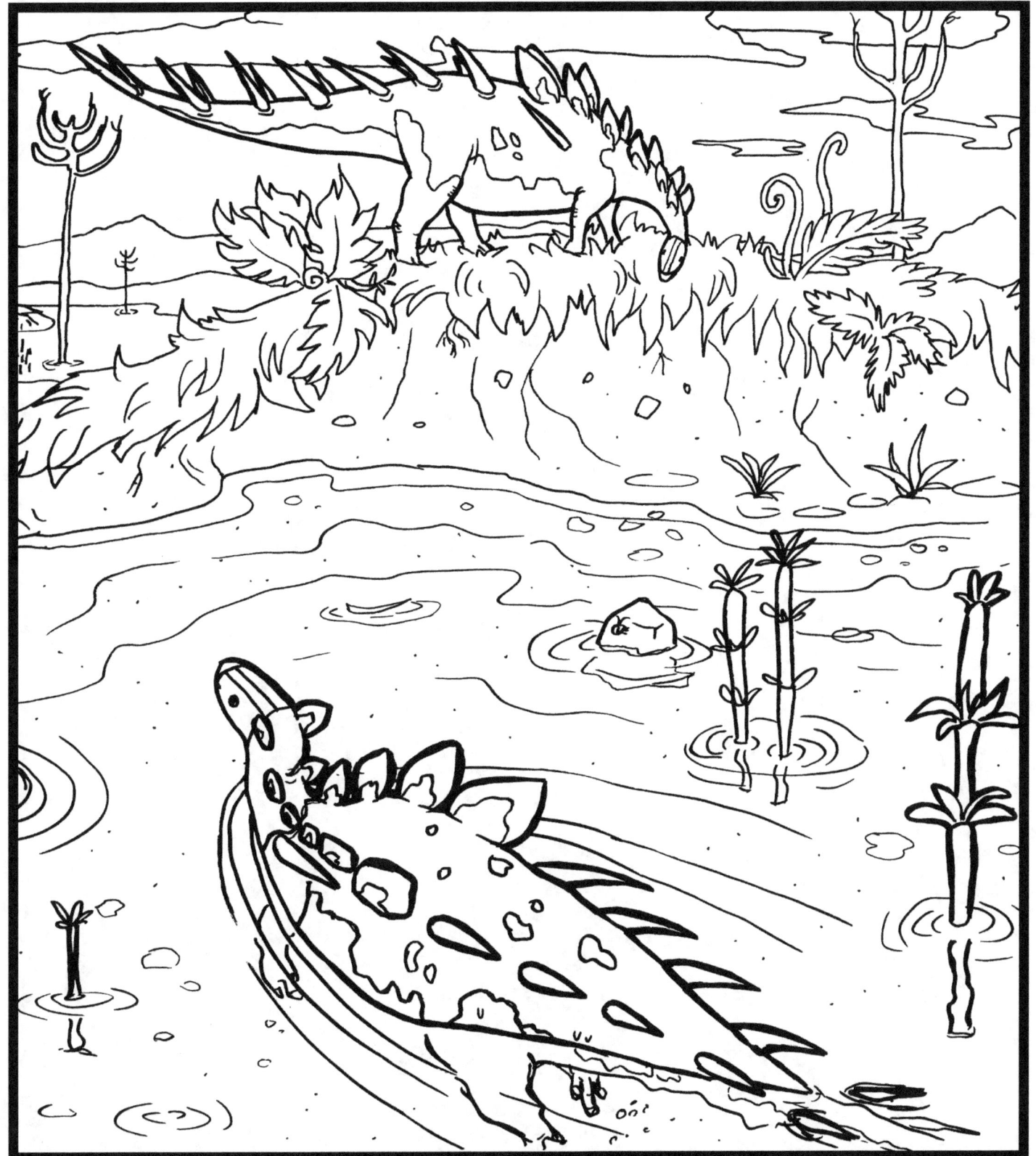

KENTROSAURUS

155 - 150 MYA, Late Jurassic, Tanzania

This 4.5 meter long stegosaur had a front half that was adorned with two rows of standing plates, and a back half that sported two rows of spikes. Being a smaller member of the group, and having a center of mass that was unusually far back for a quadruped, it was able to rear up on its hind limbs to reach food that was higher up. This trait also allowed it to quickly pivot, swinging the spiked tail at its attacker.

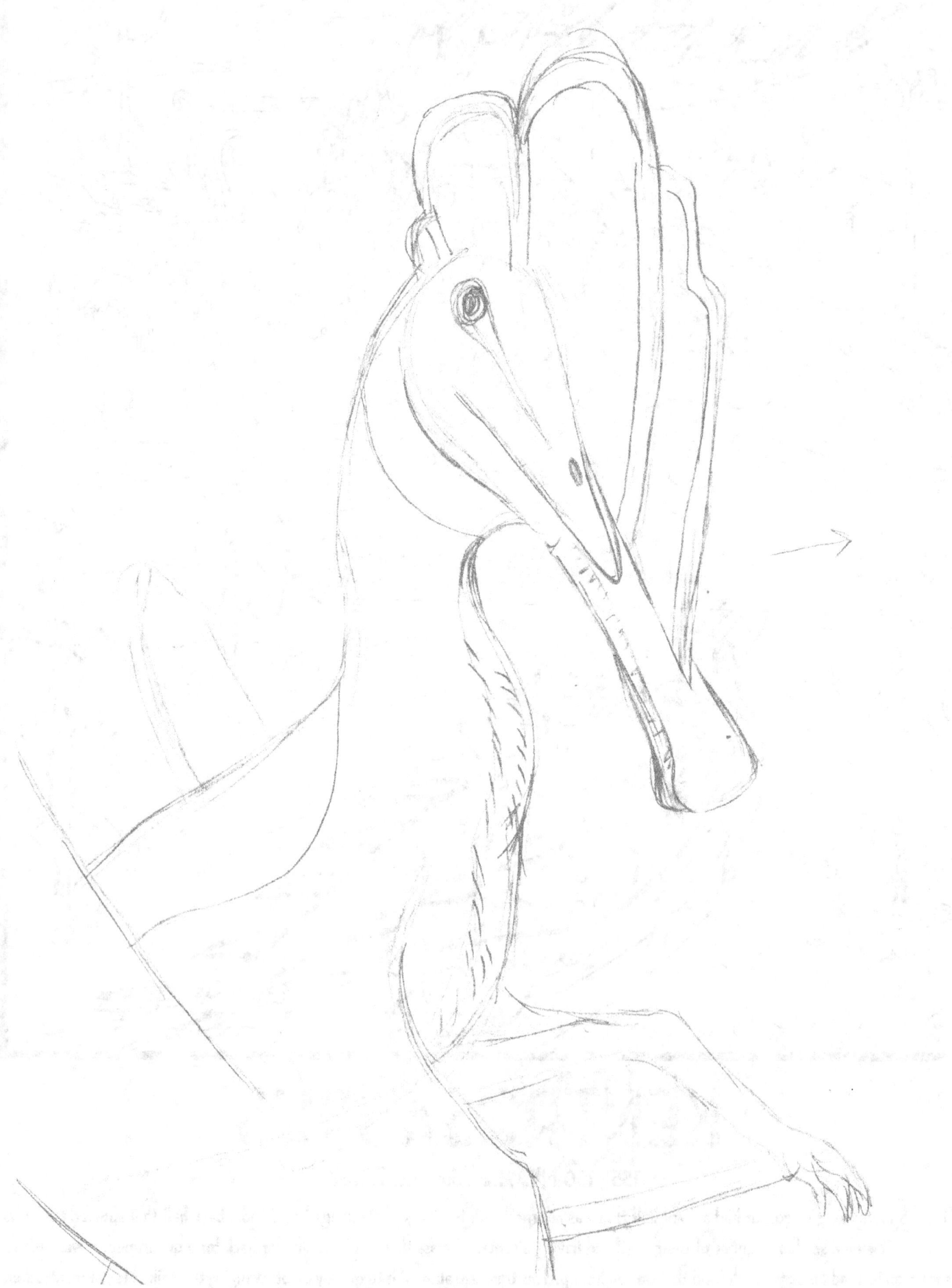

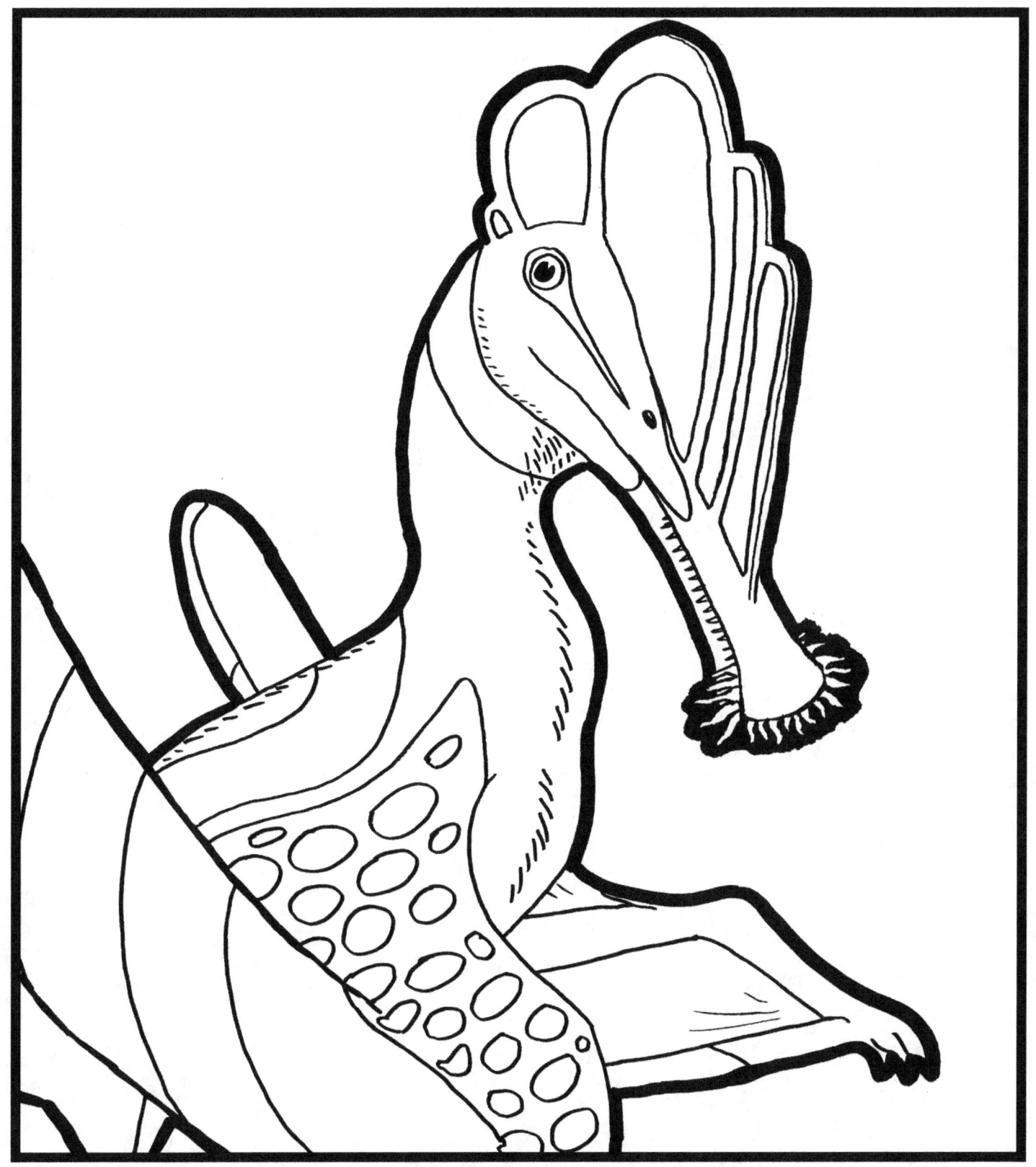

GNATHOSAURUS

152 - 145 MYA, Late Jurassic, England and Germany

This filter-feeder's wingspan was on the smaller end of the spectrum for a pterosaur, at 1.75 meters across. It lived right up to the end of the Jurassic, and was originally thought to have been a prehistoic crocodile when first discovered.

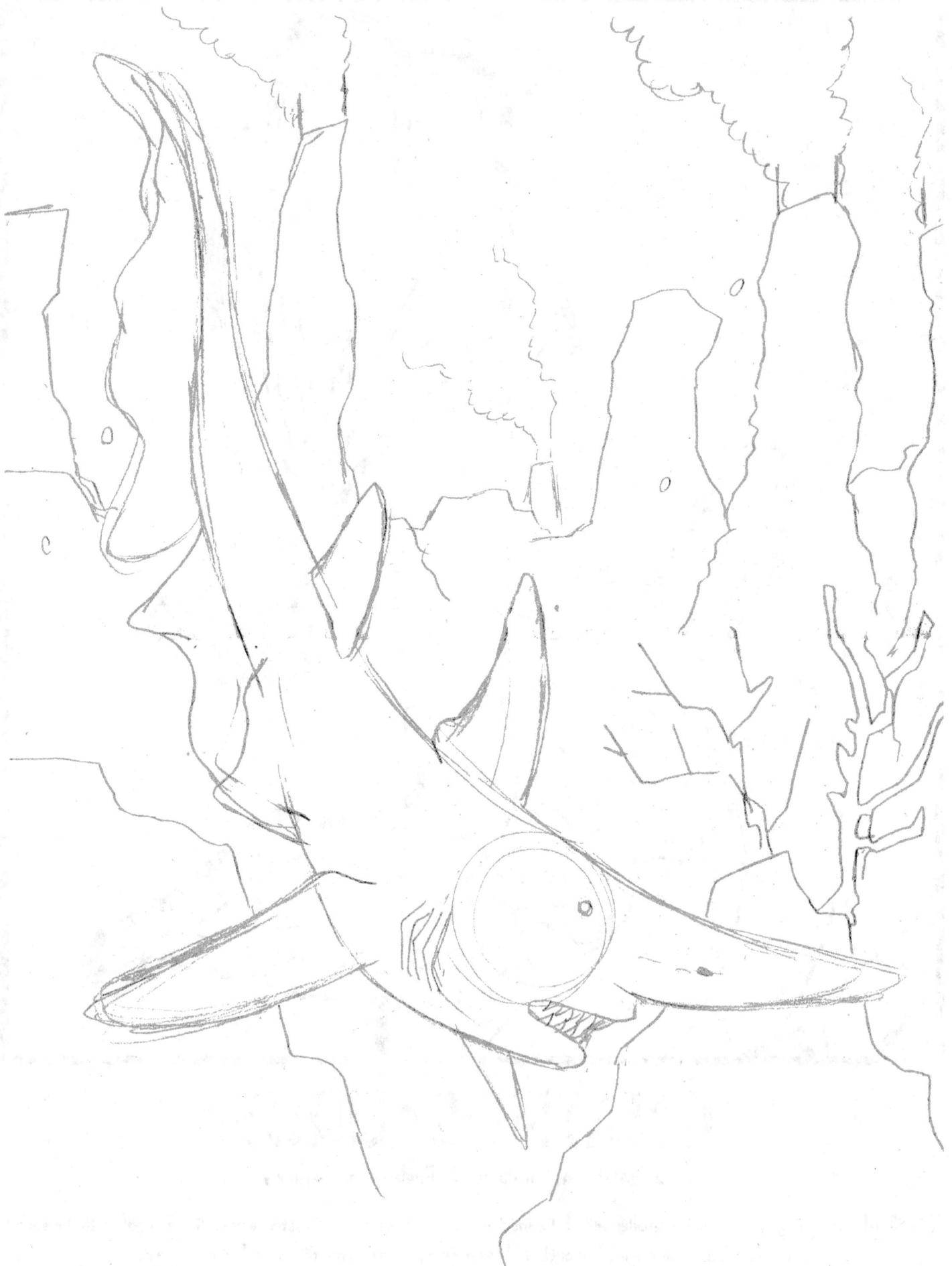

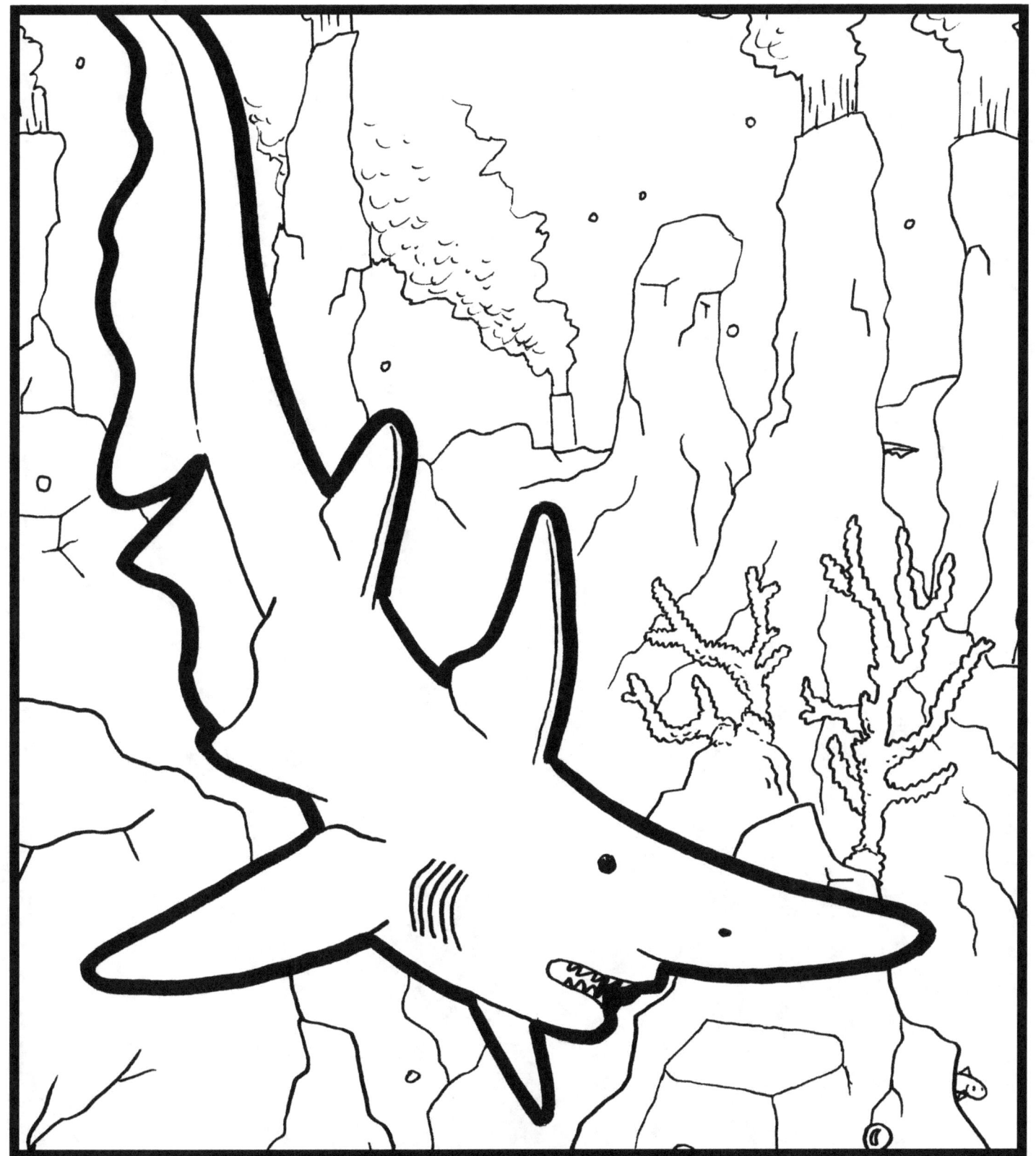

SCAPANORHYNCHUS

140 - 20 MYA, Early Cretaceous to Miocene, Worldwide

Individual species of this shark came in a range of sizes, from 65 centimeters to about 3 meters. They were similar to today's goblin shark, and were survivors, only relatively recently going extinct after they appeared in the early Cretaceous. Like the modern goblin shark, these fish dwelled in the depths of the ocean, using the electricity sensors in the snout to find their way around.

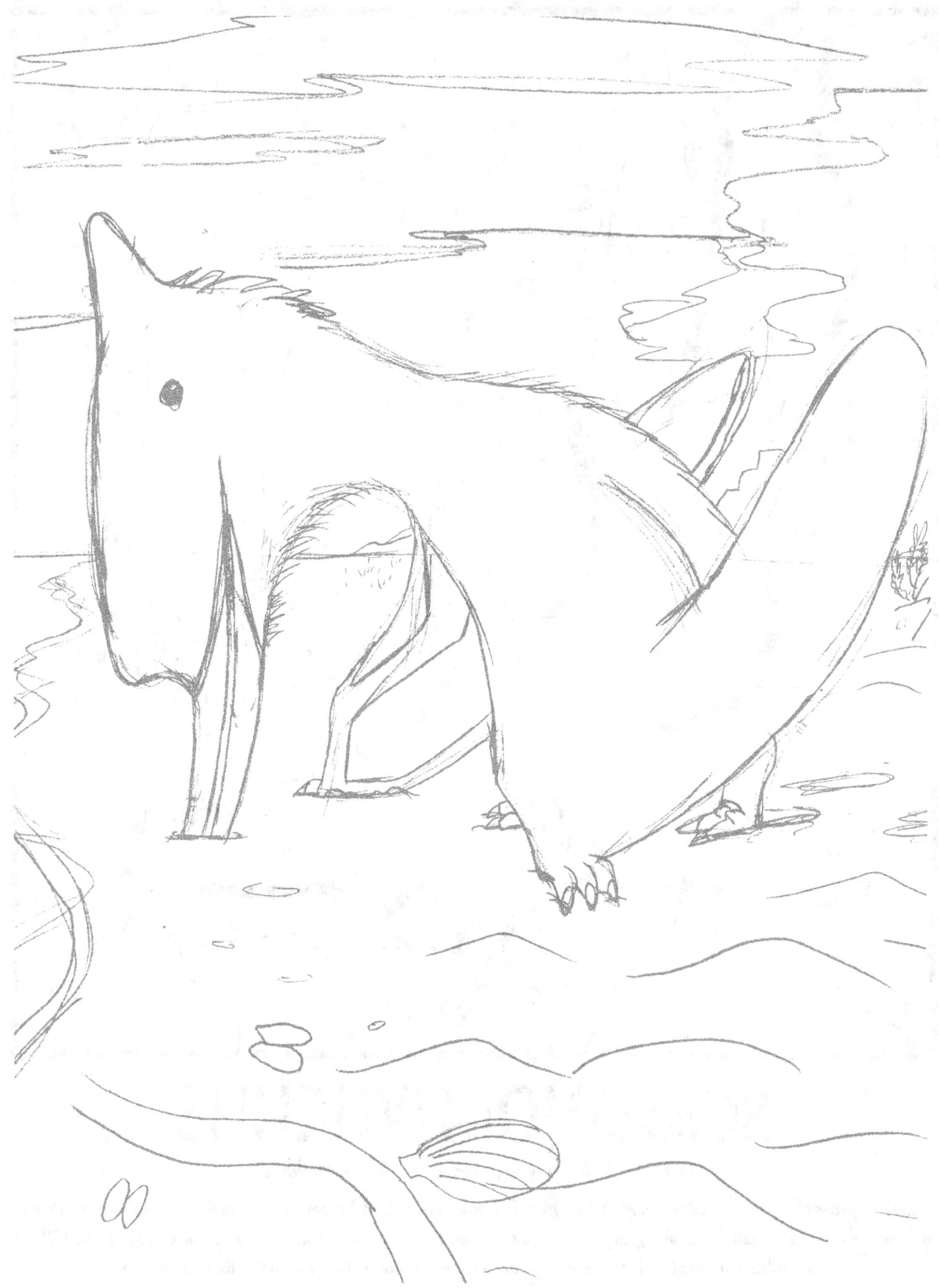

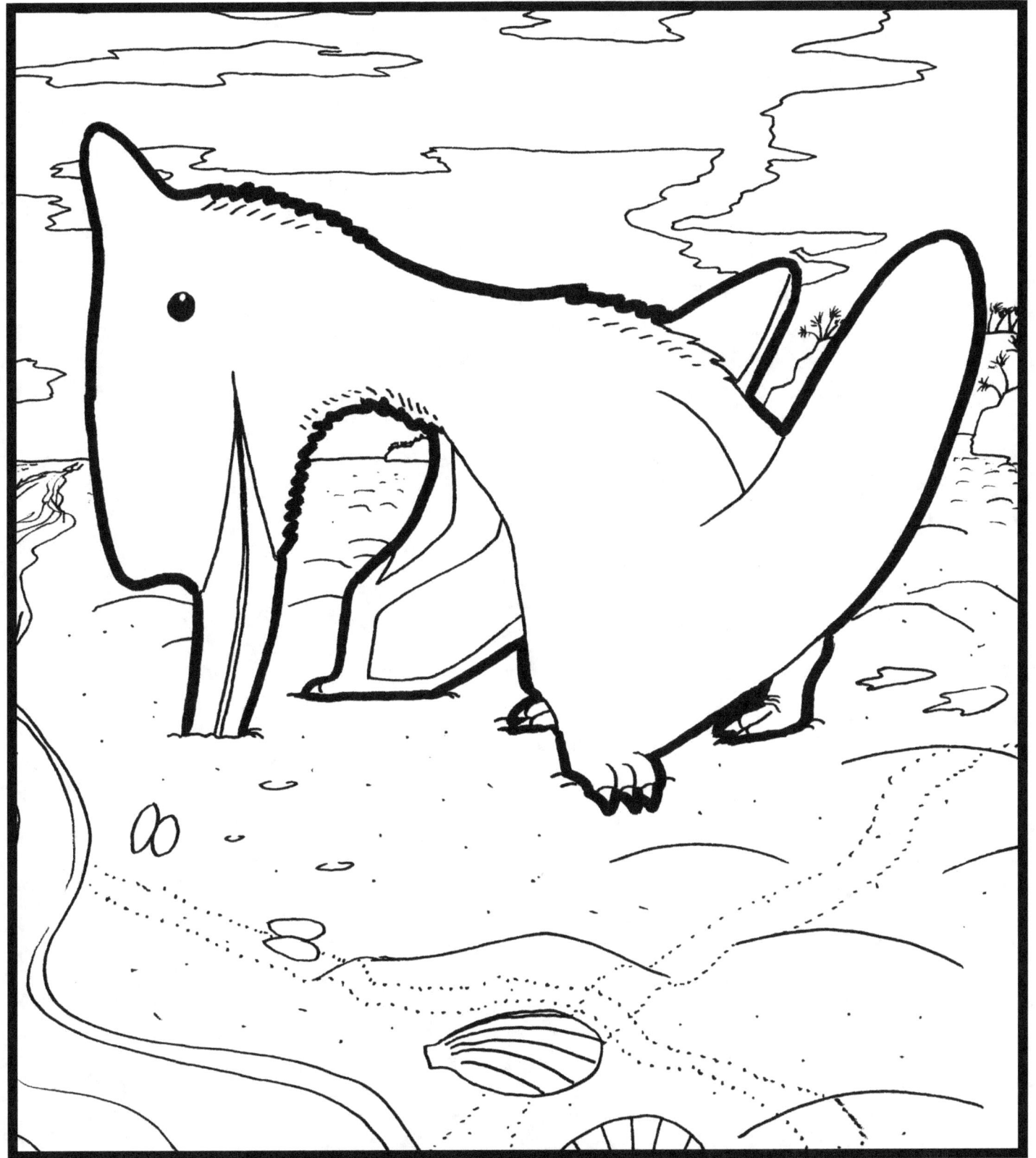

DSUNGARIPTERUS

130 MYA, Early Cretaceous, China and Mongolia

The upcurved, pointed jaws and flat back teeth of this pterosaur were ideal for feeding on mussels, snails, and even small vertebrates. It would extract the invertebrate from the mud or sand with the curve of its beak, and then crush it with the back teeth. It was a medium-sized animal, with a wingspan of around 3.5 meters.

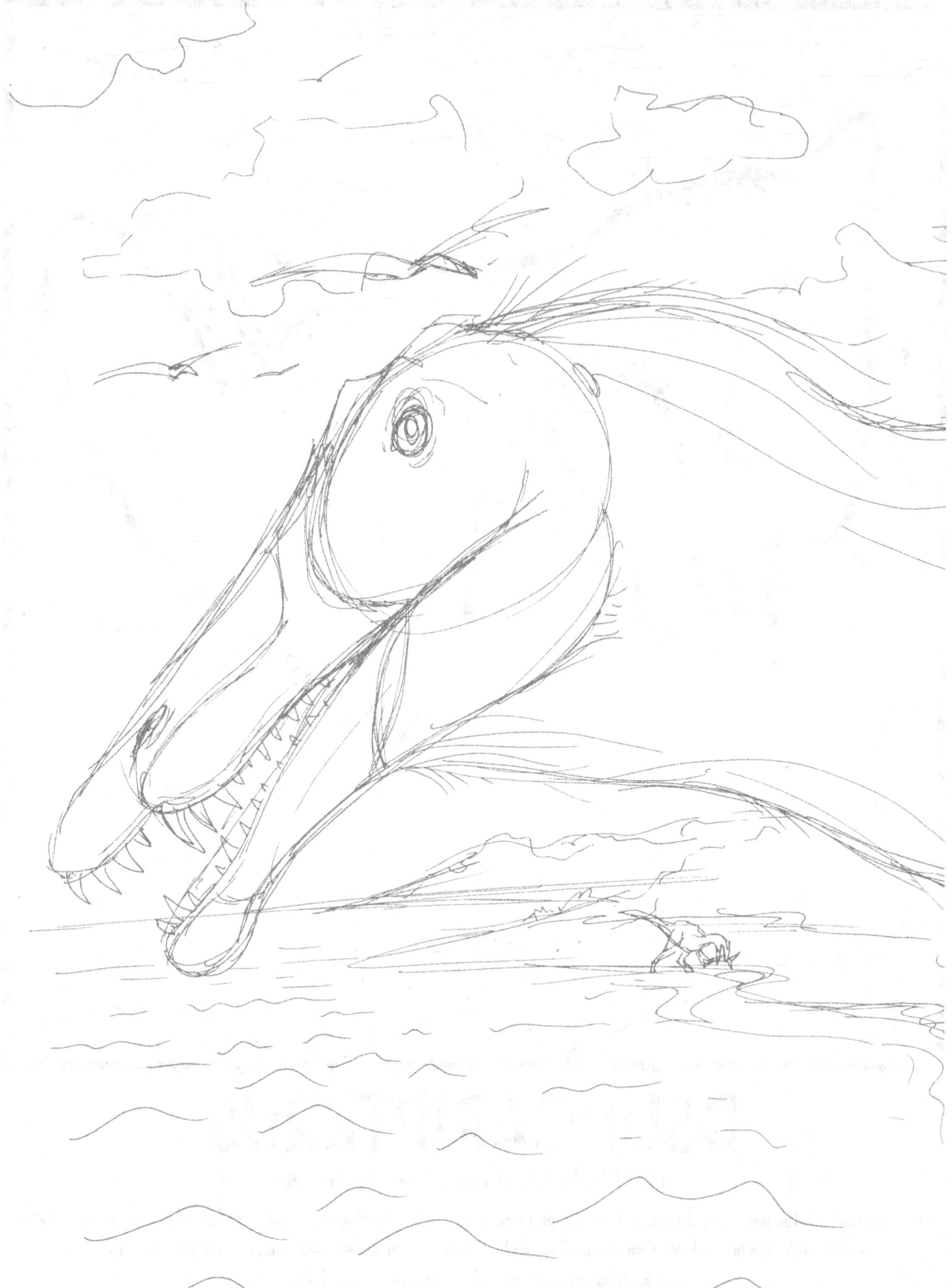

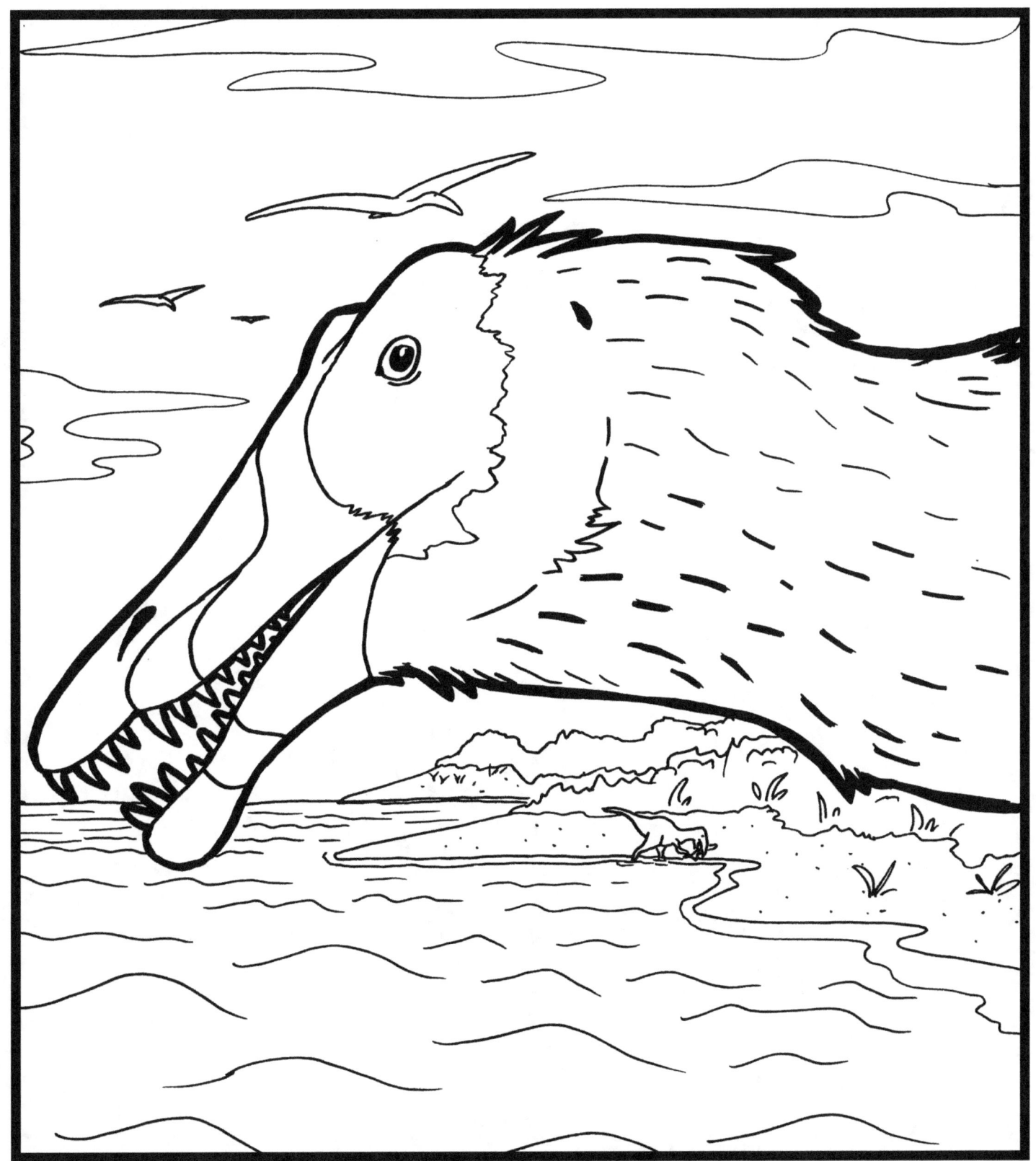

BARYONYX

130 - 125 MYA, Late Cretaceous, England, Spain, Portugal

Baryonyx were specialized in catching and eating fish, and it used both its teeth and the "heavy claw" on its hands to acquire prey.
It was probably not limited to eating fish, as pterosaur and Iguandon remains have been found with evidence of predation or scavenging.

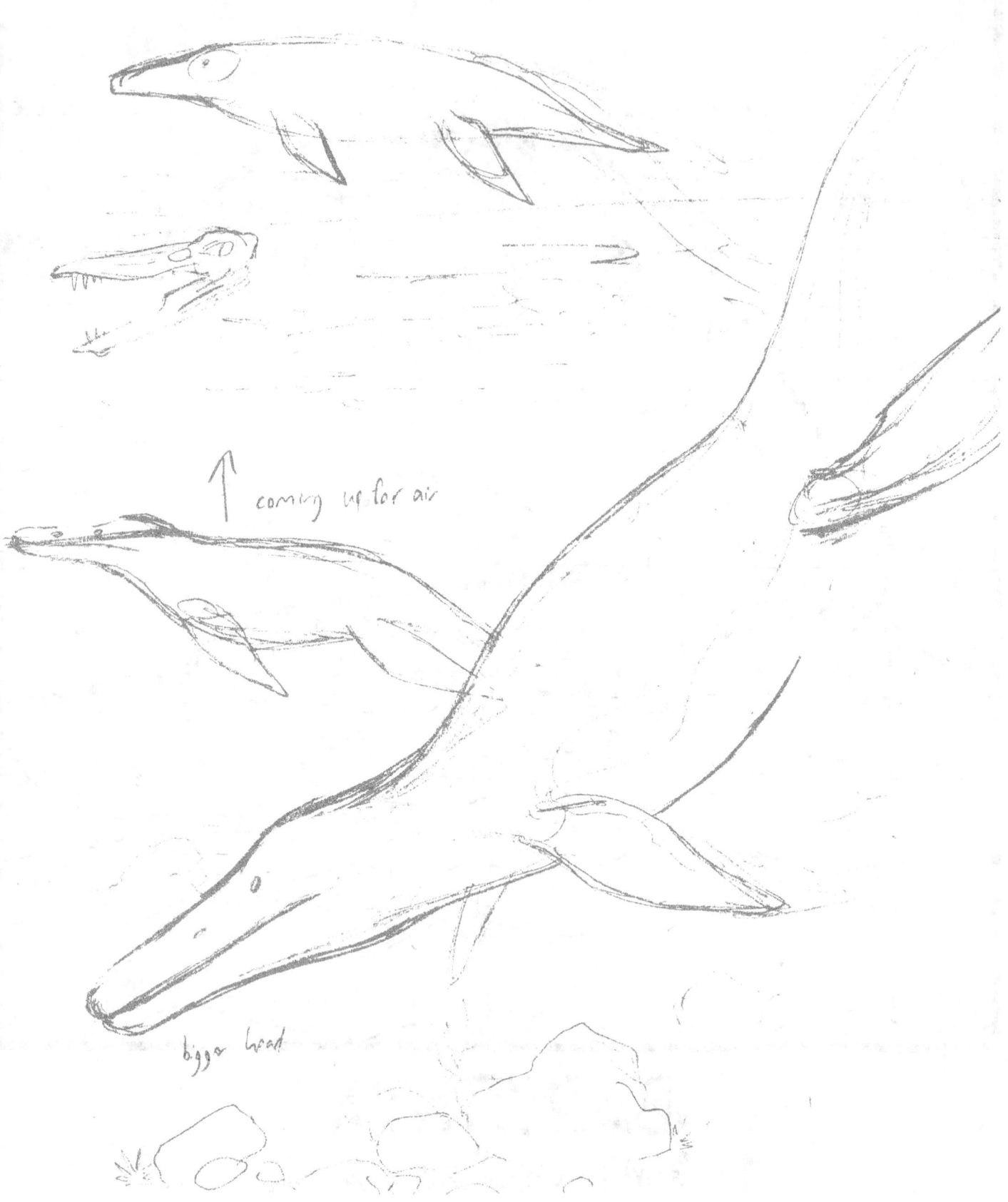

coming up for air

bigger head

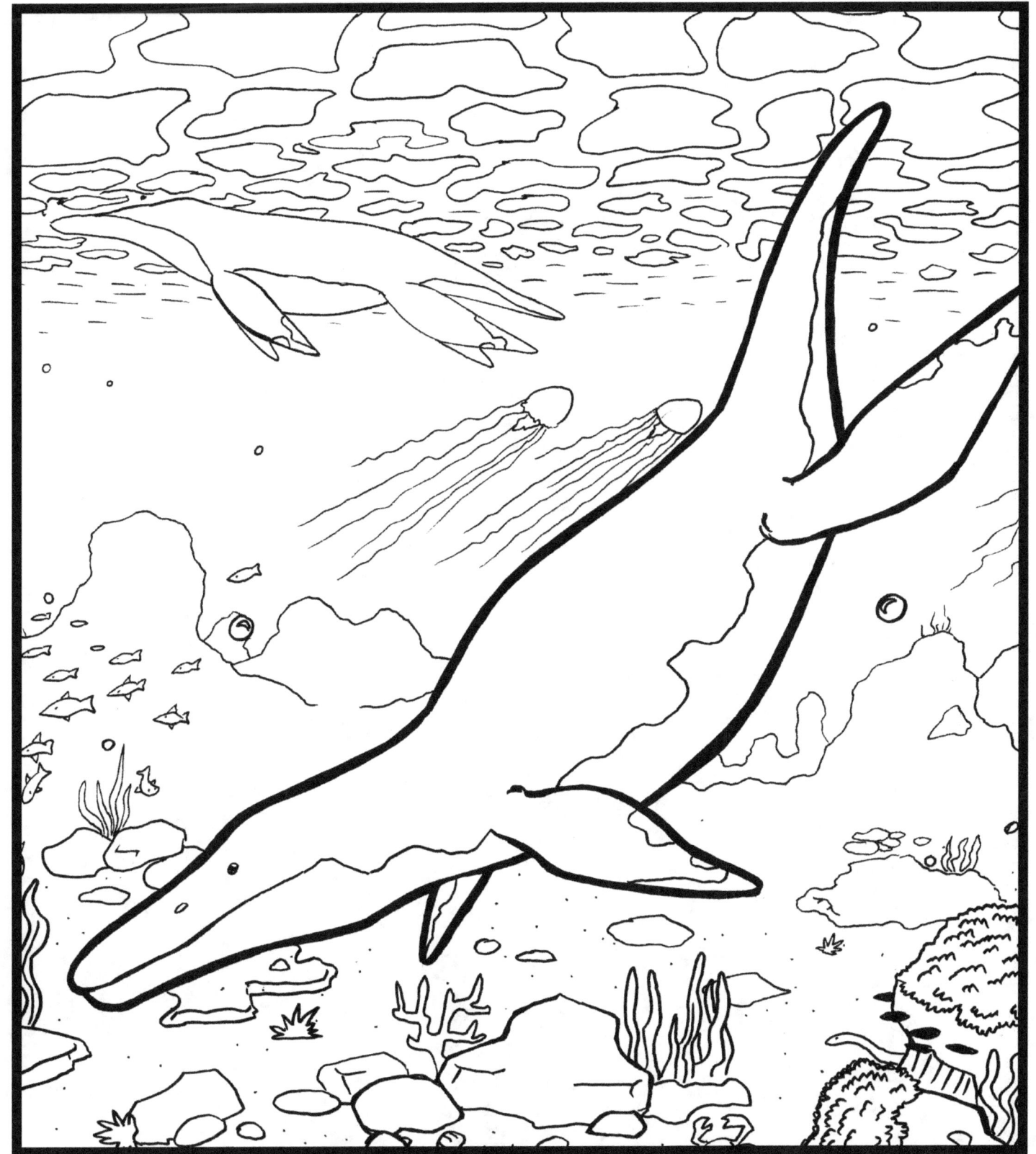

KRONOSAURUS

129 - 99 MYA, Early Cretaceous, Australia and Columbia

Kronosaurus was one of the largest pliosaurs, reaching a length of up to 10 meters. Its name means "lizard of Cronus," the figure being one of the titans from ancient Greek mythology. A fast and powerful swimmer, it preyed on plesiosaurs, turtles, fish, and squid.

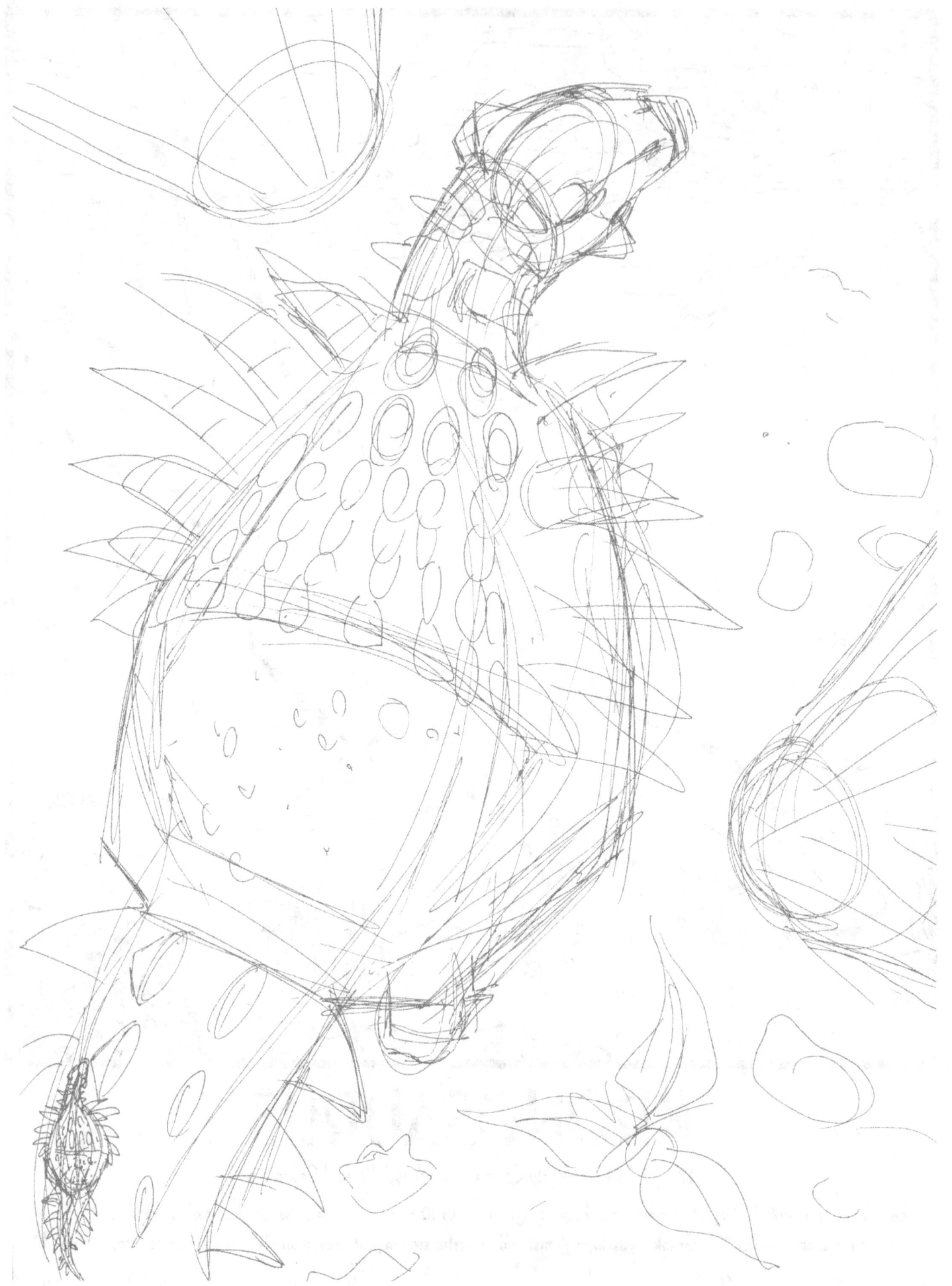

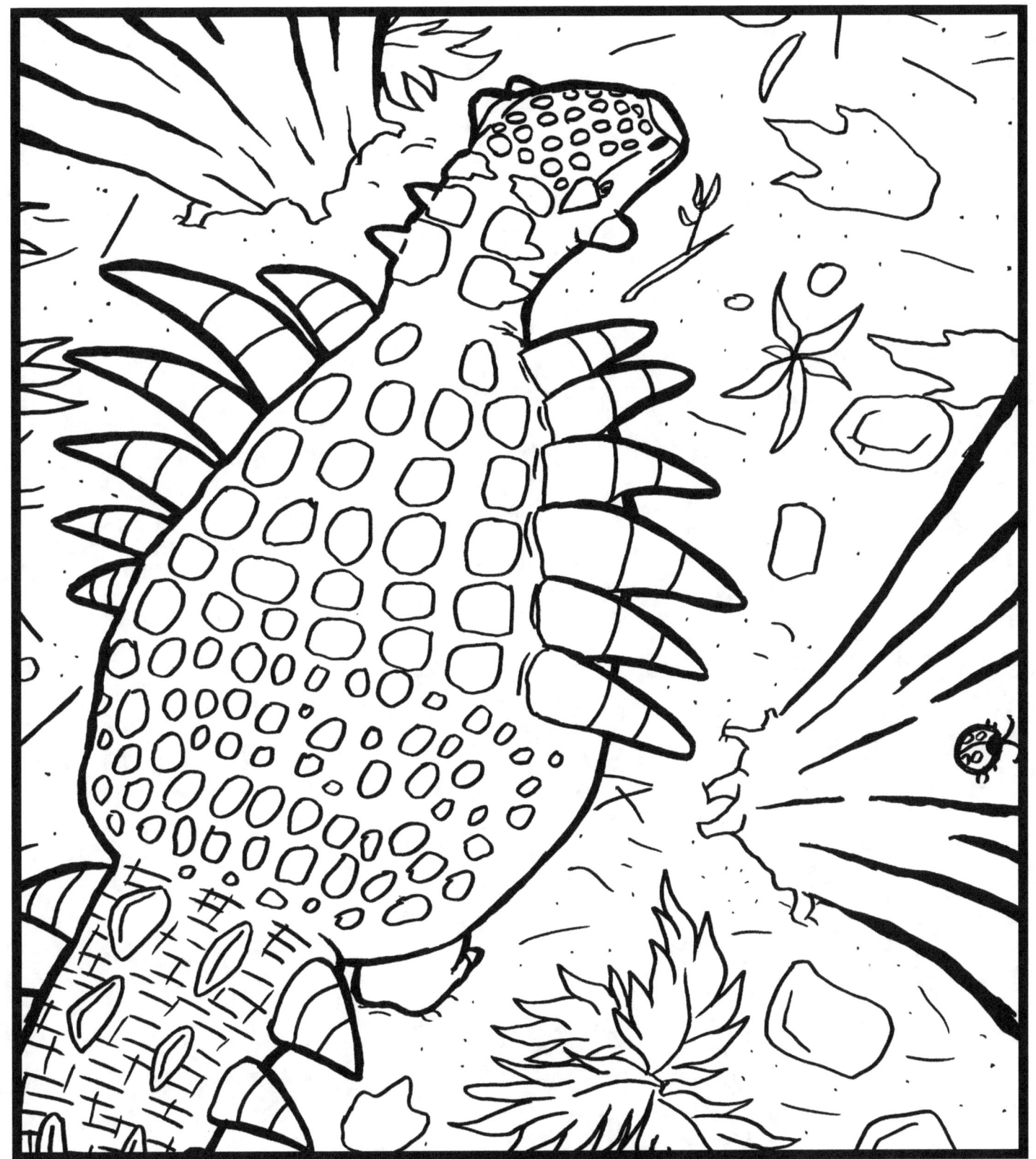

GASTONIA

126 MYA, Early Cretaceous, North America

This 5 meter-long herbivore was heavily-armored was abundant in the partly wooded areas it inhabited.

Unlike some ankylosaurs, Gastonia lacked a tail club, and instead defended itself with rows of bony spikes protruding out from the back.

Even for this group of dinosaurs, its limbs were short, and it was wide around the middle.

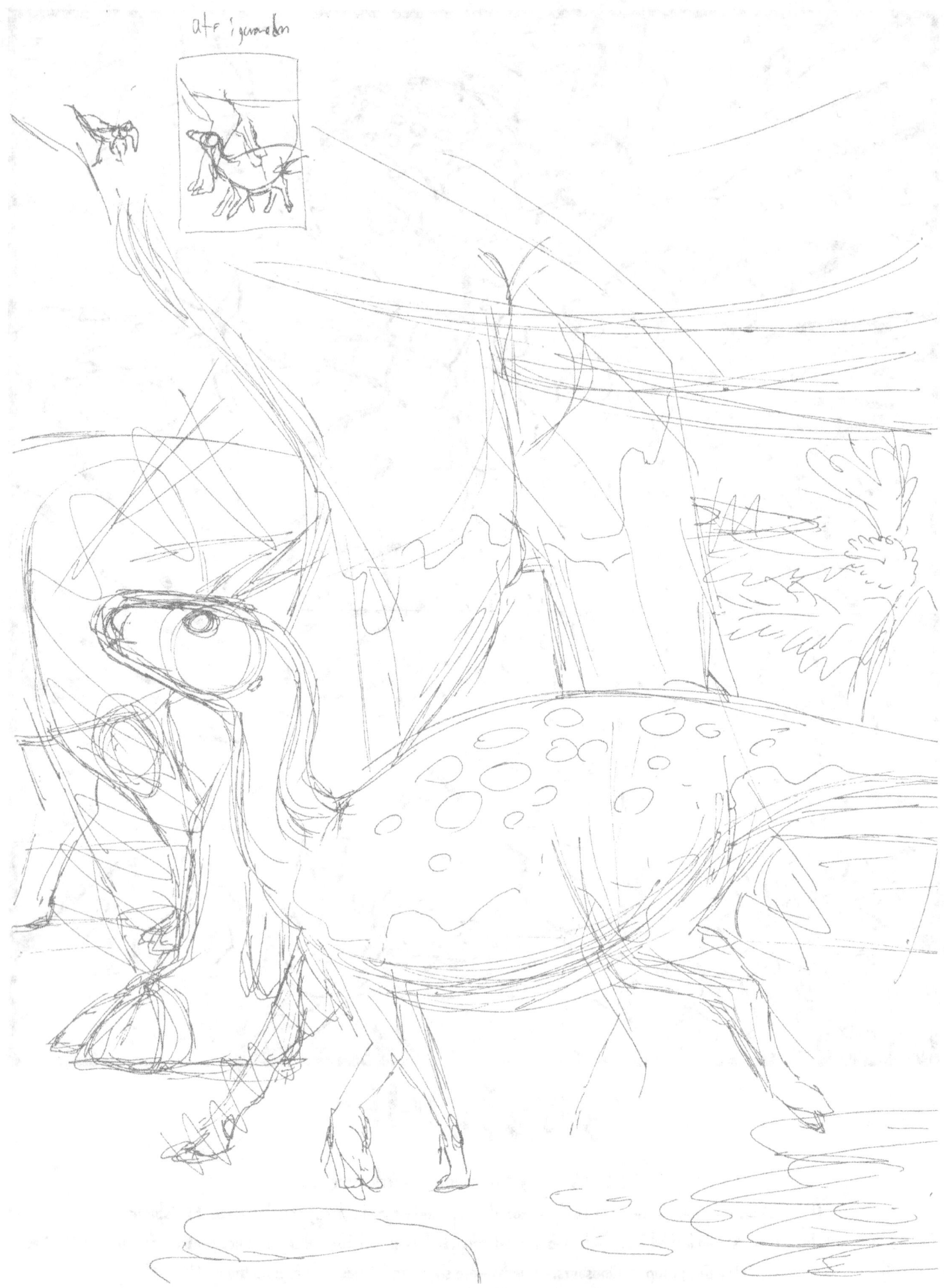

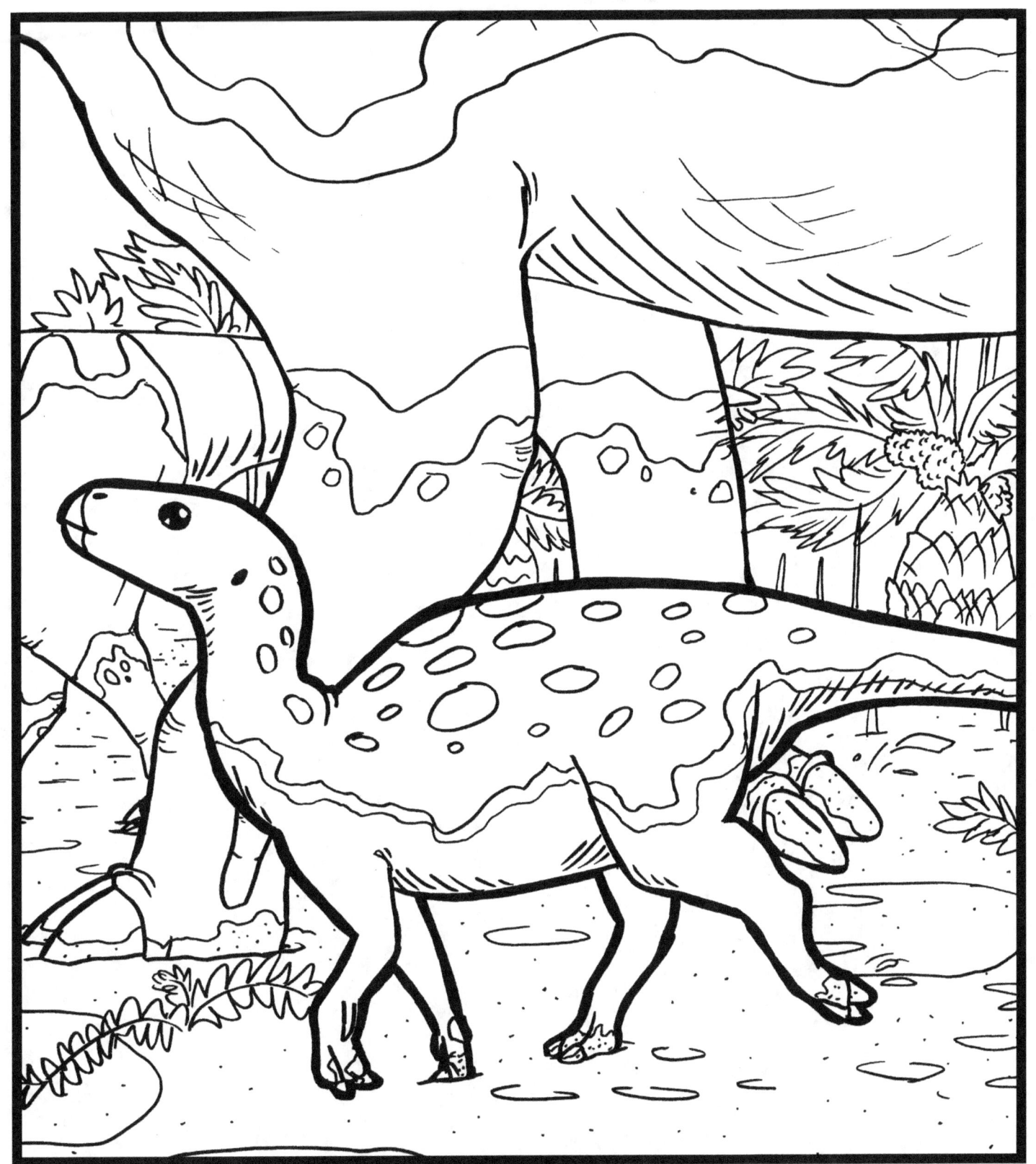

IGUANODON

126 - 125 MYA, Early Cretaceous, North Africa, Belgium, England, Germany, United States

Iguanodon were herbivores that were capable of moving in either a two-legged or four-legged way. They averaged 10 meters in length. Each front foot had a spiked thumb they could have used to defend themselves with (or for getting into seeds and nuts), while the little finger was flexible and allowed the animal to grasp and manipulate objects.

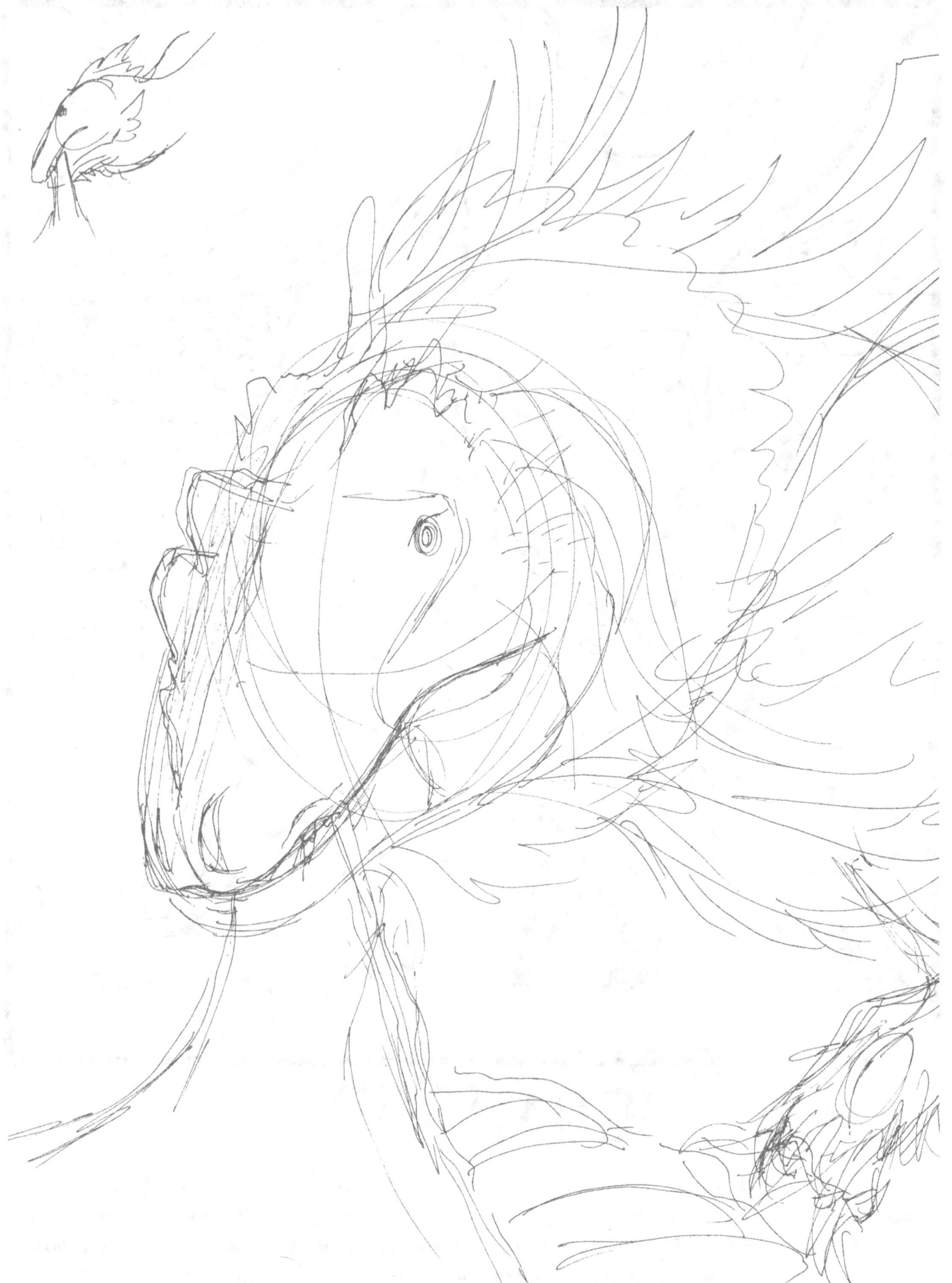

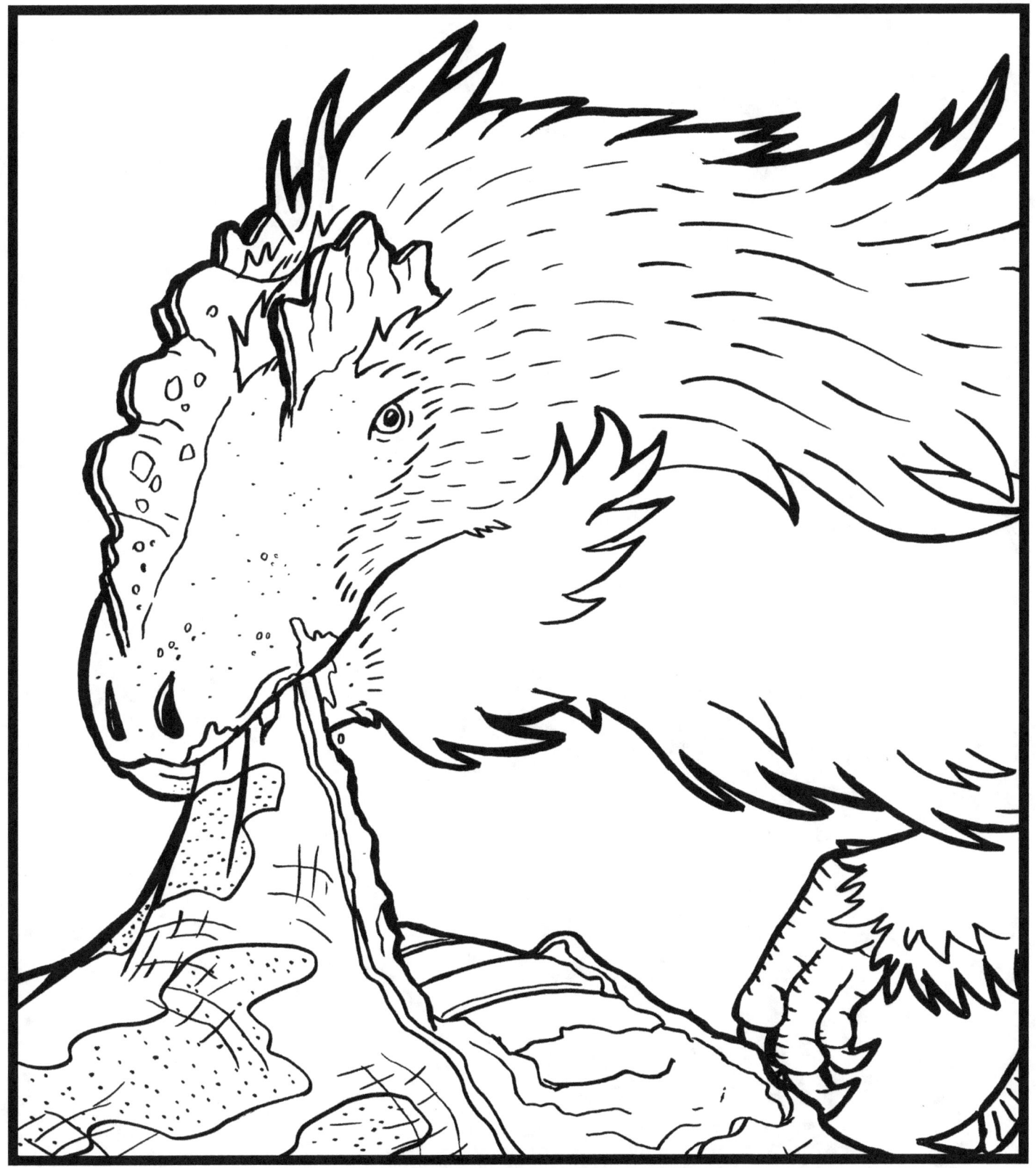

YUTYRANNUS

124 MYA, Early Cretaceous, China

This large tyrannosauroid is the largest member of the group to have been confirmed to have been covered in feathers. Because of the climate they lived in, this covering was likely used primarily as insulation against the cold environment. They were large, bipedal predators capable of reaching sizes of up to 9 meters long.

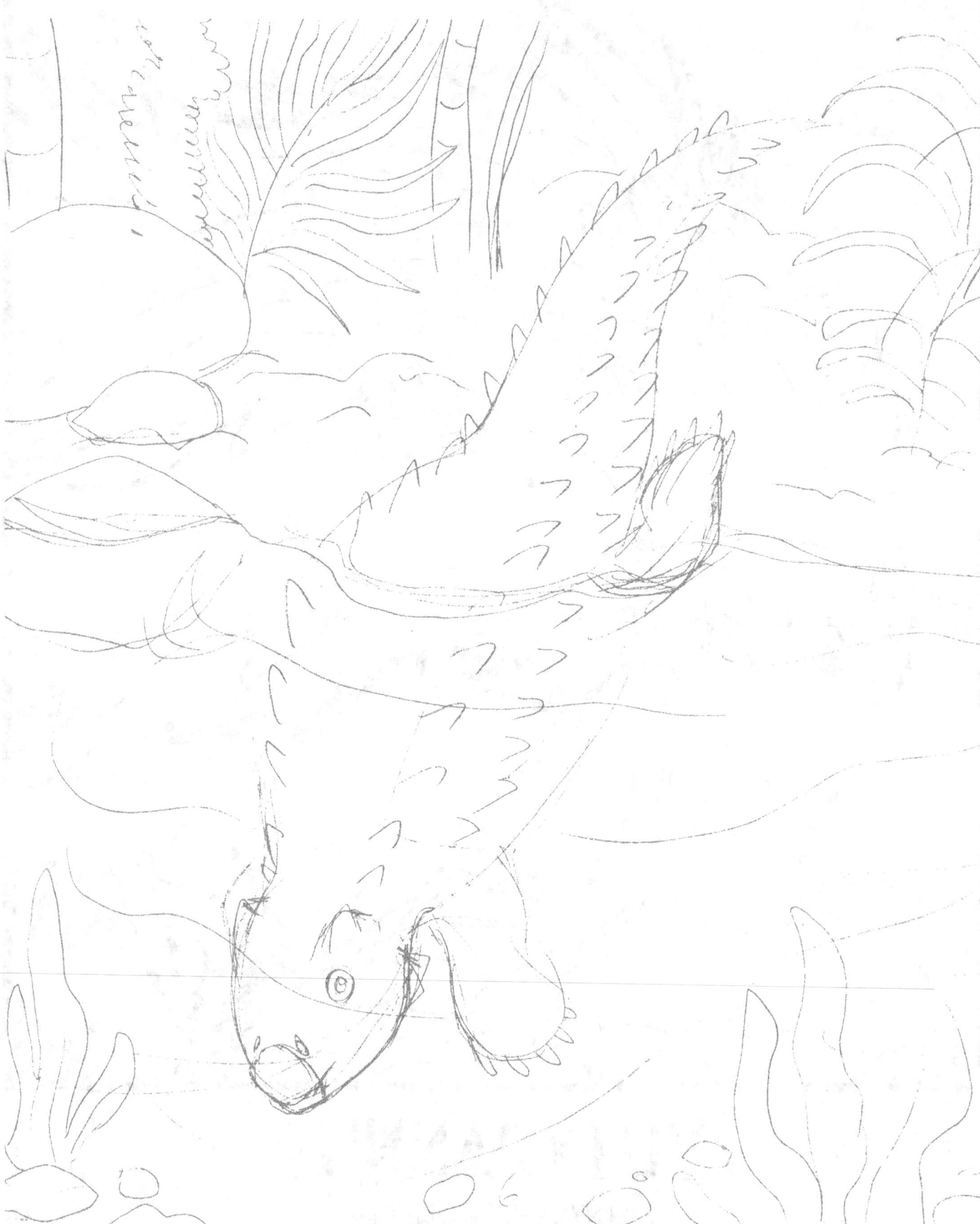

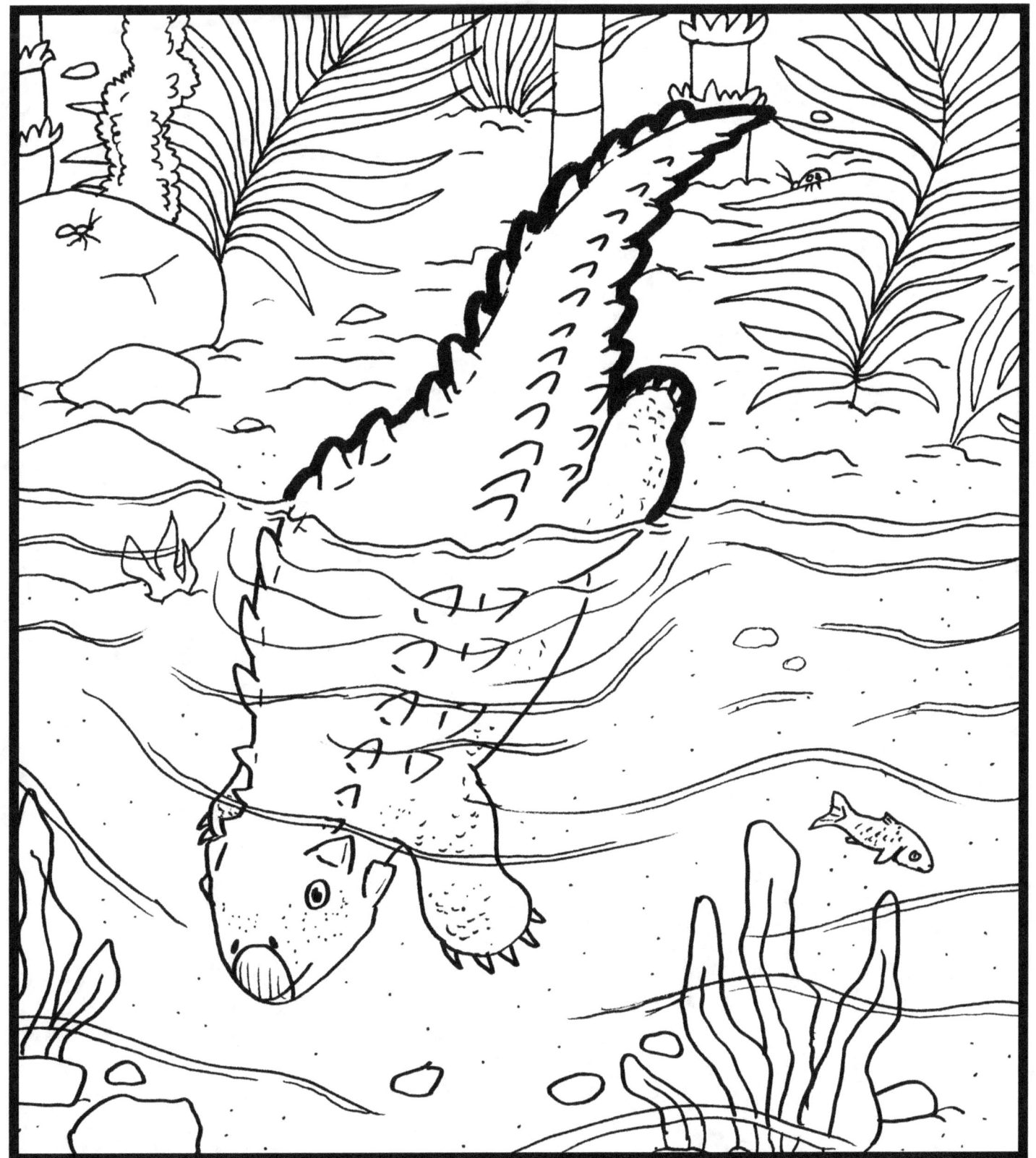

LIAONINGOSAURUS

122 MYA, Early Cretaceous, China

The smallest ankylosaur discovered to date, it was possibly aquatic. It was different than other ankylosaurs, or even other ornithischians, as fish remains have been found inside of it, indicating a hunting or scavenging behavior. As of this writing, it is the only "bird-hipped" dinosaur known to science to have been carnivorous.

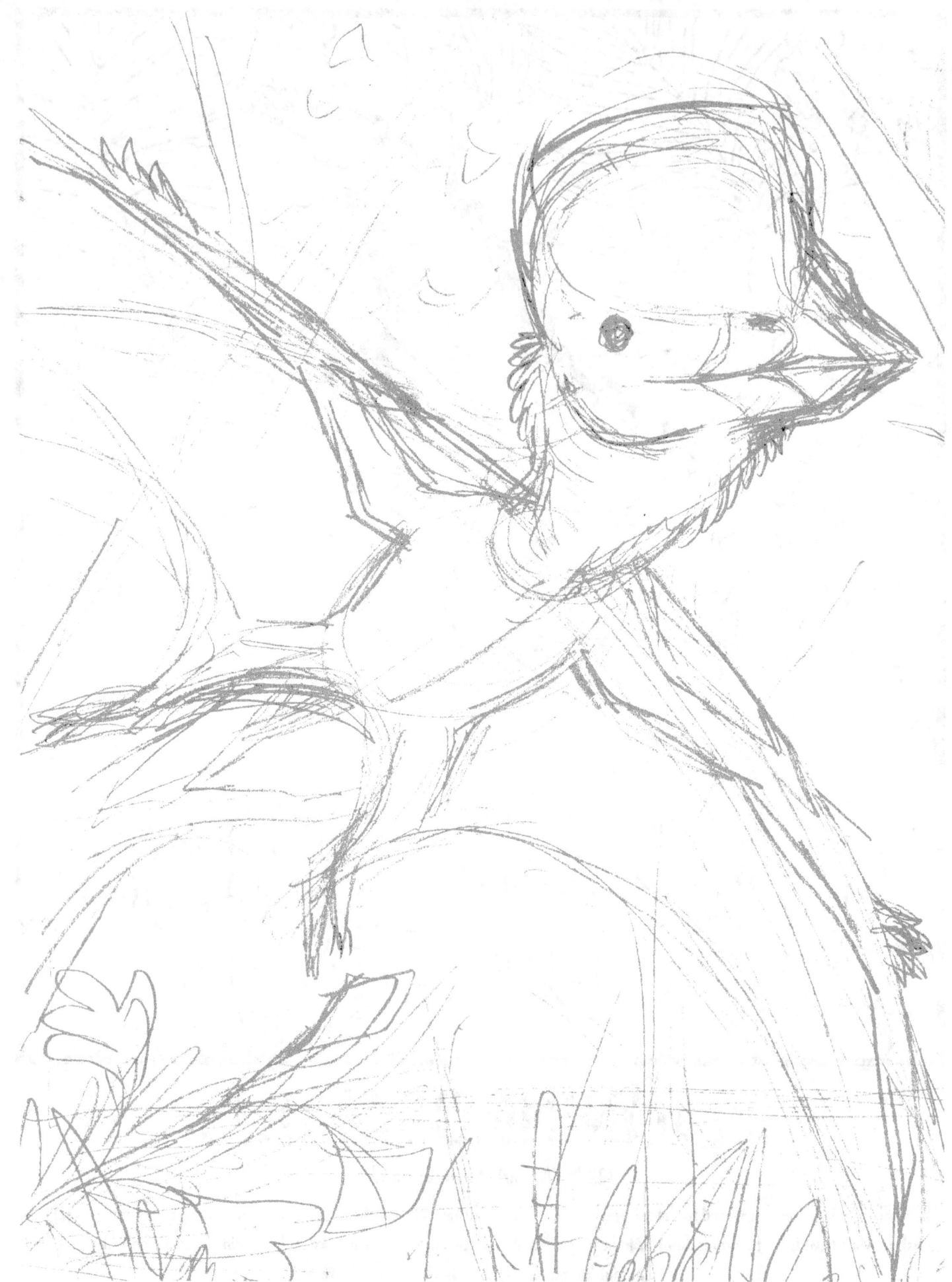

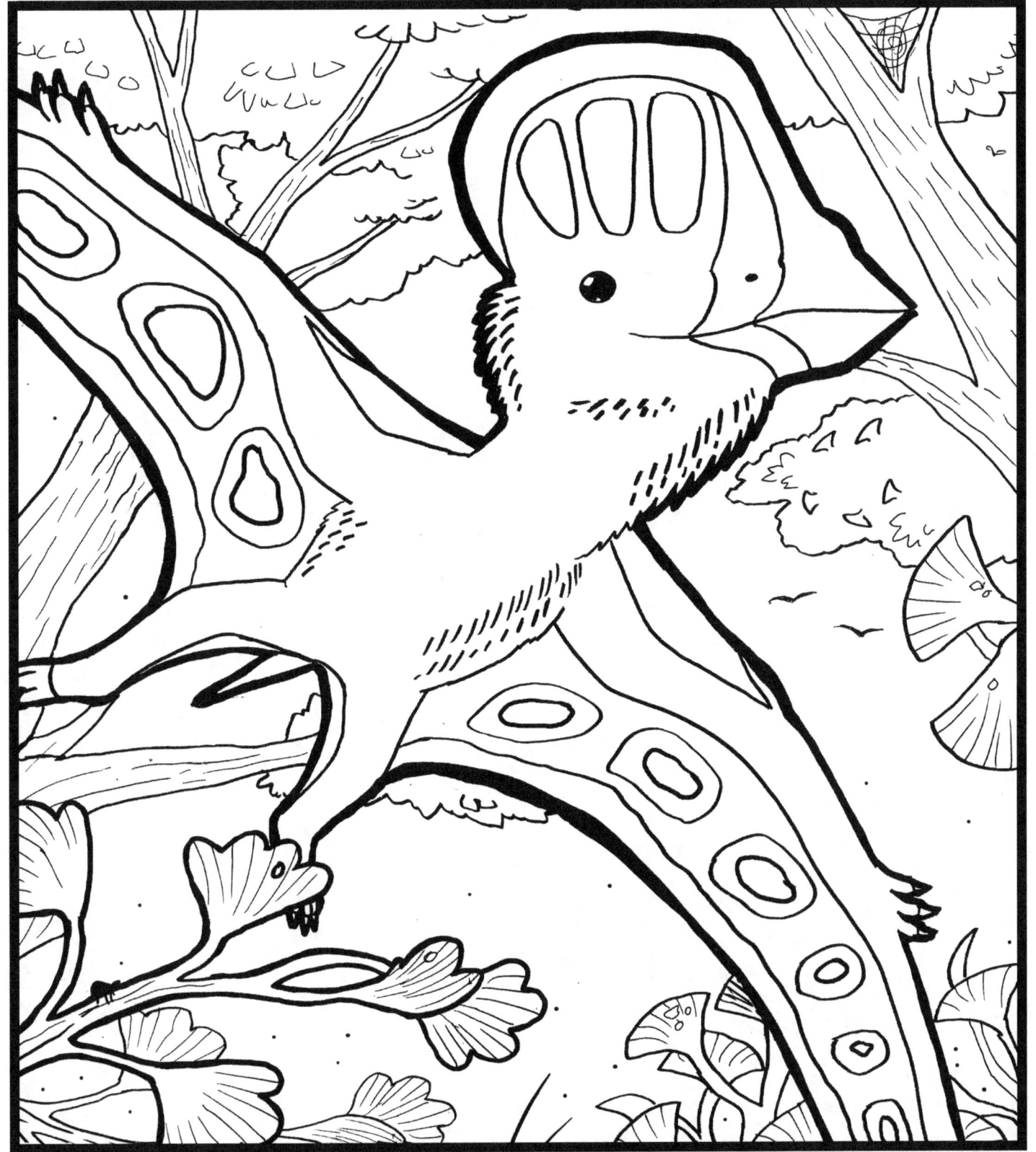

HUAXIAPTERUS

120 MYA, Early Cretaceous, China

This small tapejarid pterosaur shared a few traits with the earlier rhamphorhynchoids, and spent much of its time on the ground.
It has been found with much of the skeleton intact, allowing for more accurate, and more confident reconstructions.
In addition, the area it was discovered (the Jiufotang Formation) contained many other pterosaur fossils, making it a species-rich site.

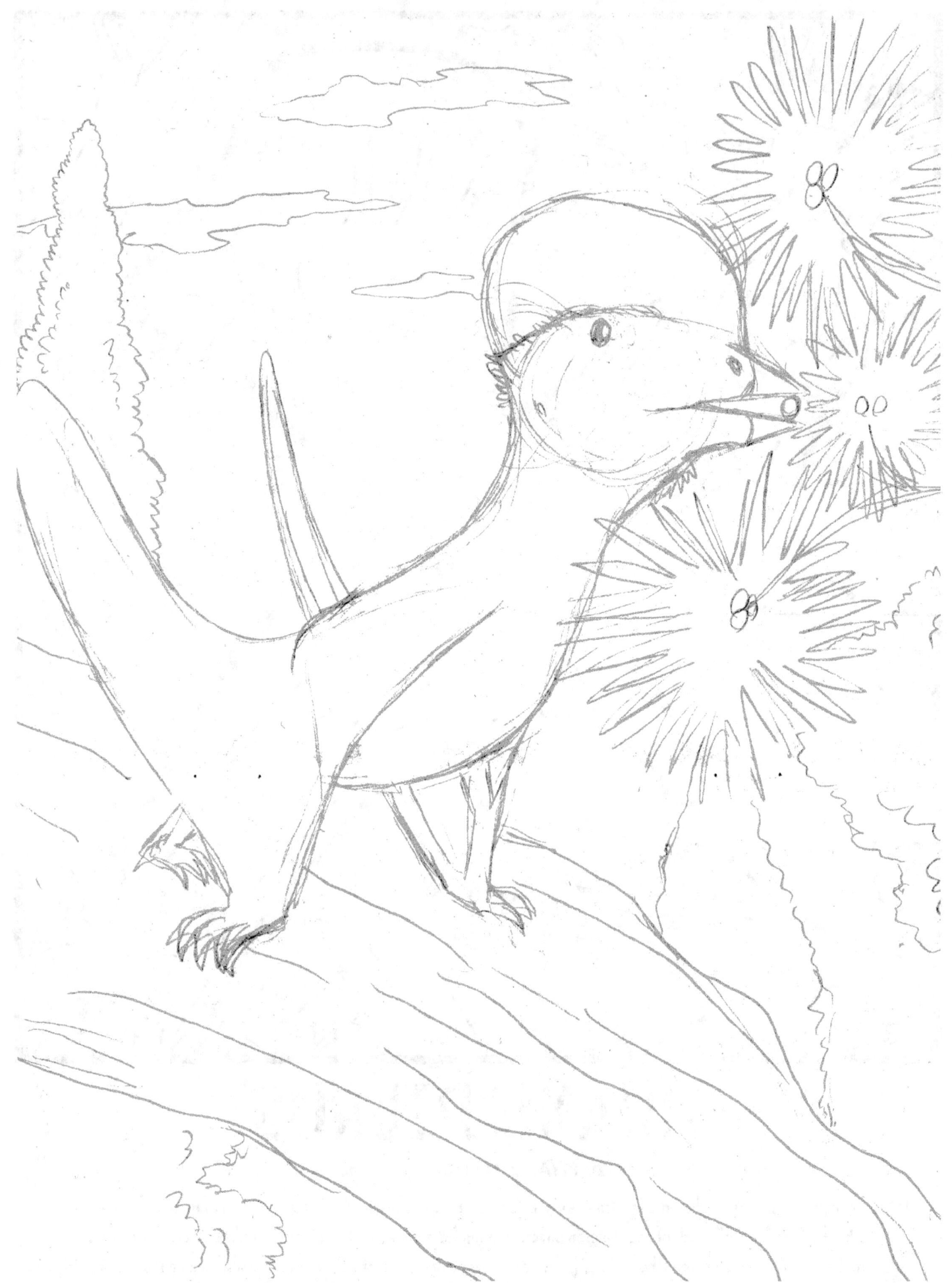

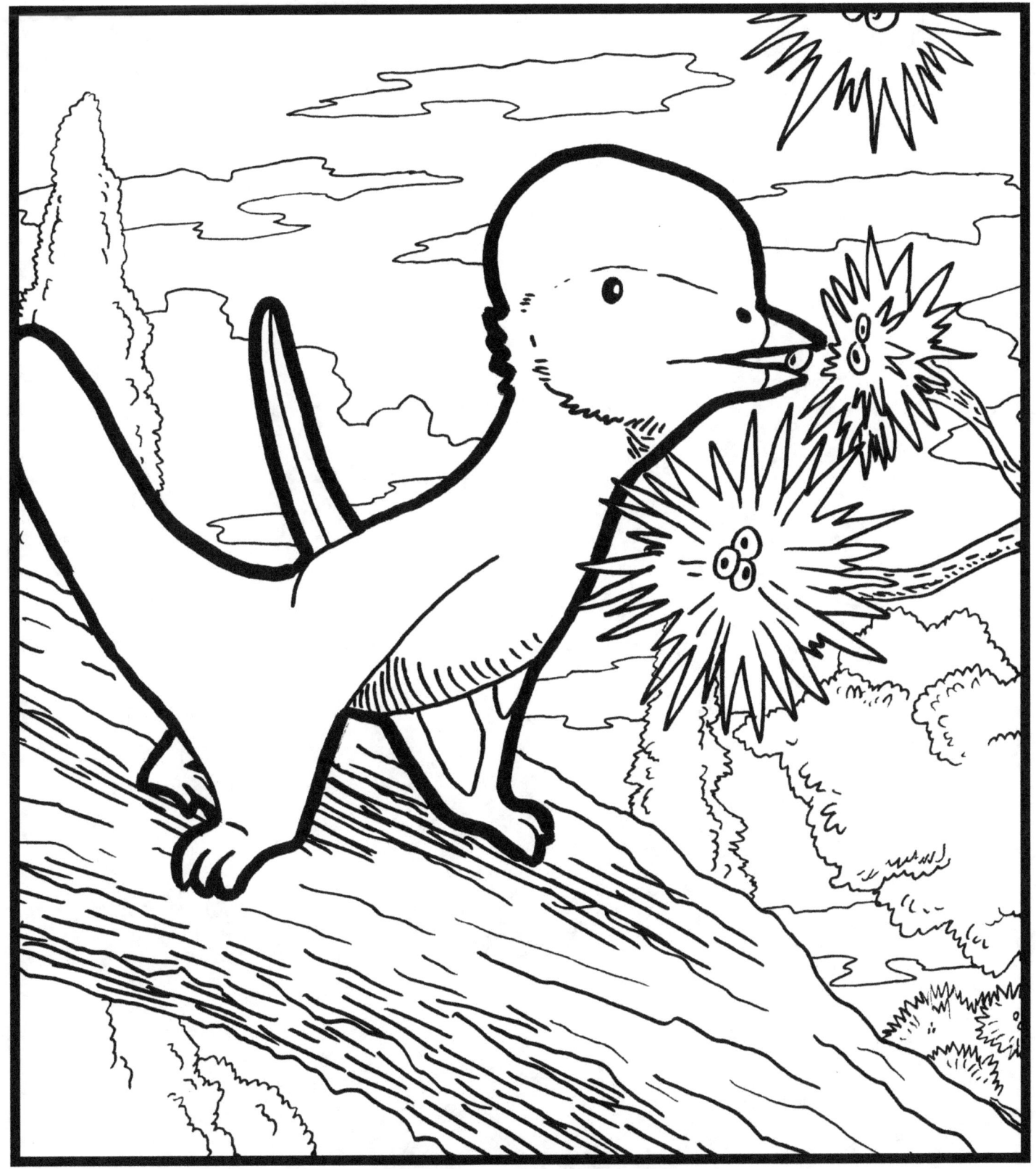

TAPEJARA

112 MYA, Mid Cretaceous, Brazil

A smaller pterosaur with a wingspan of around 3.5 meters, its head supported a long bony crest, which itself supported an even larger crest made of softer tissue. It likely lived off of fruits, insects, and maybe vertebrates, living as an omnivore or herbivore.

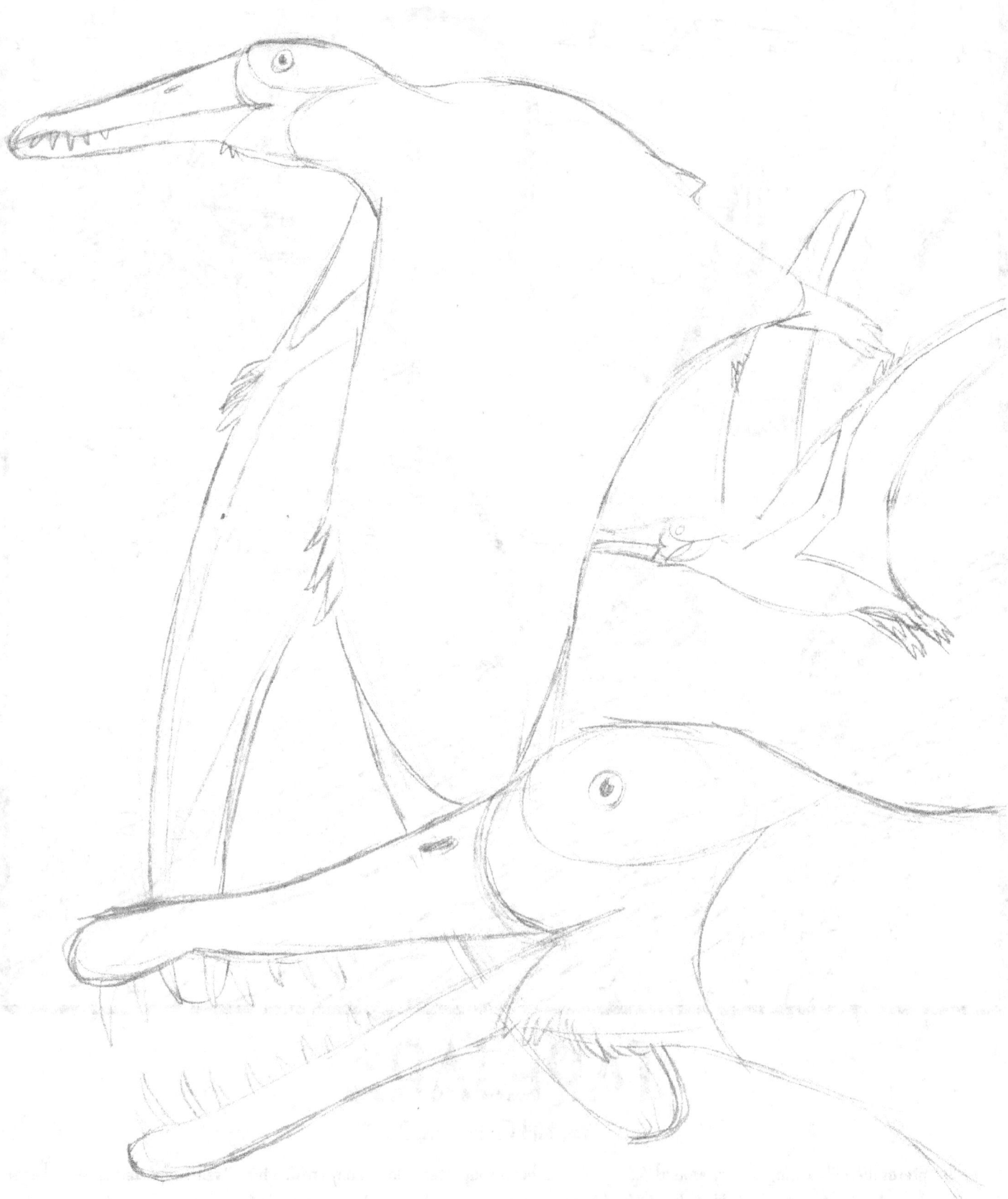

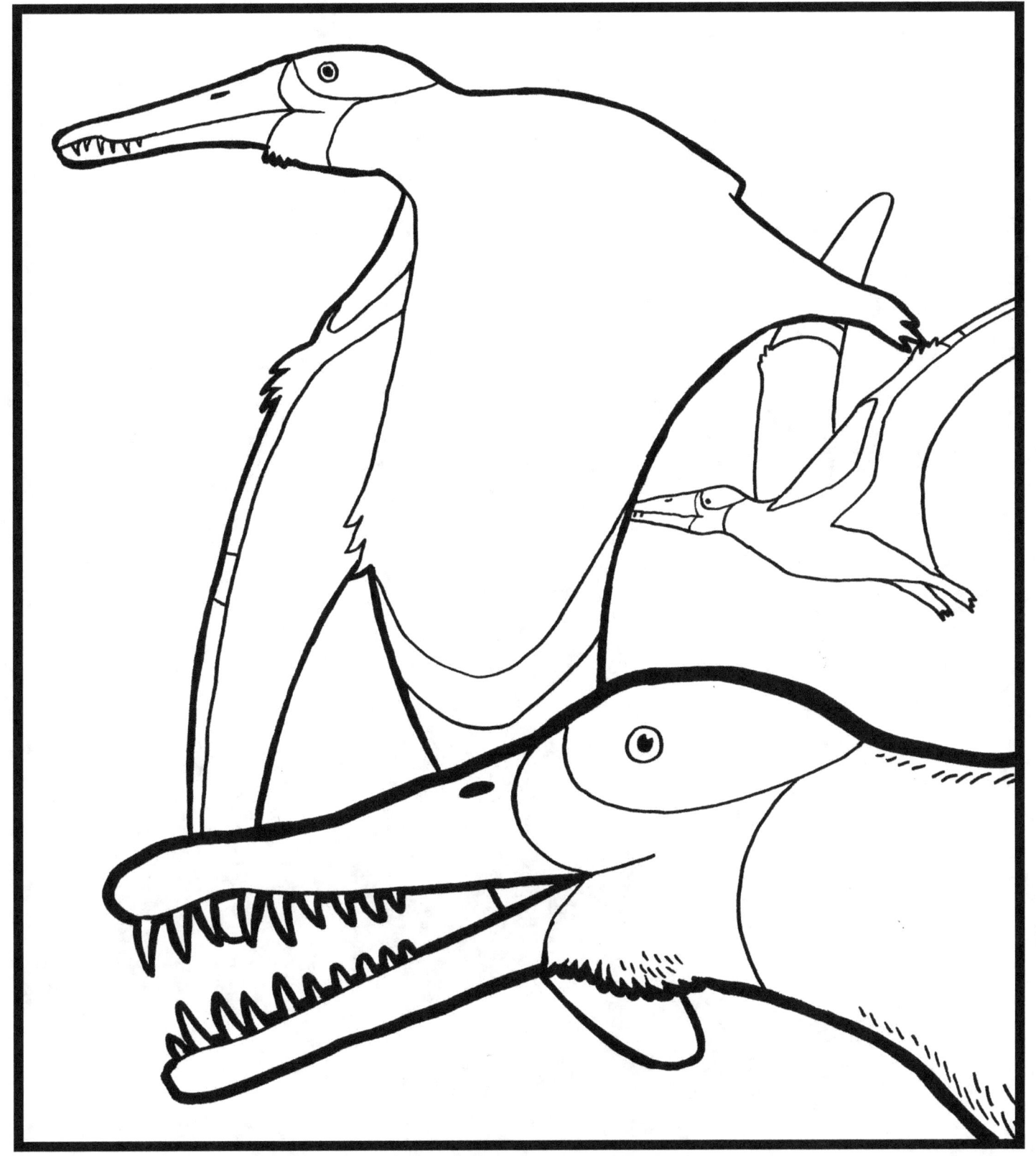

CEARADACTYLUS

112 MYA, Early Cretaceous, Brazil

This pterosaur's long, cone-shaped teeth were suitable for grabbing and holding onto slippery fish. It was a mid-size pterosaur, with a wingspan about 5.5 meters across. Living on the coast, it probably spent its days soaring over waves, and catching fish in mid-flight.

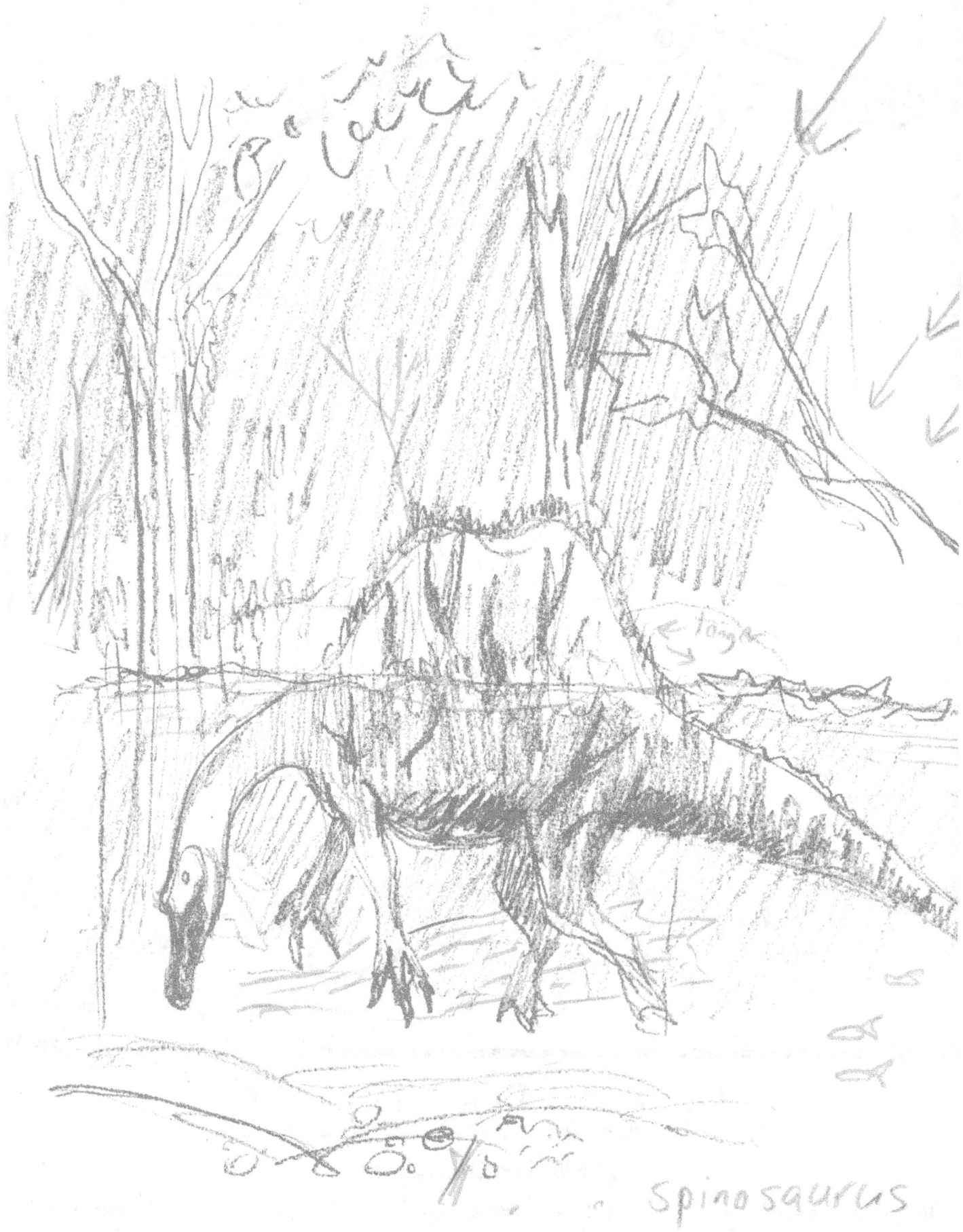

spinosaurus

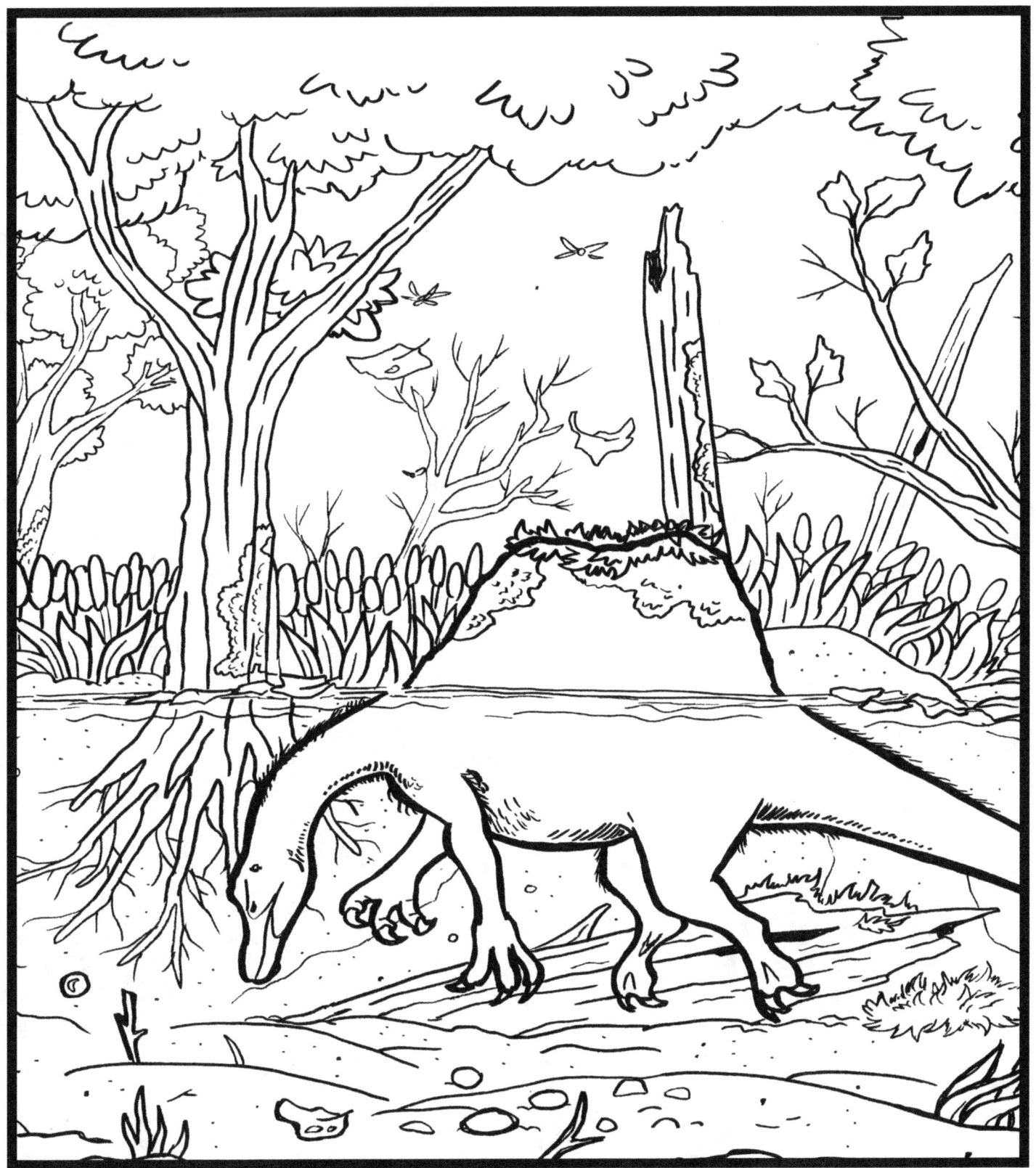

SPINOSAURUS

112 - 93.5 MYA, Late Jurassic to Early Cretaceous, Egypt

A theropod that is as controversial as it is popular, Spinosaurus was a fish-eating dinosaur that lived in and around shorelines and mudflats. Though its mouth and teeth were shaped to best grasp slippery fish, it would probably not have turned up its nose at other prey, much like the alligators and crocodiles of today.

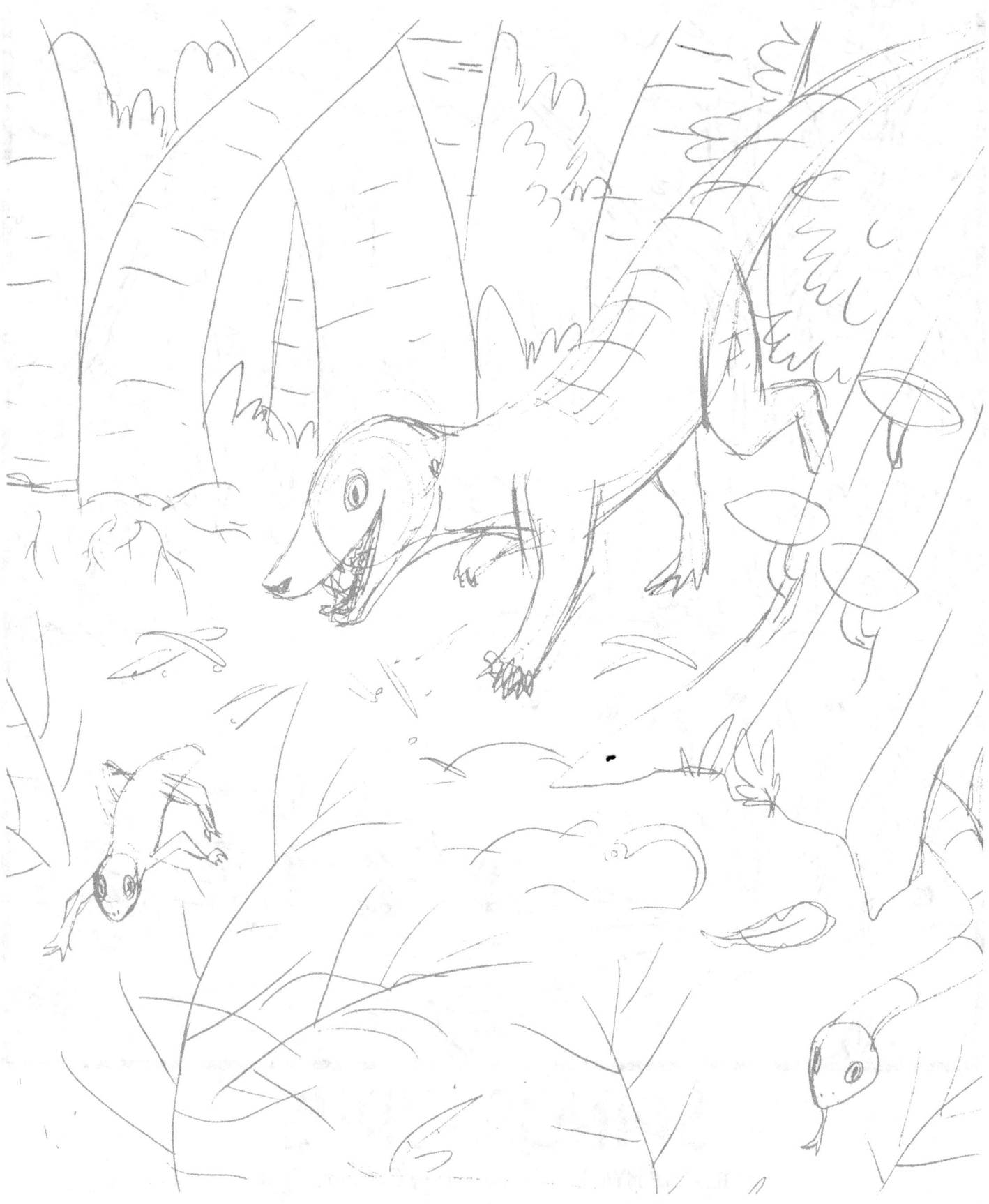

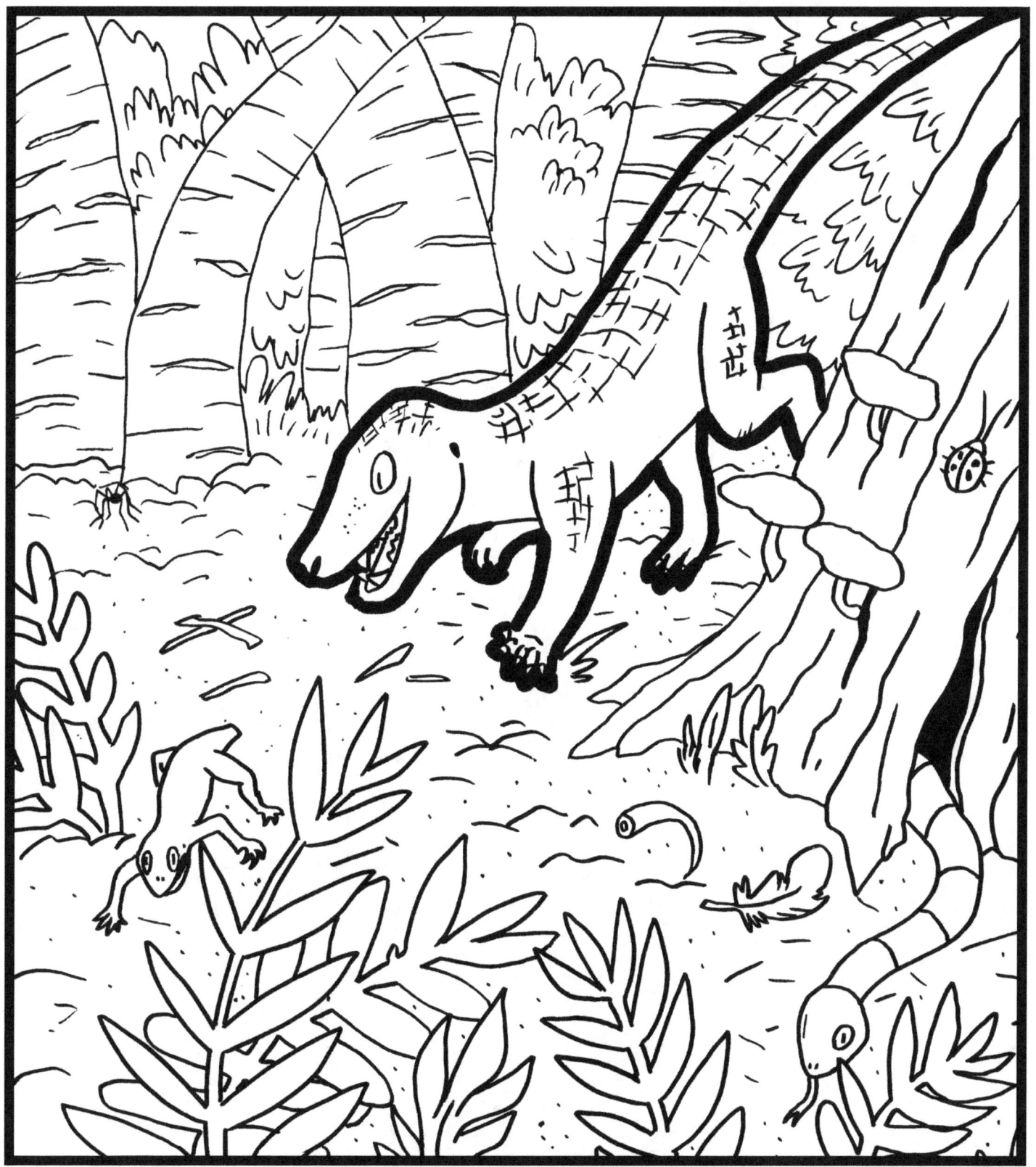

PAKASUCHUS

105 MYA, Early Cretaceous, Tanzania

This notosuchian's tooth remains are said to be able to pass as those of a mammal's. Its teeth have varied shapes, unlike the more uniform teeth other crocodyliforms sport. It was a swift runner, and having fewer osteoderms made it even lighter on its feet. It also had a doglike nose, and a slender build. Its name means "cat crocodile," referencing its similarity to a modern cat.

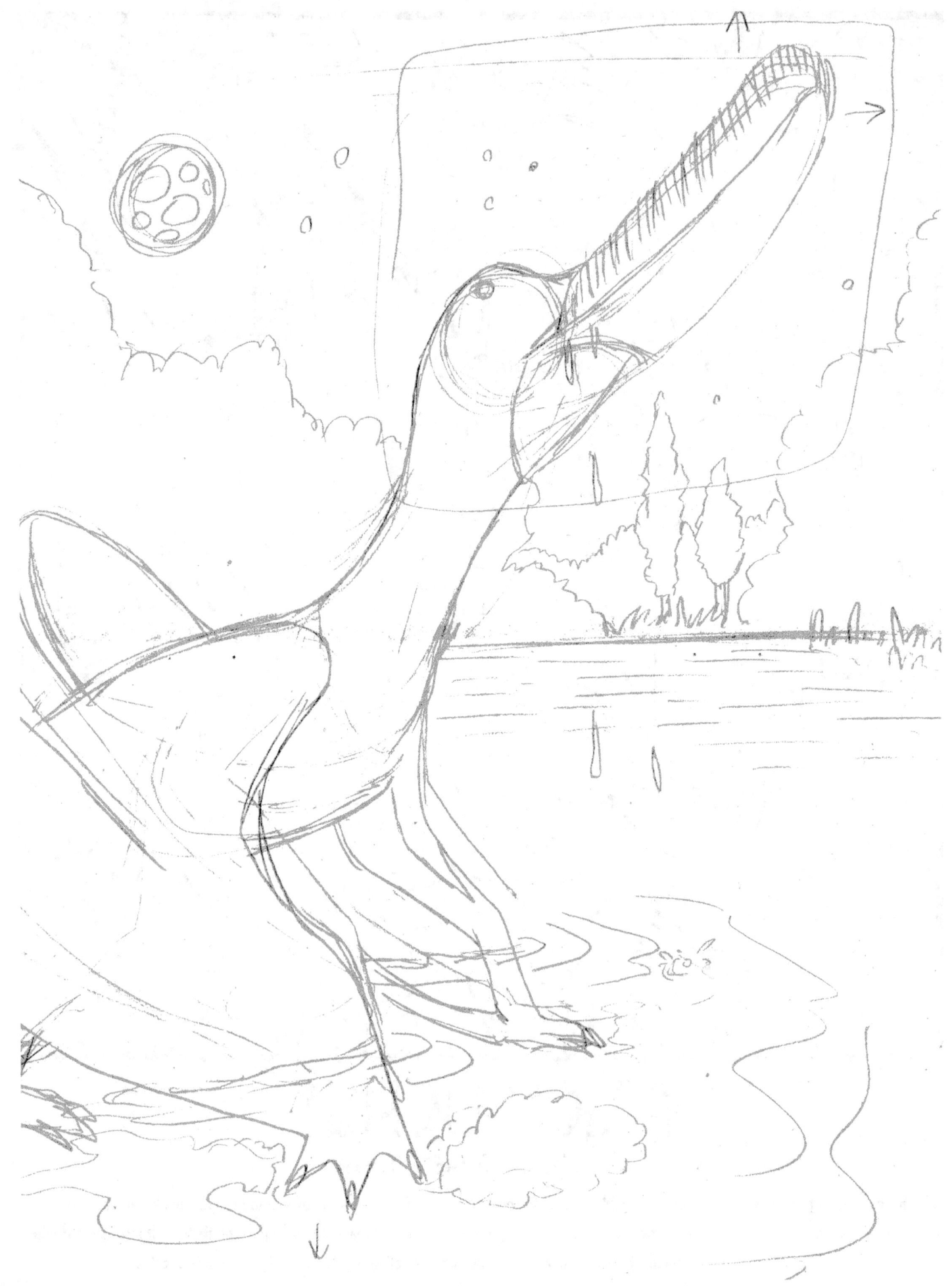

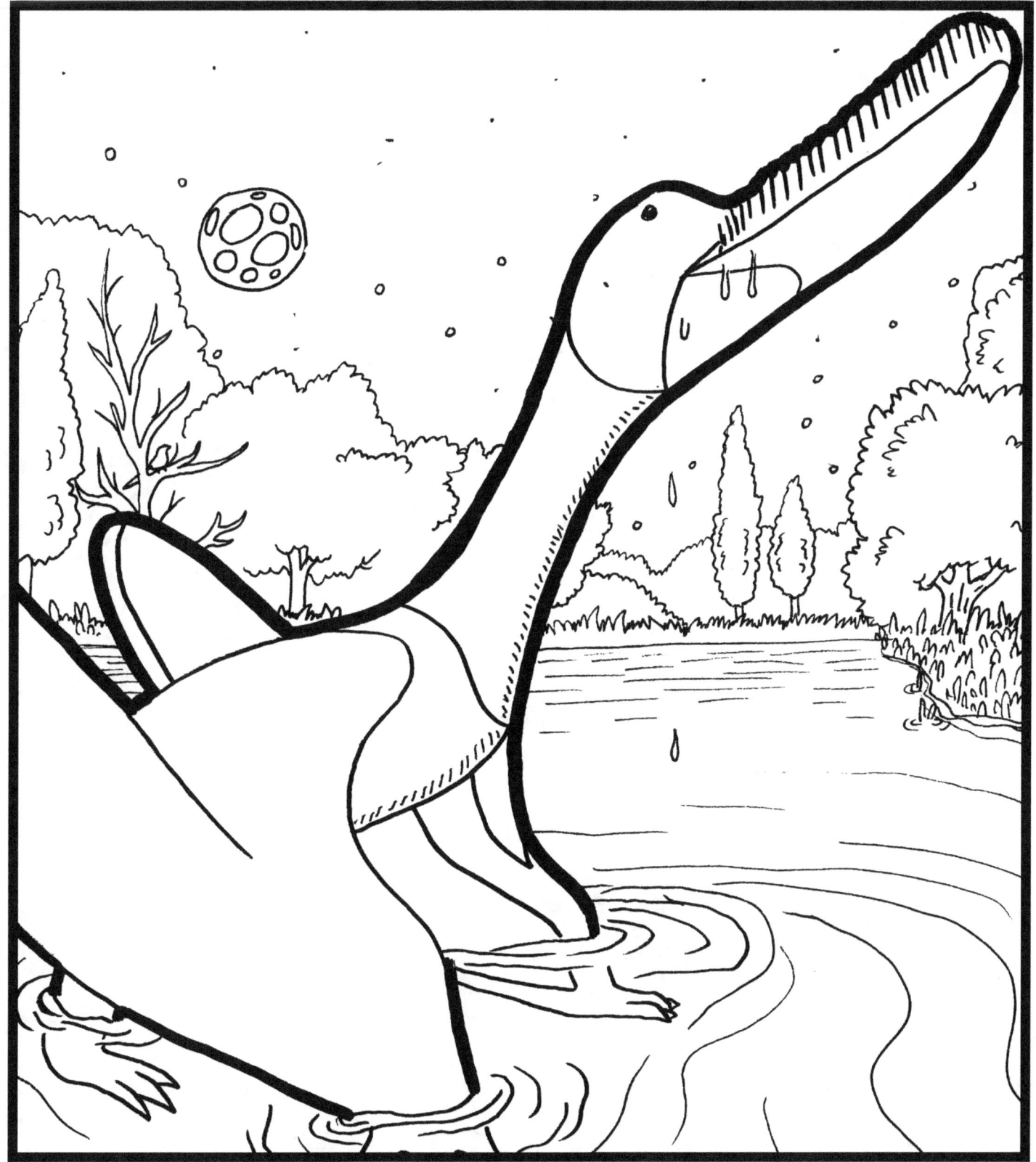

PTERODAUSTRO

105 MYA, Early Cretaceous, Argentina, Chile

A rather small pterosaur, its long jaws were filled with bristle-like teeth that were well-suited for sifting through water for tiny crustaceans. Looking at the scleral rings (a ring-shaped bone that supports the eye), and comparing them to the scleral rings of birds today, it was found to have been nocturnal.

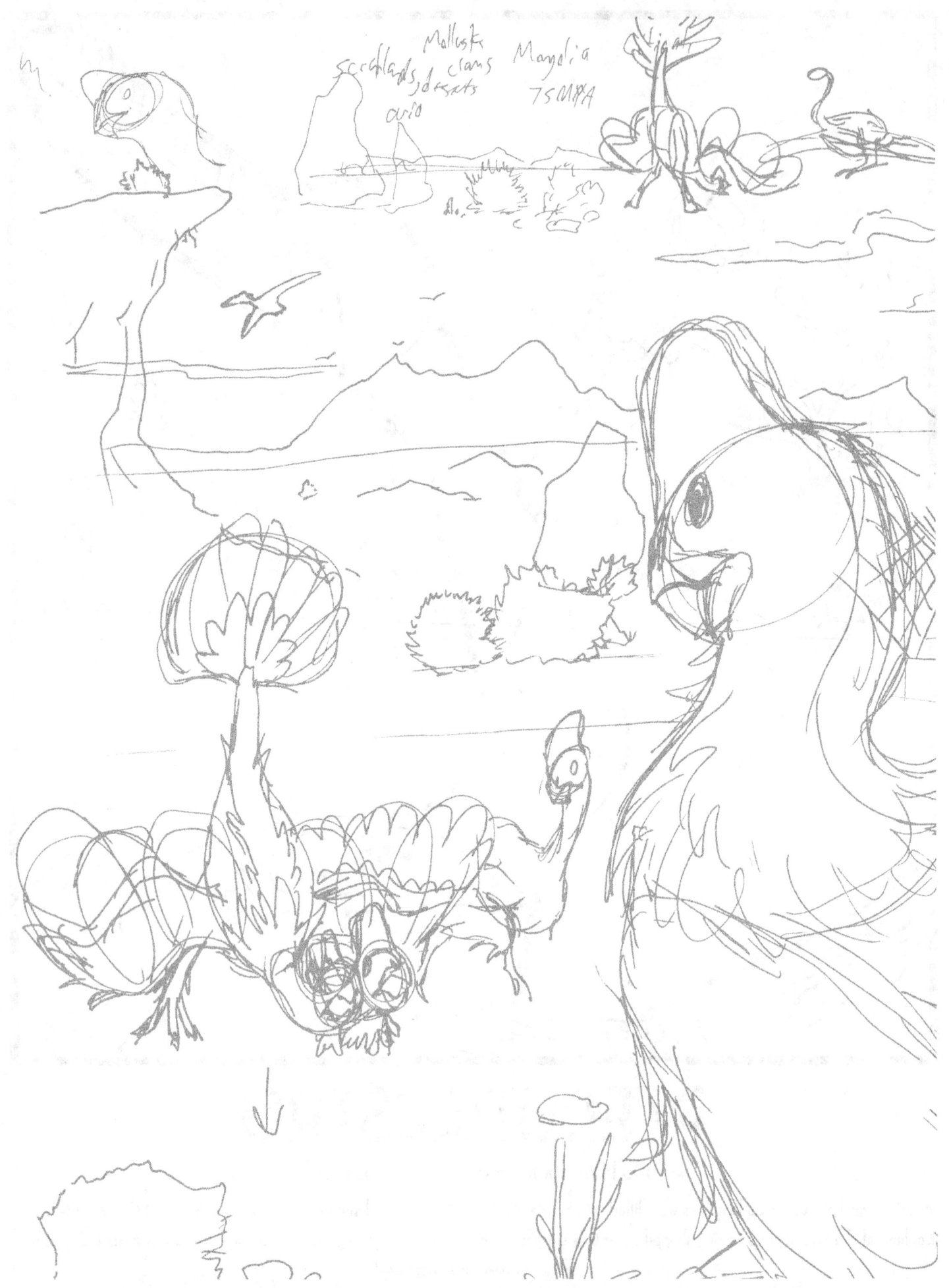

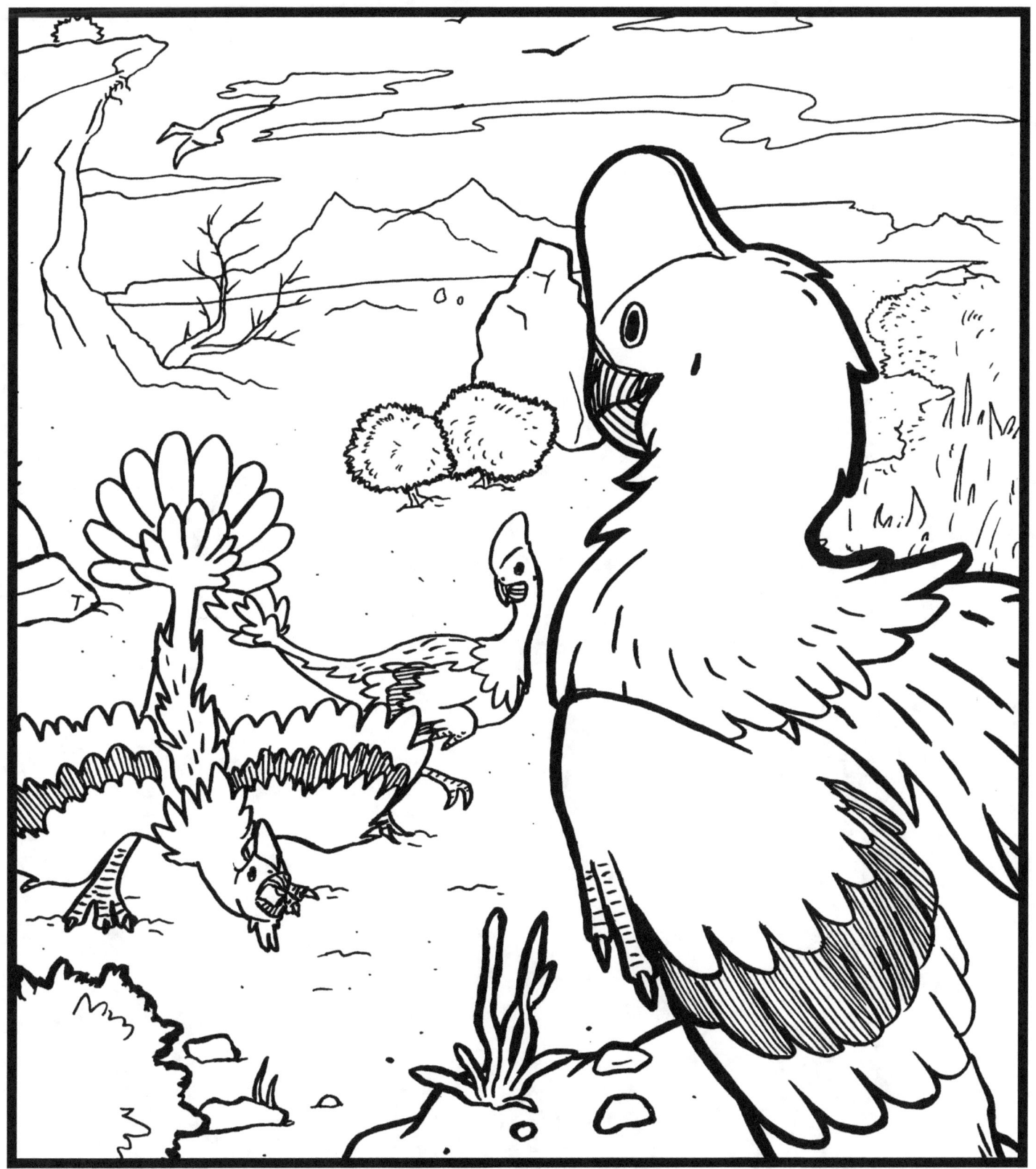

CITIPATI

84 - 75 MYA, Late Cretaceous, Mongolia

These oviraptorids are commonly used to reconstruct members of the poorly preserved Oviraptors, since more and better quality fossils have been documented. The first Citipati were discovered crouching over their own nests, arms spread across on either side over the eggs. This posture is often found in modern birds, and it was decided that Citipati and other oviraptorids were also covered in feathers.

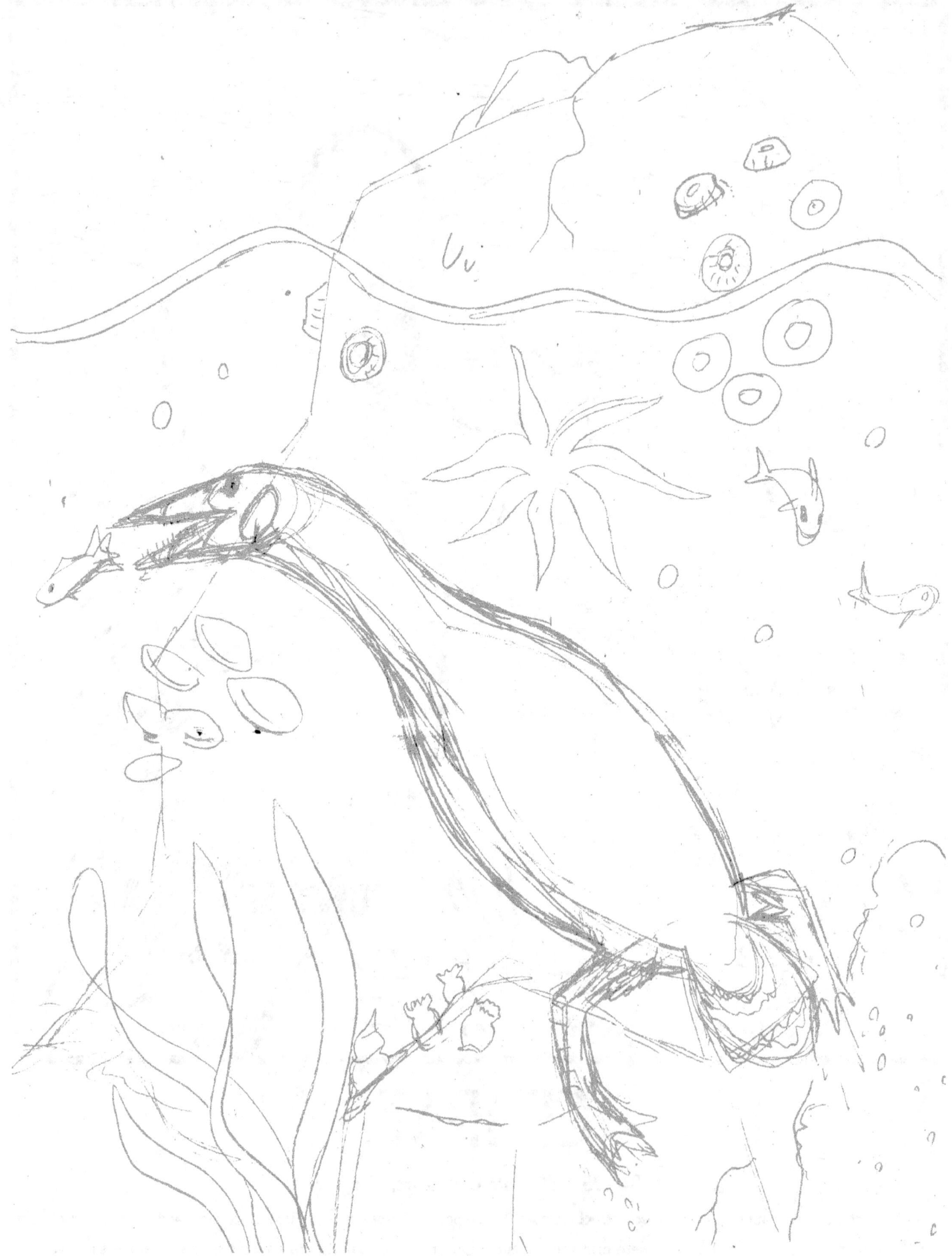

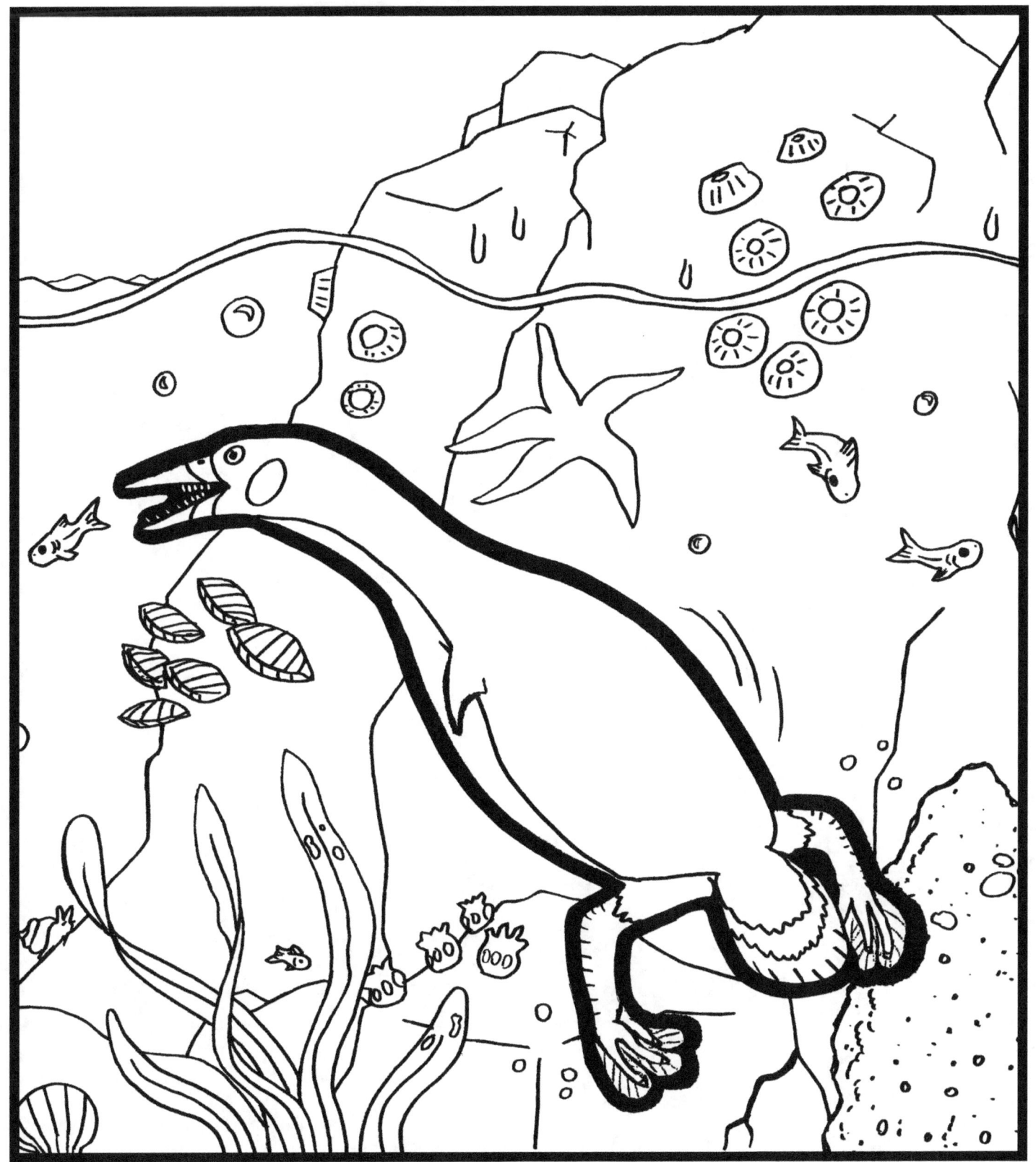

HESPERORNIS

83.5 - 78 MYA, Late Cretaceous, Canada, Russia, United States

This 1.8 meter-long flightless aquatic bird was certainly as elegant in the water as it was ungainly on land. Its wings had been reduced to tiny flippers, and it retained teeth in its beak, which helped grip the slippery prey it would encounter. Hesperornis itself fell prey to the mosasaurs, which have been discovered with this bird's remnants inside of them.

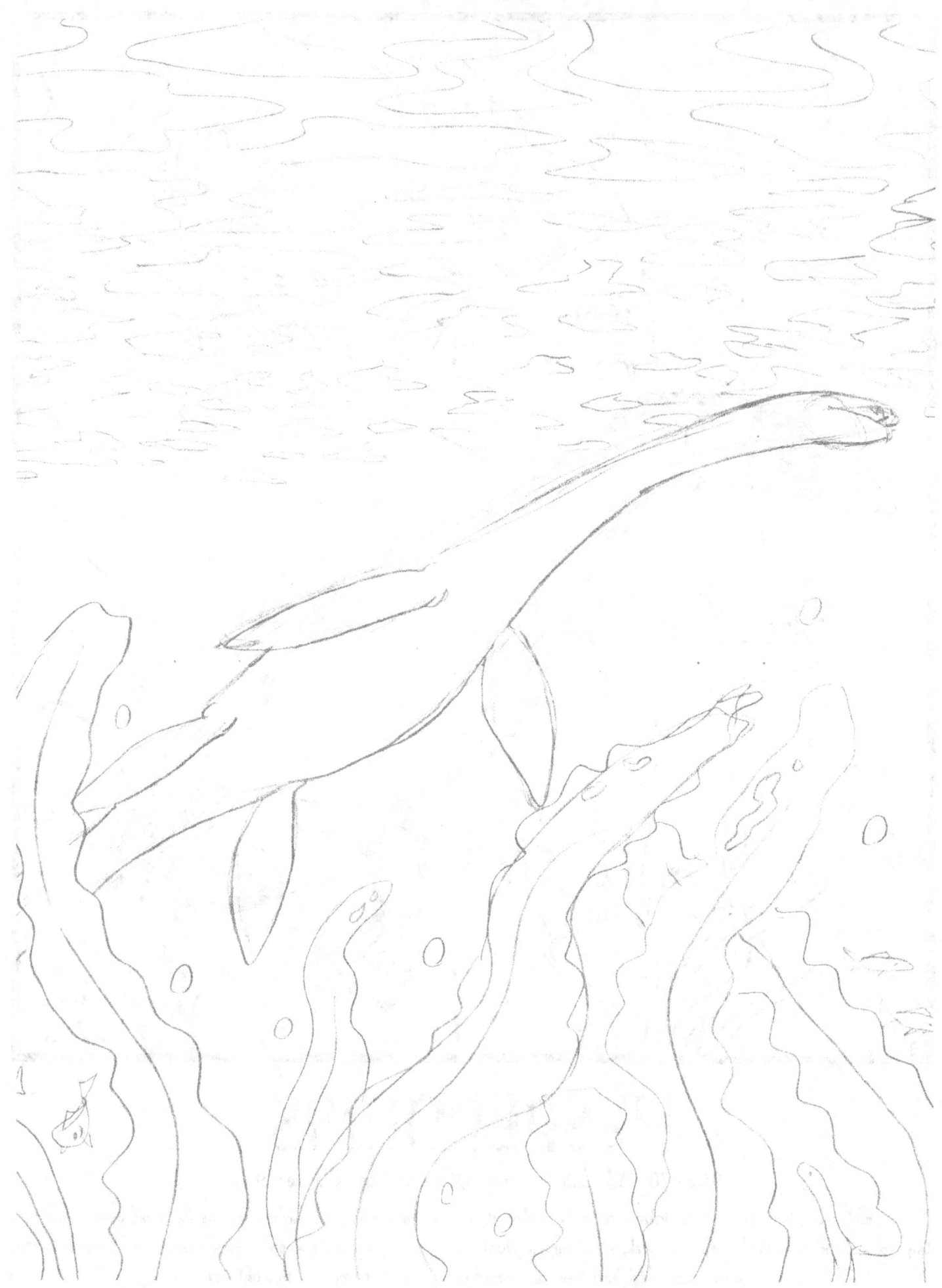

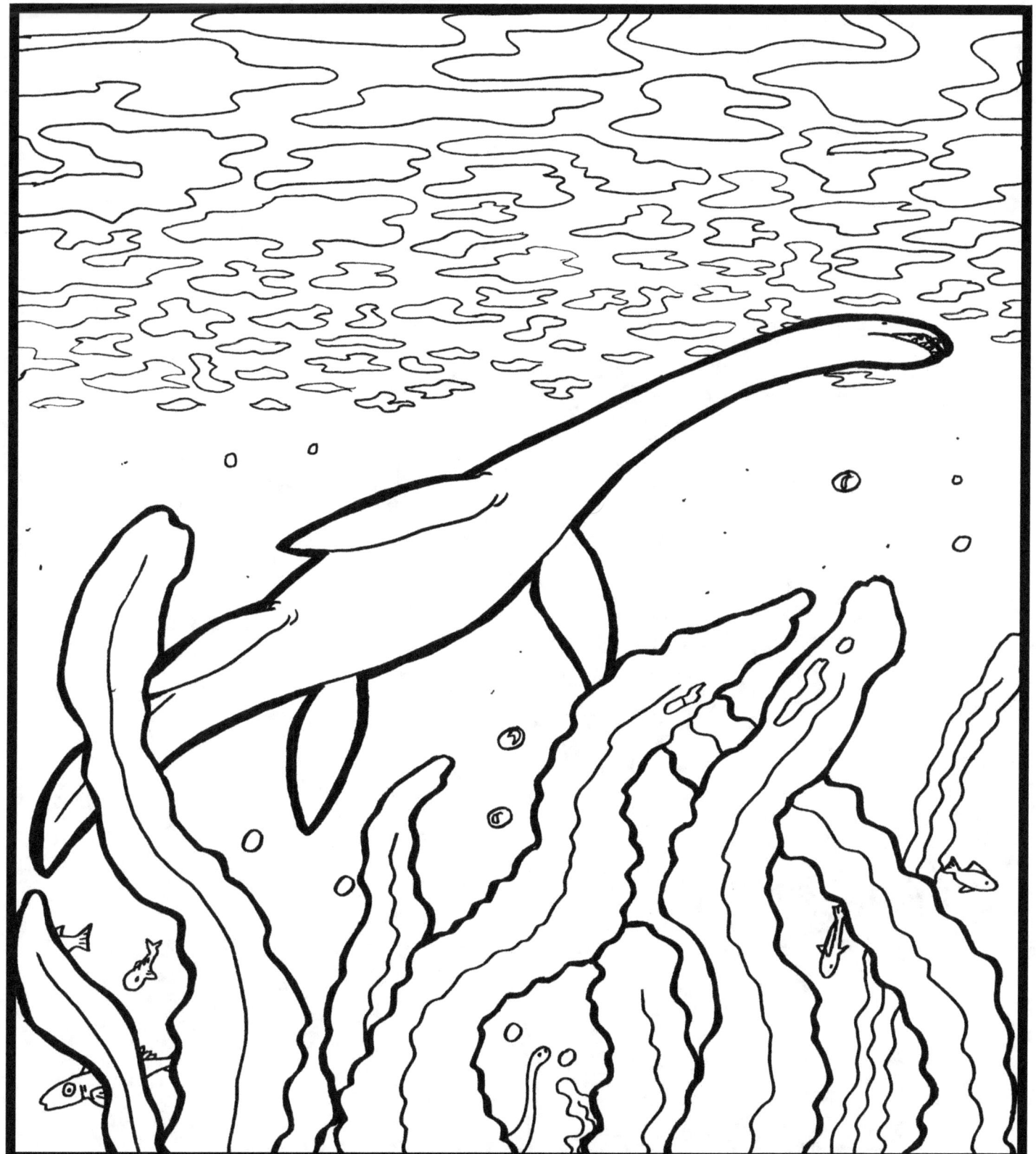

ELASMOSAURUS

80 MYA, Late Cretaceous, United States and Canada

At over 10 meters long, this was one of the largest plesiosaurs. Plesiosaurs generally had long necks, equally sized paddles, short tails, and gave birth to live young. Elasmosaurus had 72 neck vertebrae, and such a long neck helped it get close to the small fish and squid on which it preyed. Stones have also been found in the stomach area, indicating it had swallowed them to help with digestion.

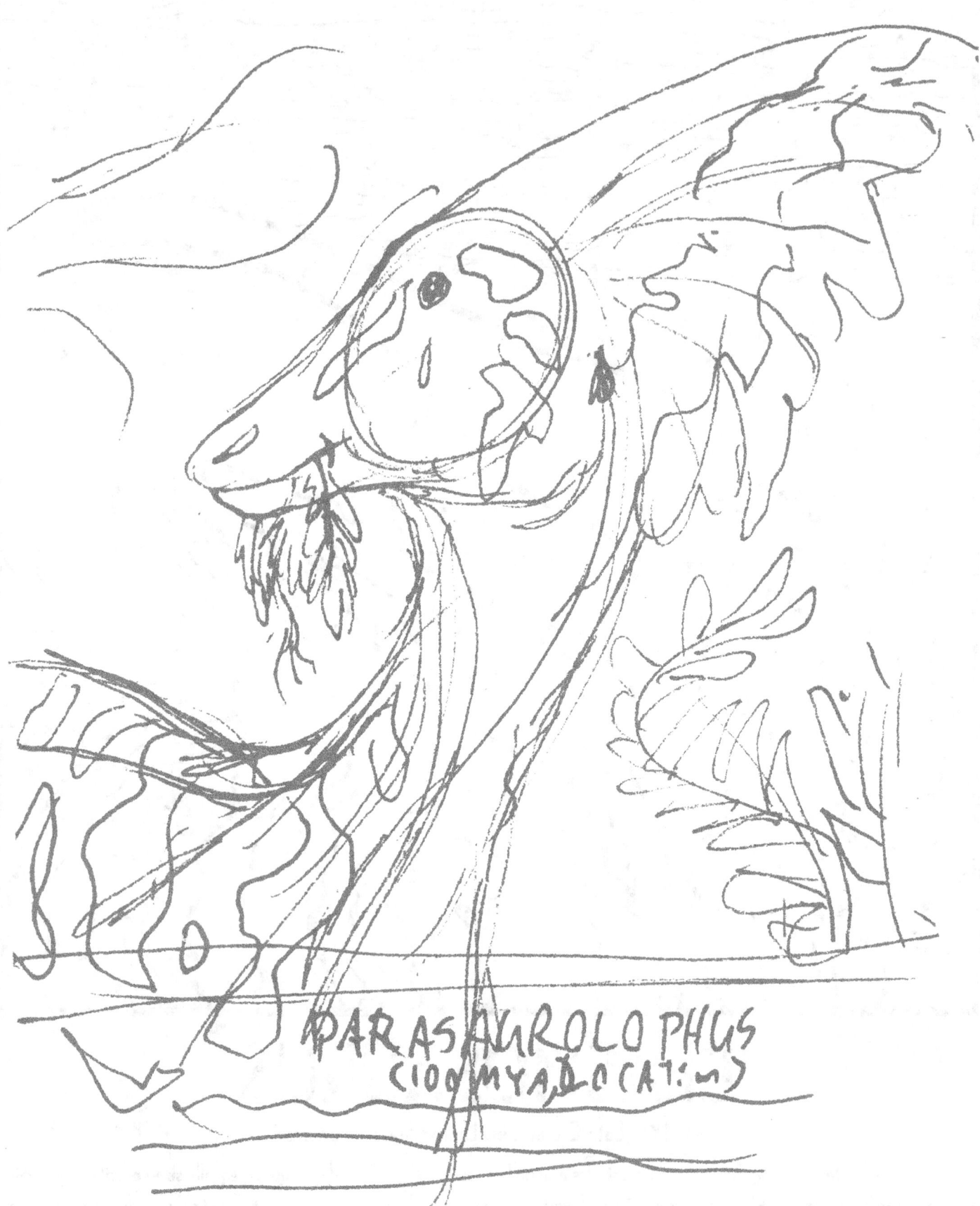

PARASAUROLOPHUS
(100 MYA, LOCATION)

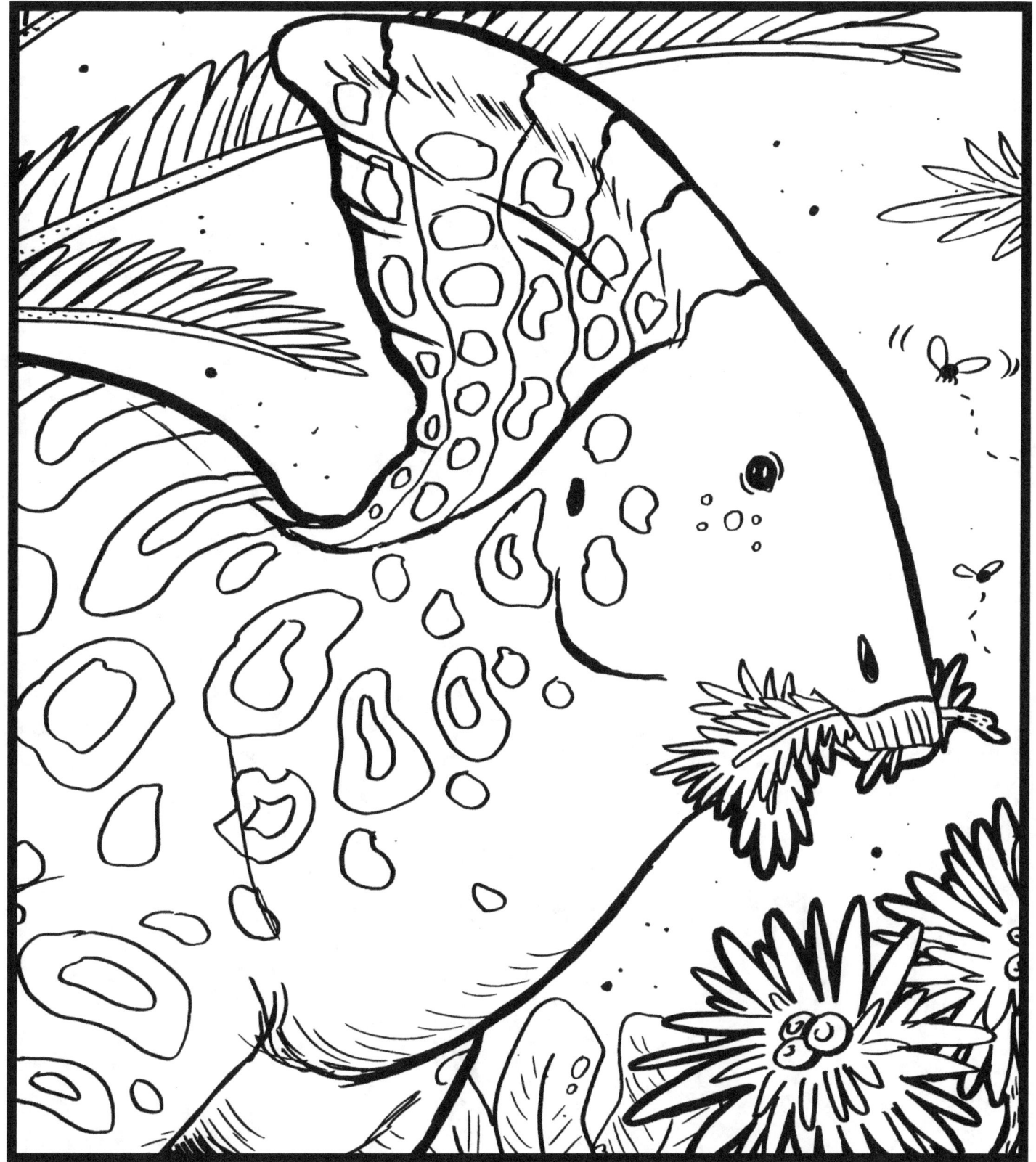

PARASAUROLOPHUS

76 - 74 MYA, Late Cretaceous, United States and Canada

Like other hadrosaurs, this dinosaur could be bipedal or quadrupedal. These animals were rare in the fossil record. Their dental battery was suited for grinding up tough plants, and worn down teeth were simply replaced. The crest has been given a variety of uses over the years, including display, sound production, a snorkle-like function, and scent detection, though most are outdated.

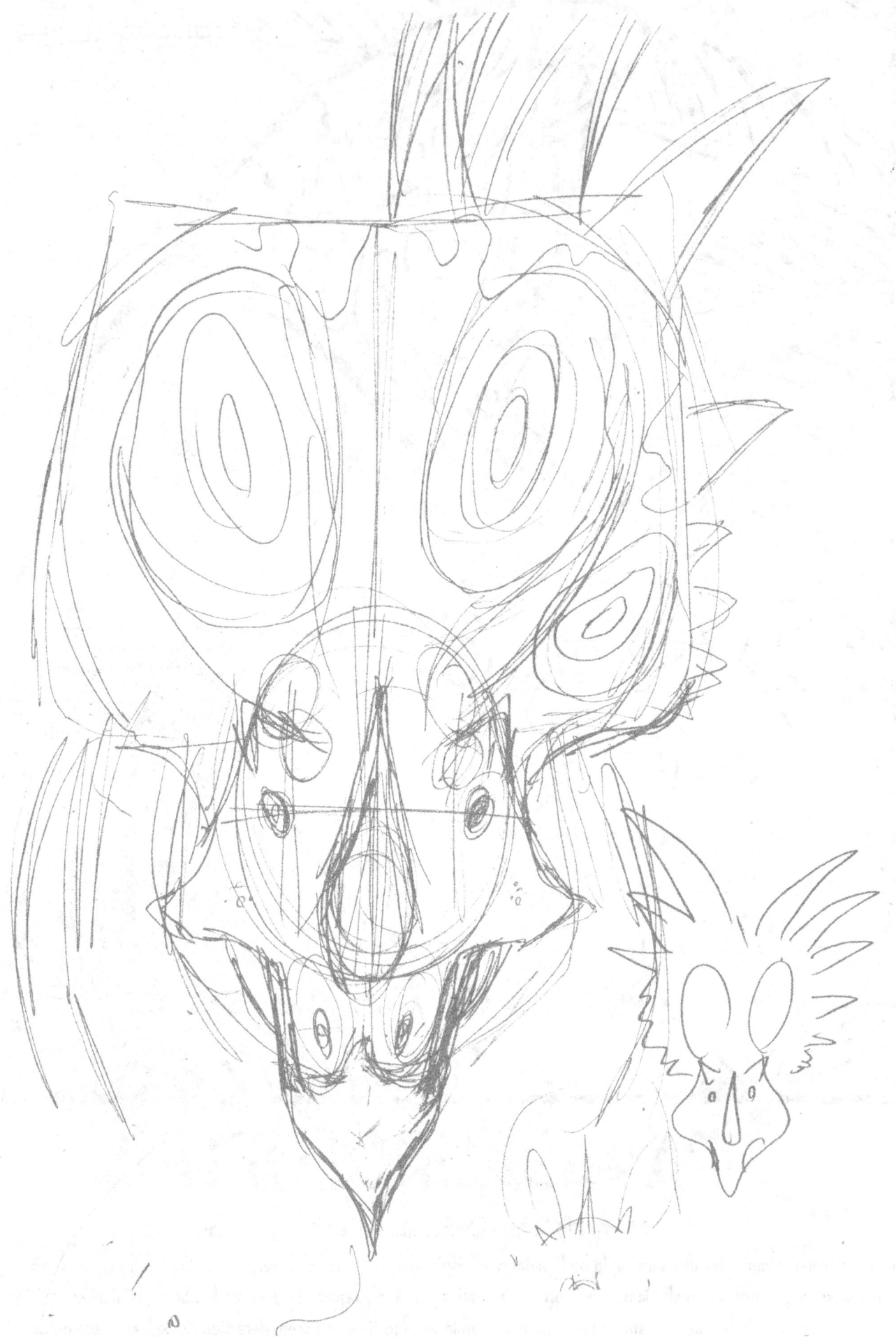

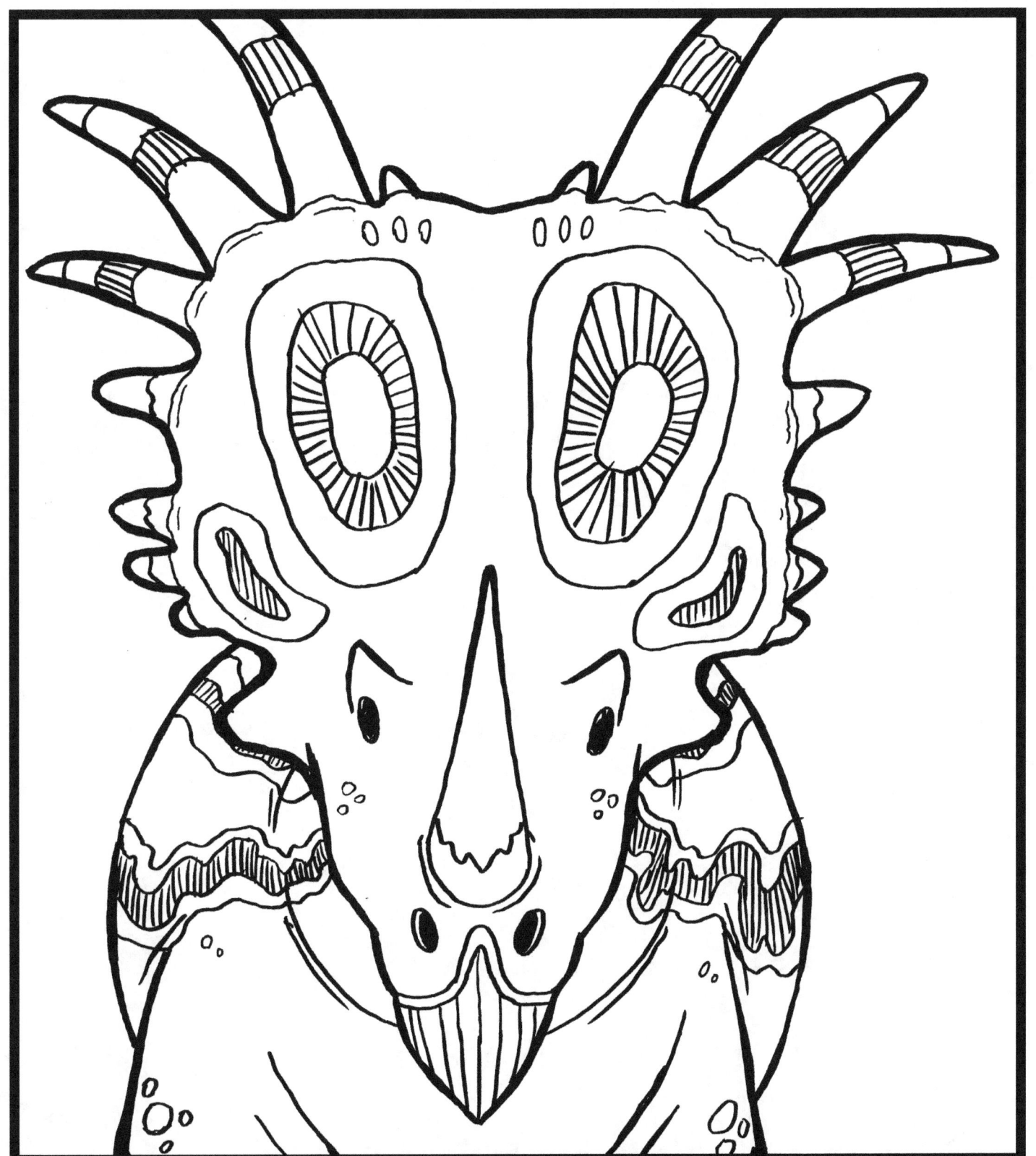

STYRACOSAURUS

75 MYA, Late Cretaceous, United States and Canada

This "spiked lizard" lived in herds, and used its beak to eat tough vegetation. Its mouth was filled with a dental battery of teeth that were constantly replaced as they got worn down by slicing vegetable matter. Because of its relative fragility, the extensive horn and frills might have been used in display rather than defense, though they might have made the animal appear larger to potential predators.

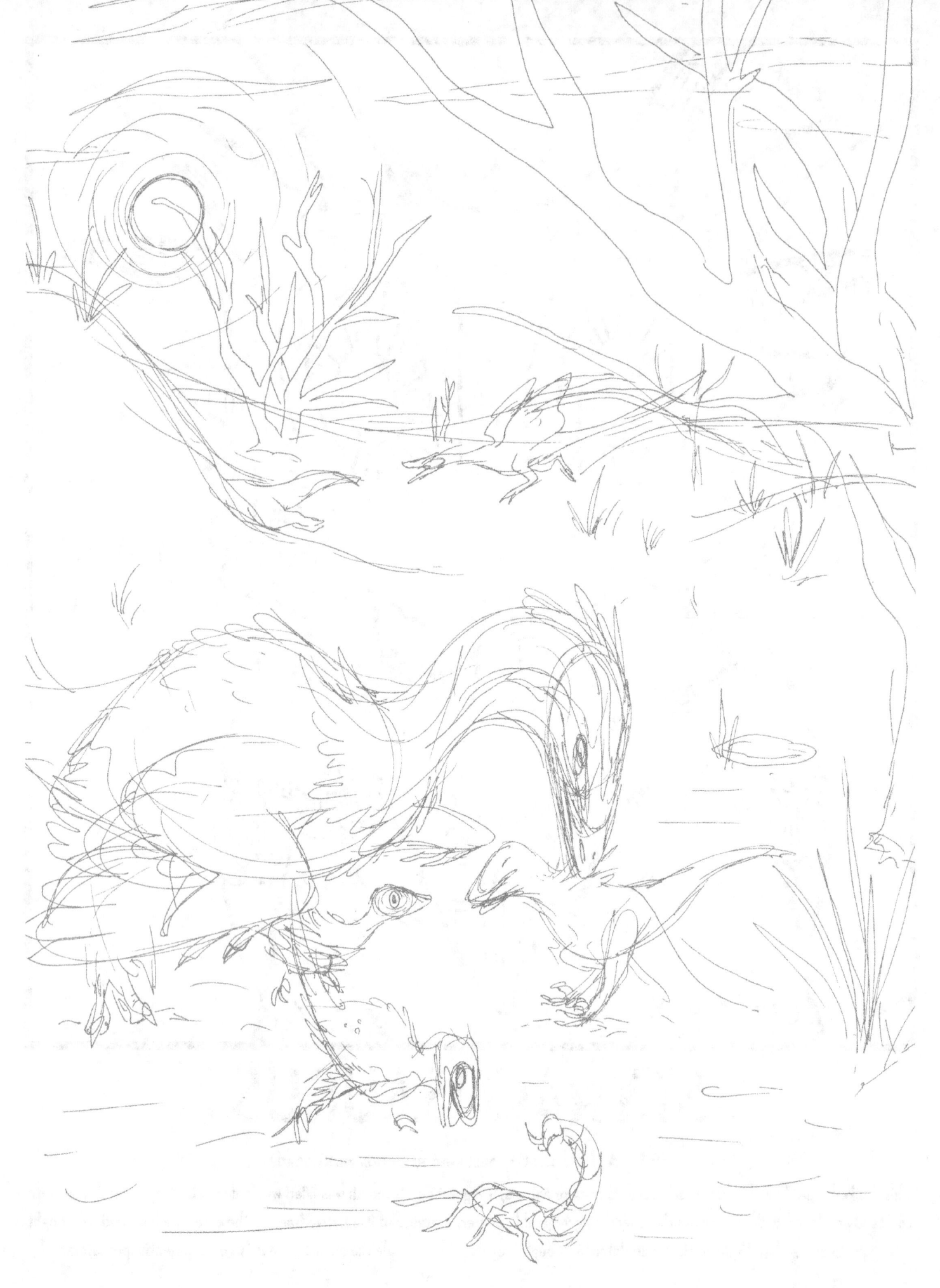

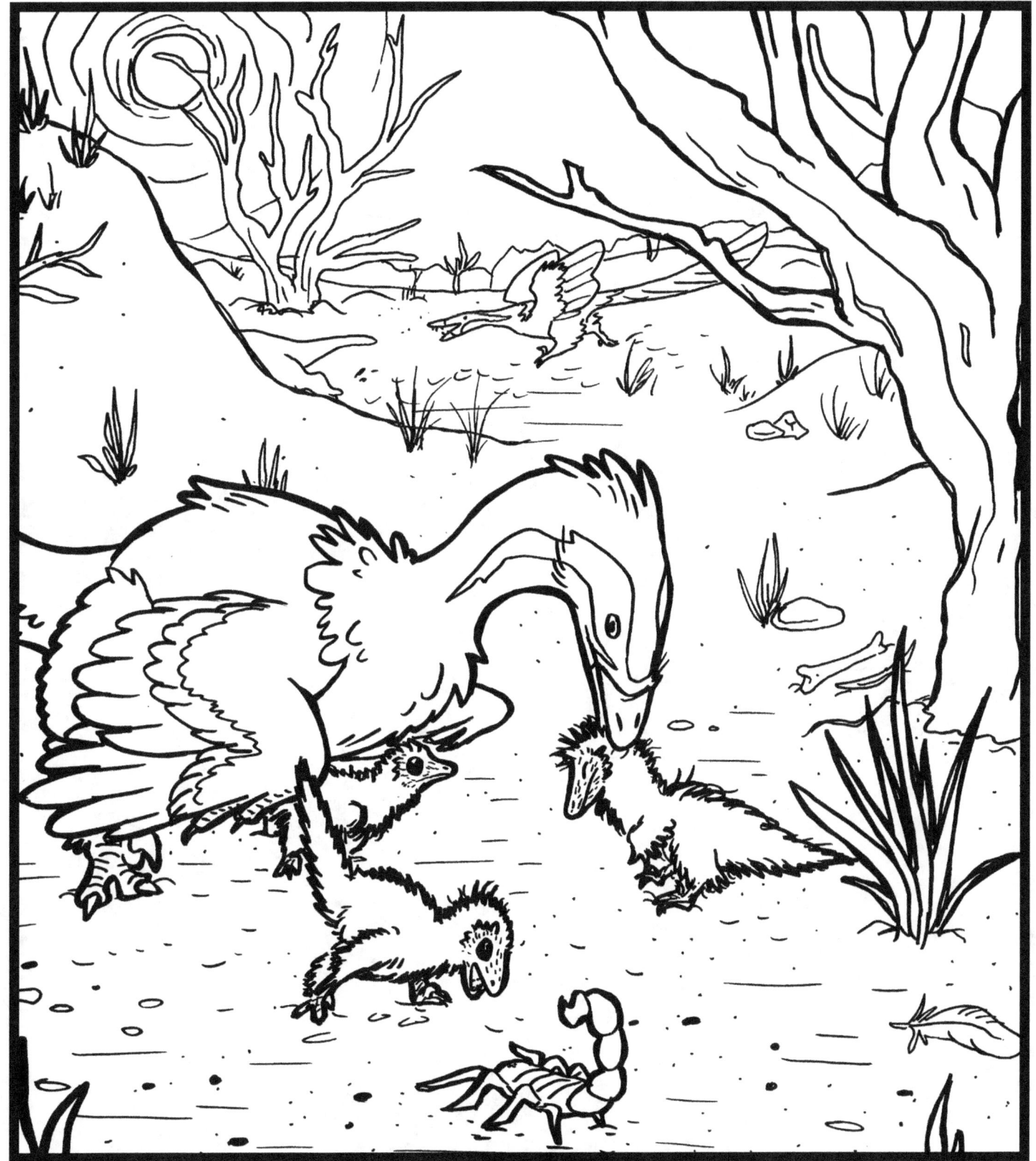

VELOCIRAPTOR

75 - 71 MYA, Late Cretaceous, Mongolia

This dinosaur was just under two meters long, and, like many other raptors, had a long sickle-shaped claw on each foot, held off the ground. They had complex feather structures rivaling those found in many modern birds, evidenced by quill knobs discovered on the forelimbs showing where the feather attachments would have been.

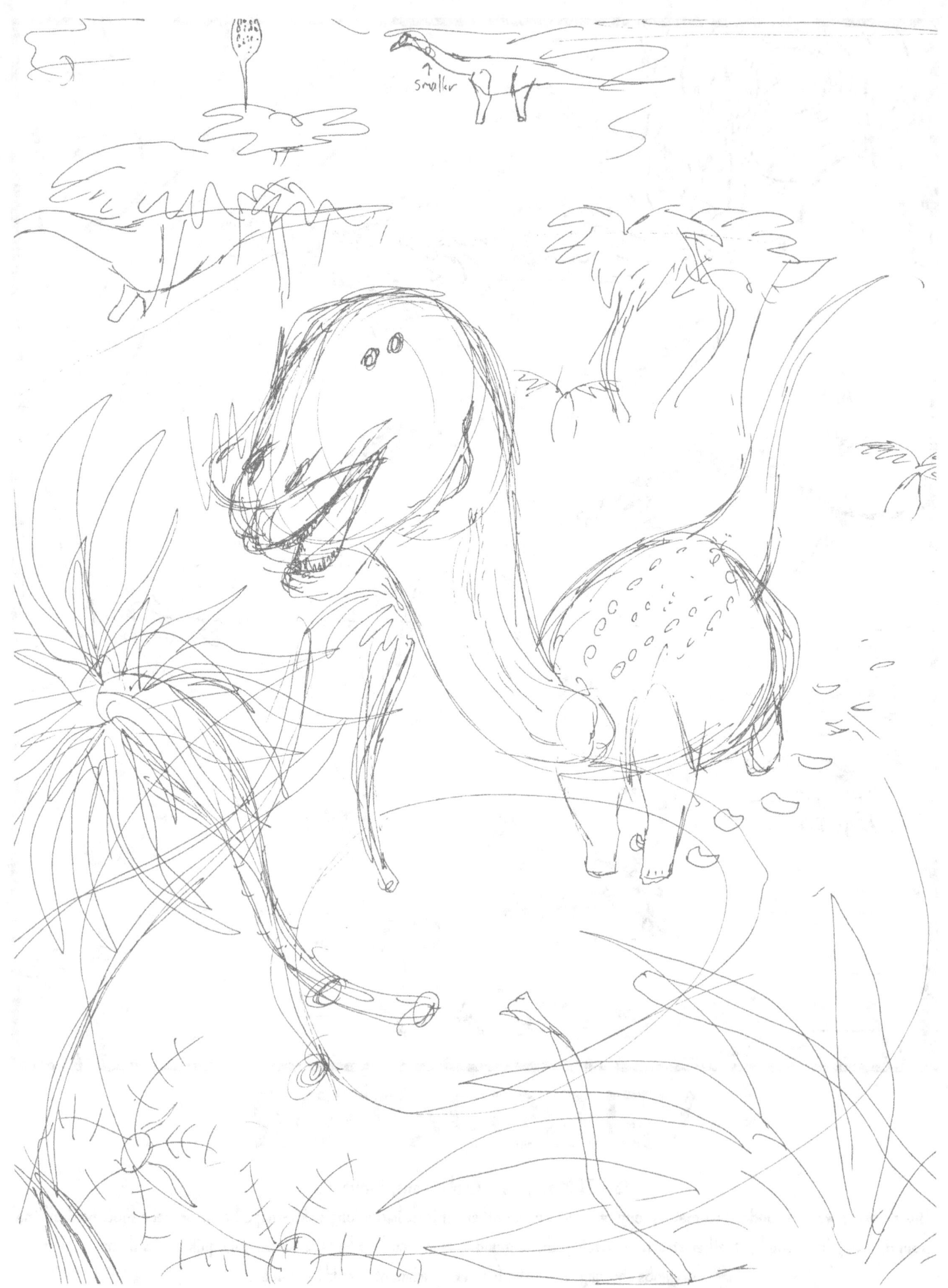

smaller

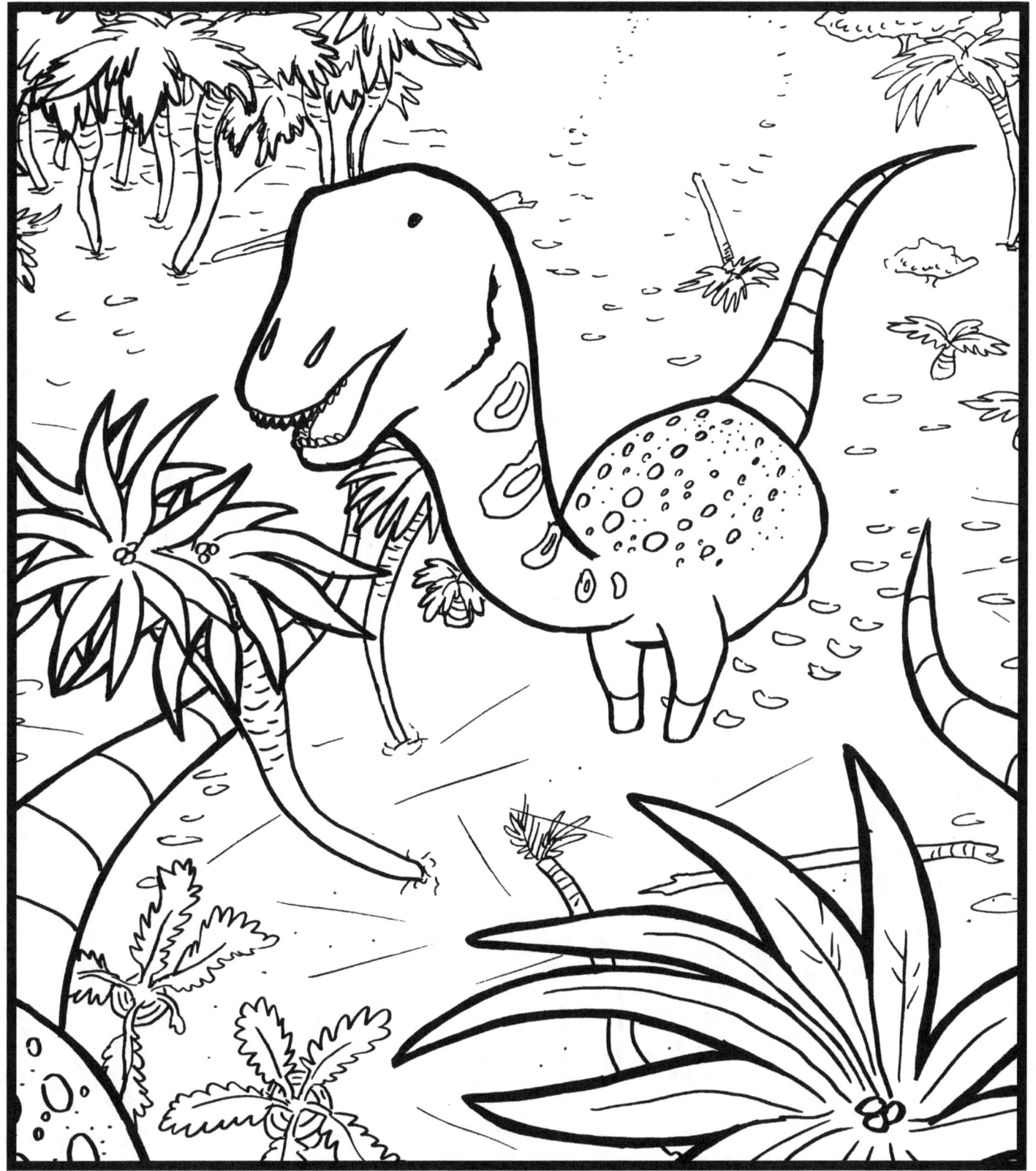

SALTASAURUS

70 MYA, Late Cretaceous, Argentina

This dinosaur was the first to be discovered bearing osteoderms, bony plates of armor embedded in the skin. Earlier, sauropods were thought to have been protected from predators due to their sheer size, but it seems that further defenses were needed for the survival of the smaller ones. Though they were estimated to have been between 8.5 and 12 meters long, this is still very small for a sauropod.

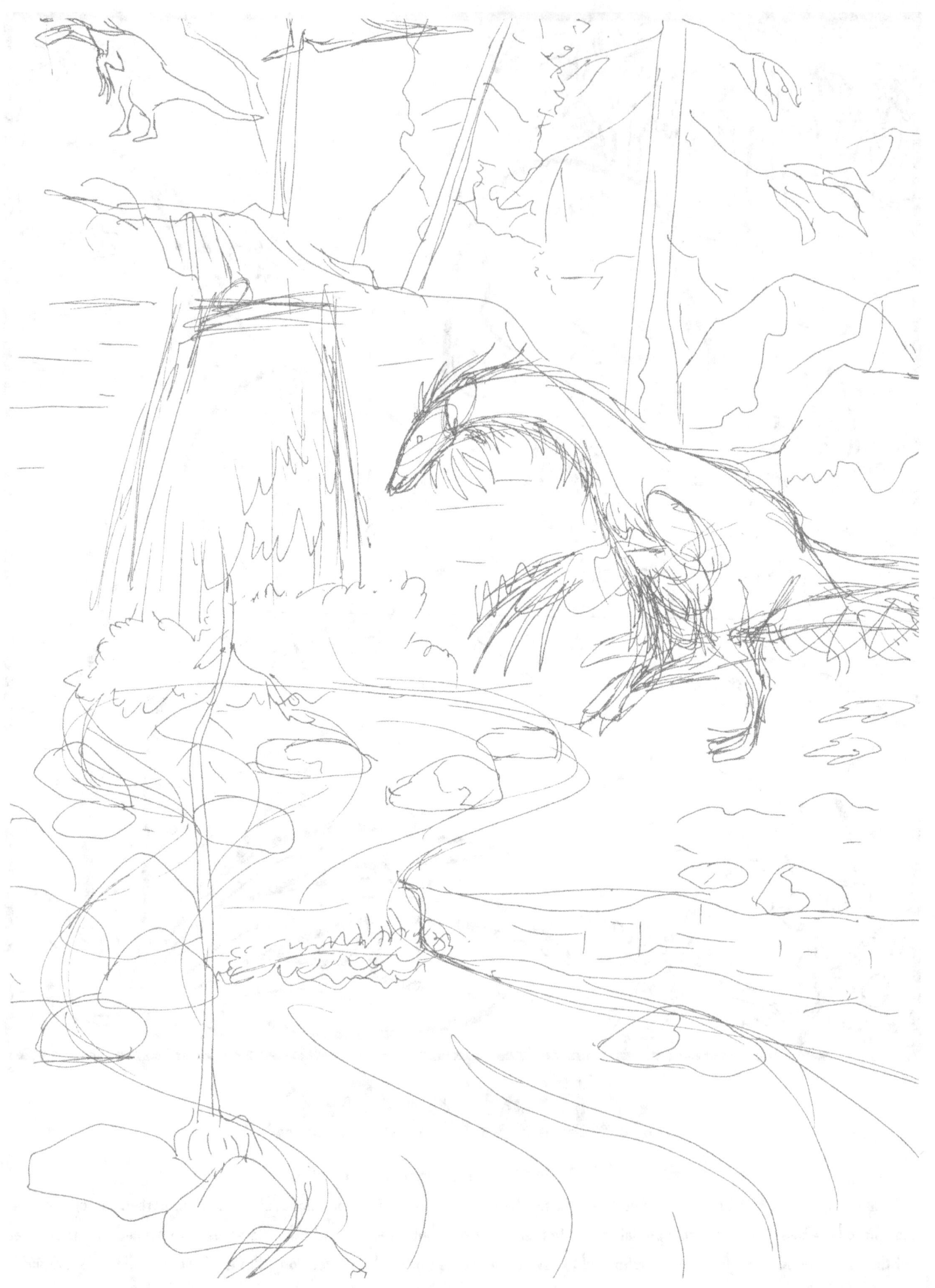

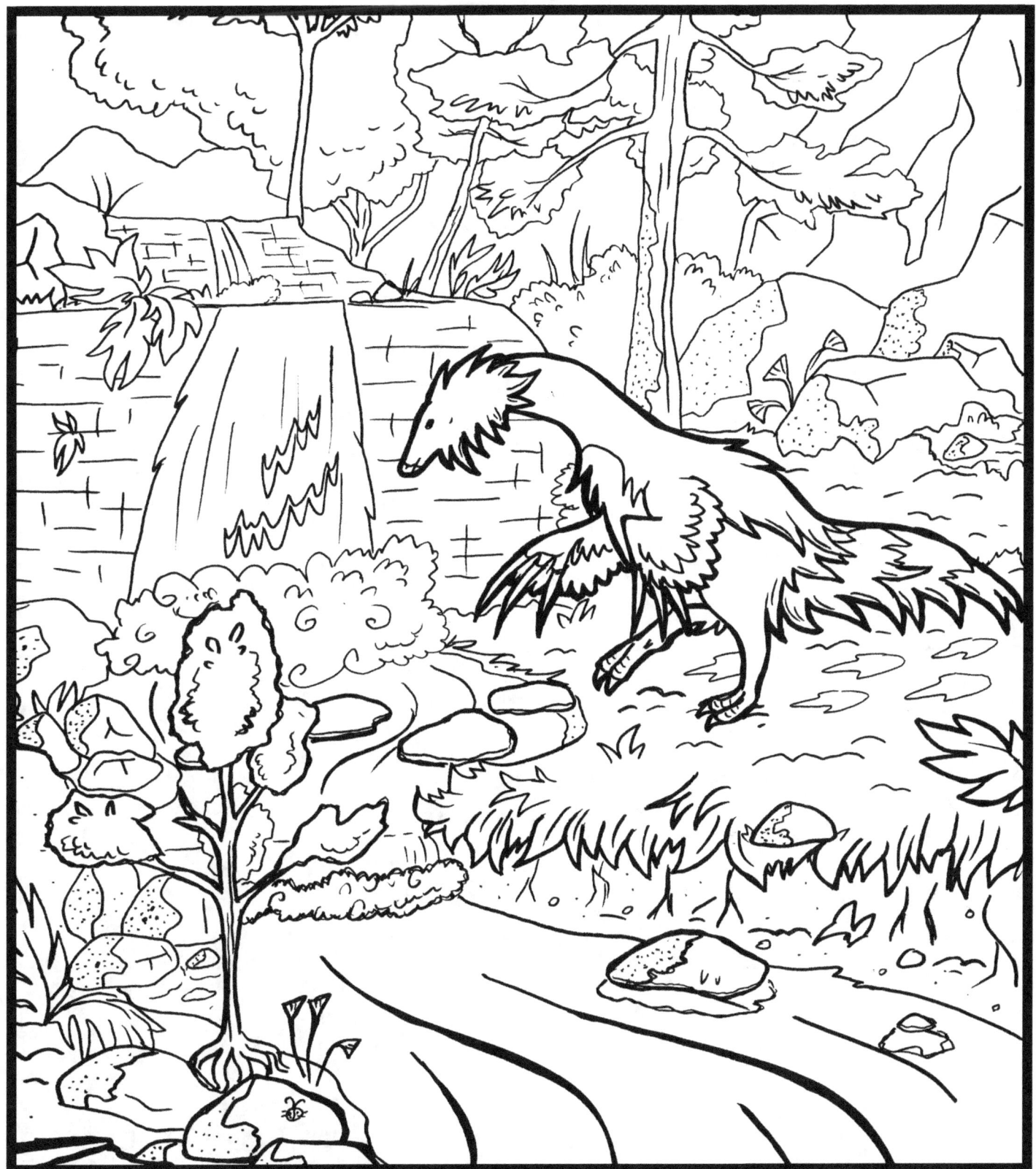

THERIZINOSAURUS

70 MYA, Late Cretaceous, Mongolia

Therizinosaurs were the first fossil theropods confirmed to be herbivores, not carnivores. At almost a meter long, their claws are the longest of any animal known. The exact purpose of these claws is still unknown, though several ideas have been suggested. These creatures were also the largest of the maniraptorans, reaching lengths of up to 10 meters.

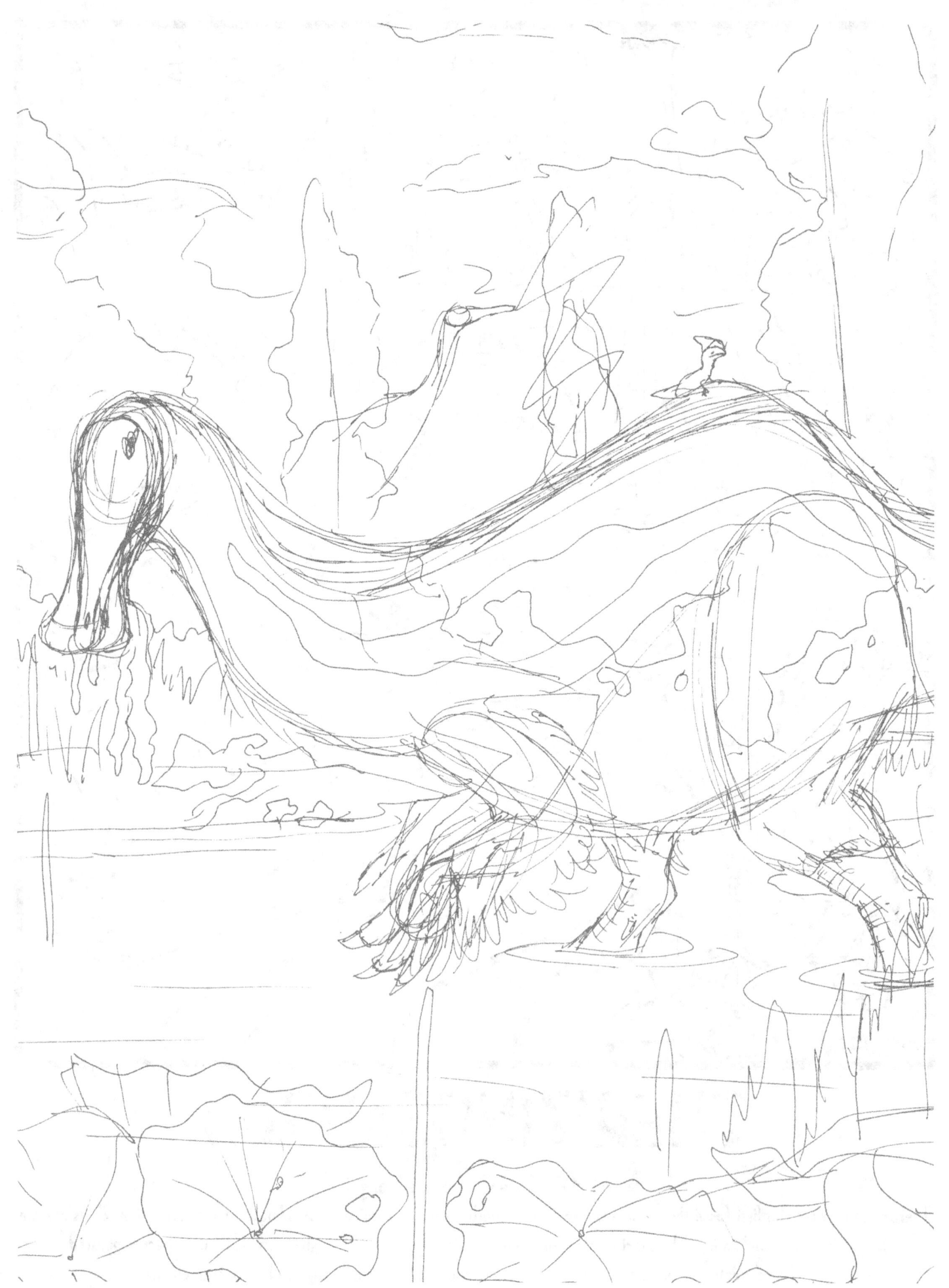

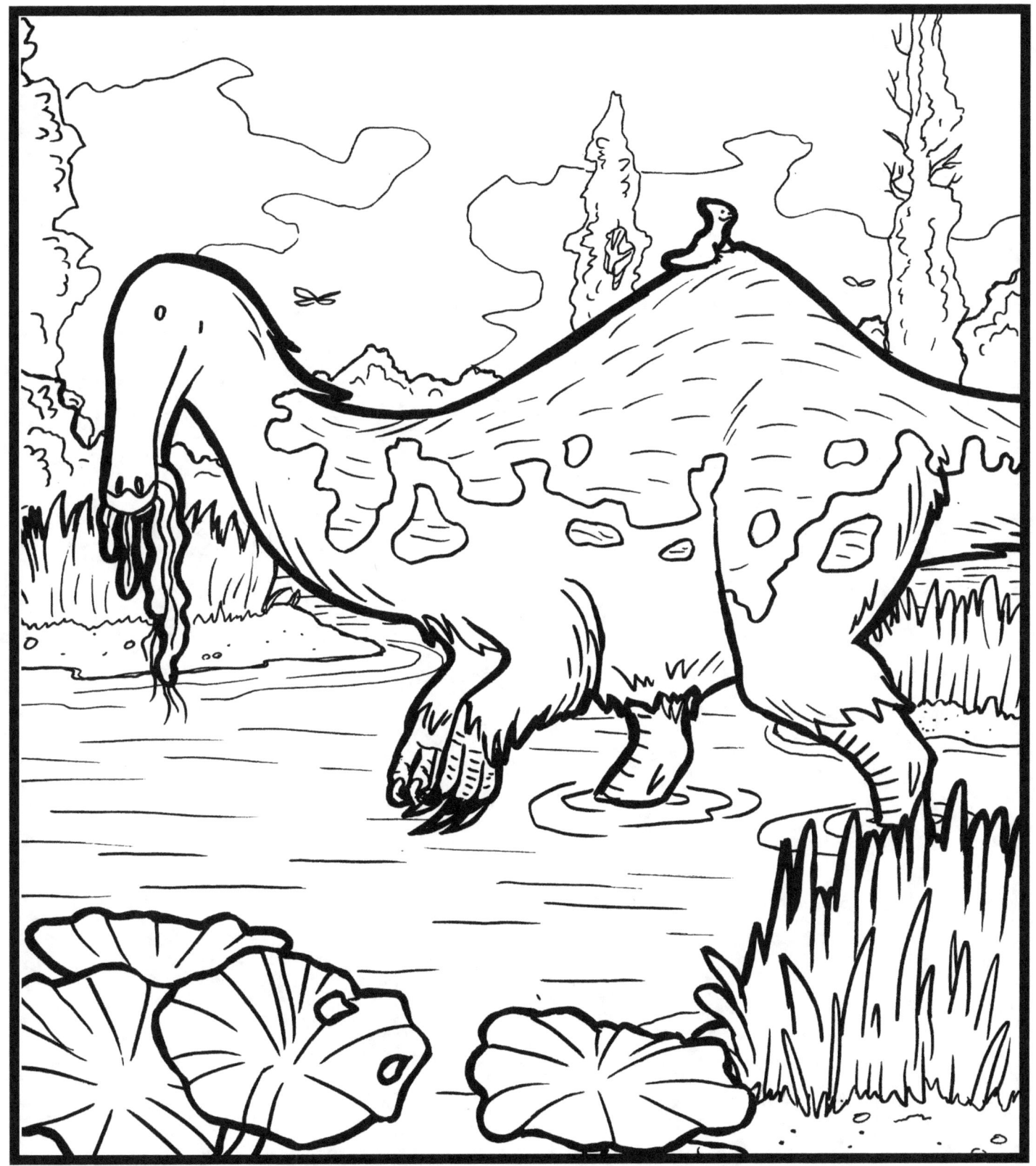

DEINOCHEIRUS

70 MYA, Late Cretaceous, Mongolia

A large ornithomimid herbivore, this dinosaur lived in swamps, deltas, and floodplains, and fed on soft plant matter. Studies of its brain cavity revealed that it had comparable intelligence to a sauropod. Growing up to 11 meters long, its sheer size helped protect it from predation. Despite its large size, several bones were pneumatized to lighten the load.

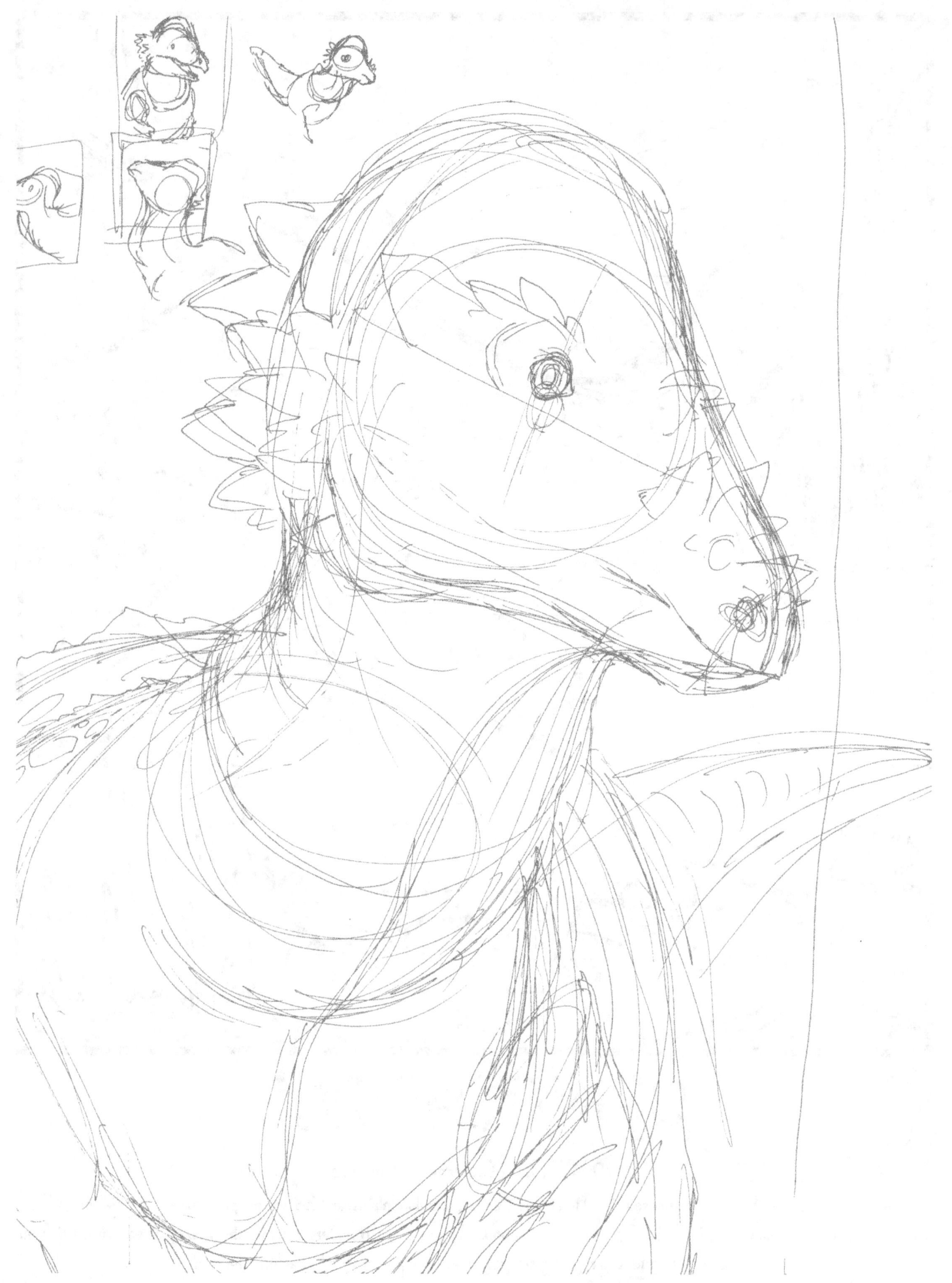

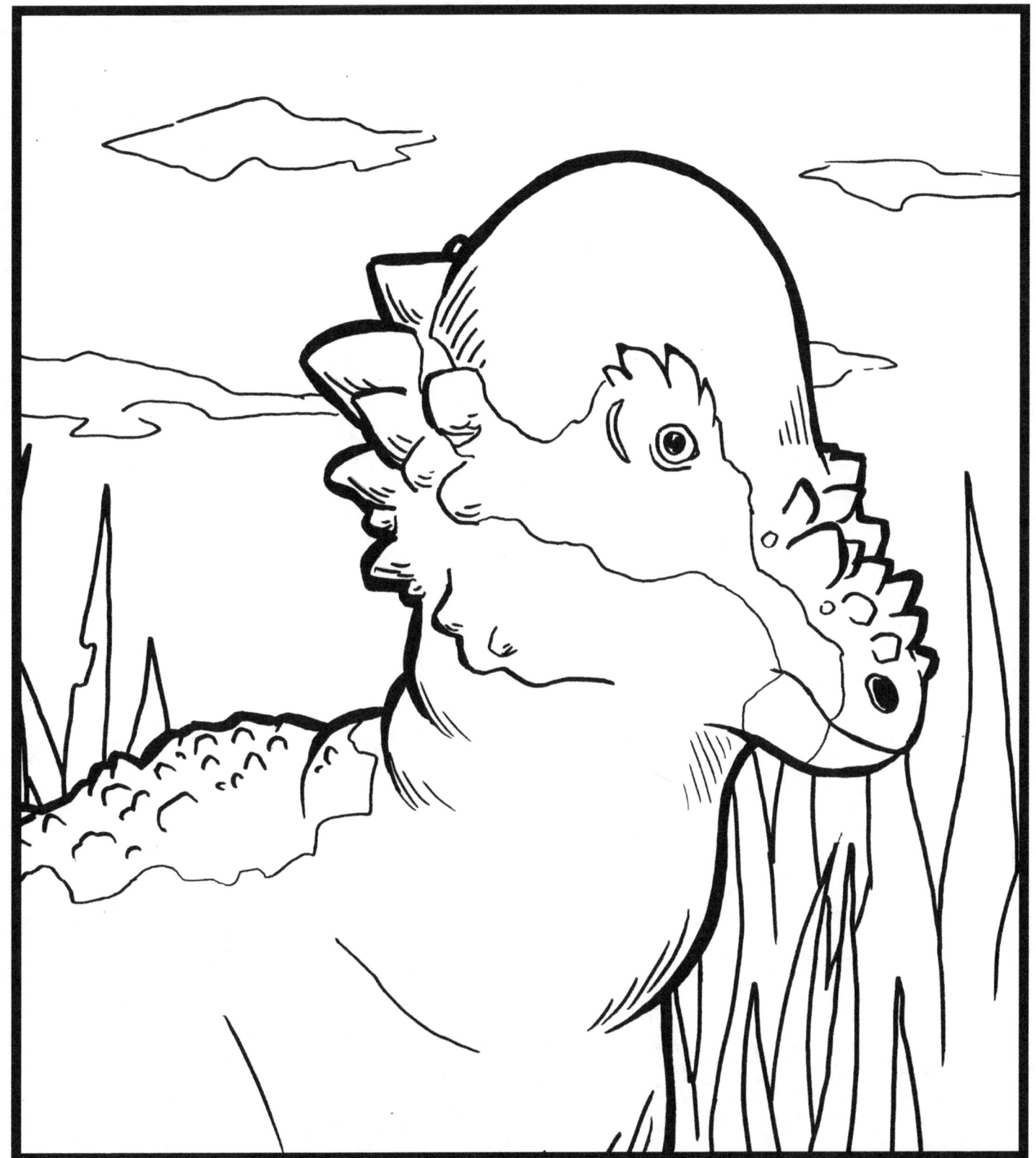

PACHYCEPHALOSAURUS

70 - 66 MYA, Late Cretaceous, United States amd Canada

One of the last non-avian dinosaurs before the Cretaceous mass-extinction, this "thick-headed lizard" was a medium-sized herbivore.
As the animal aged, the horns and spikes would gradually reduce in size, while the dome top and rounded skull knobs grew.
The dome has been thought to have been used in display, defense, and/or same-species head-butting, but more research is needed.

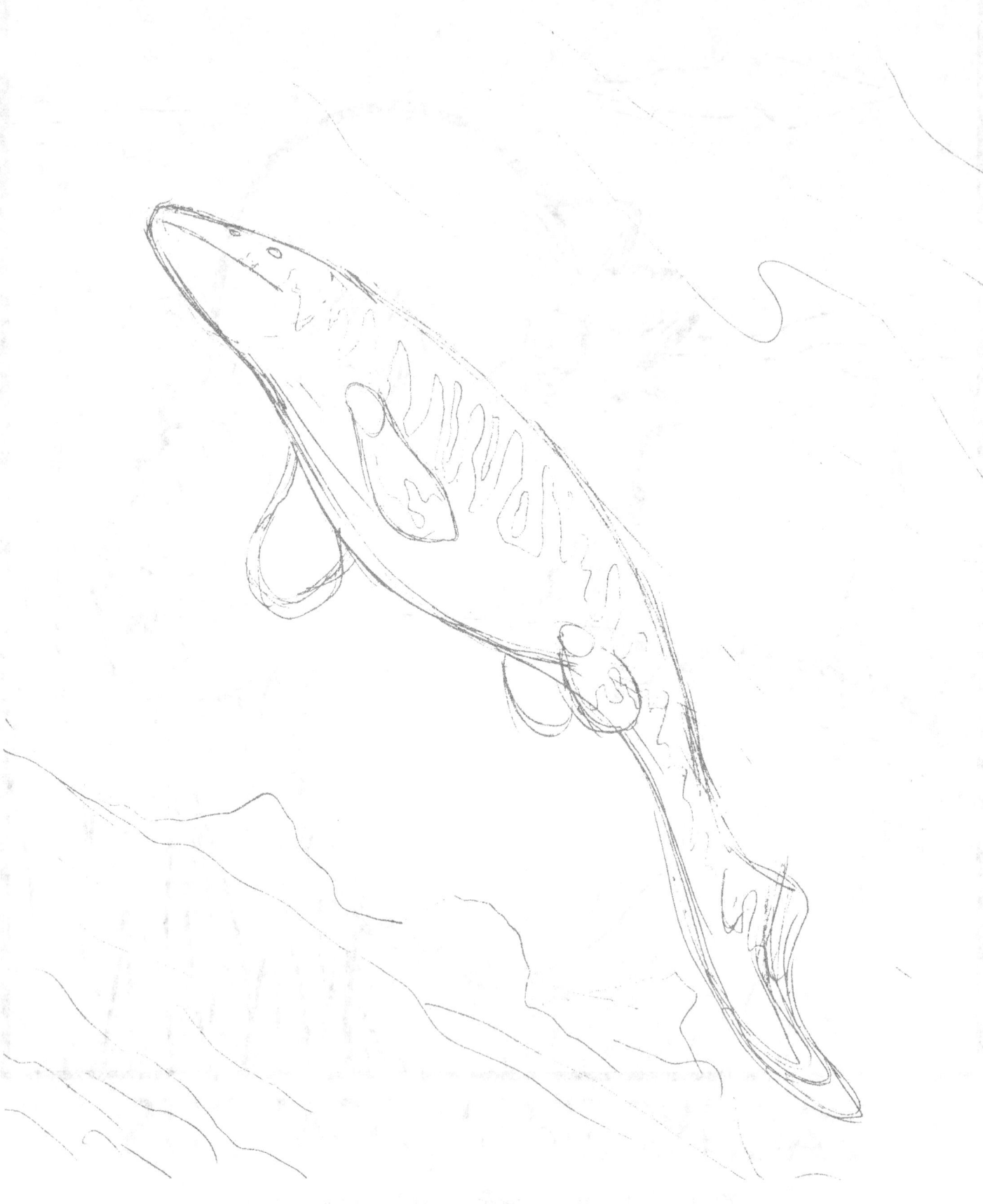

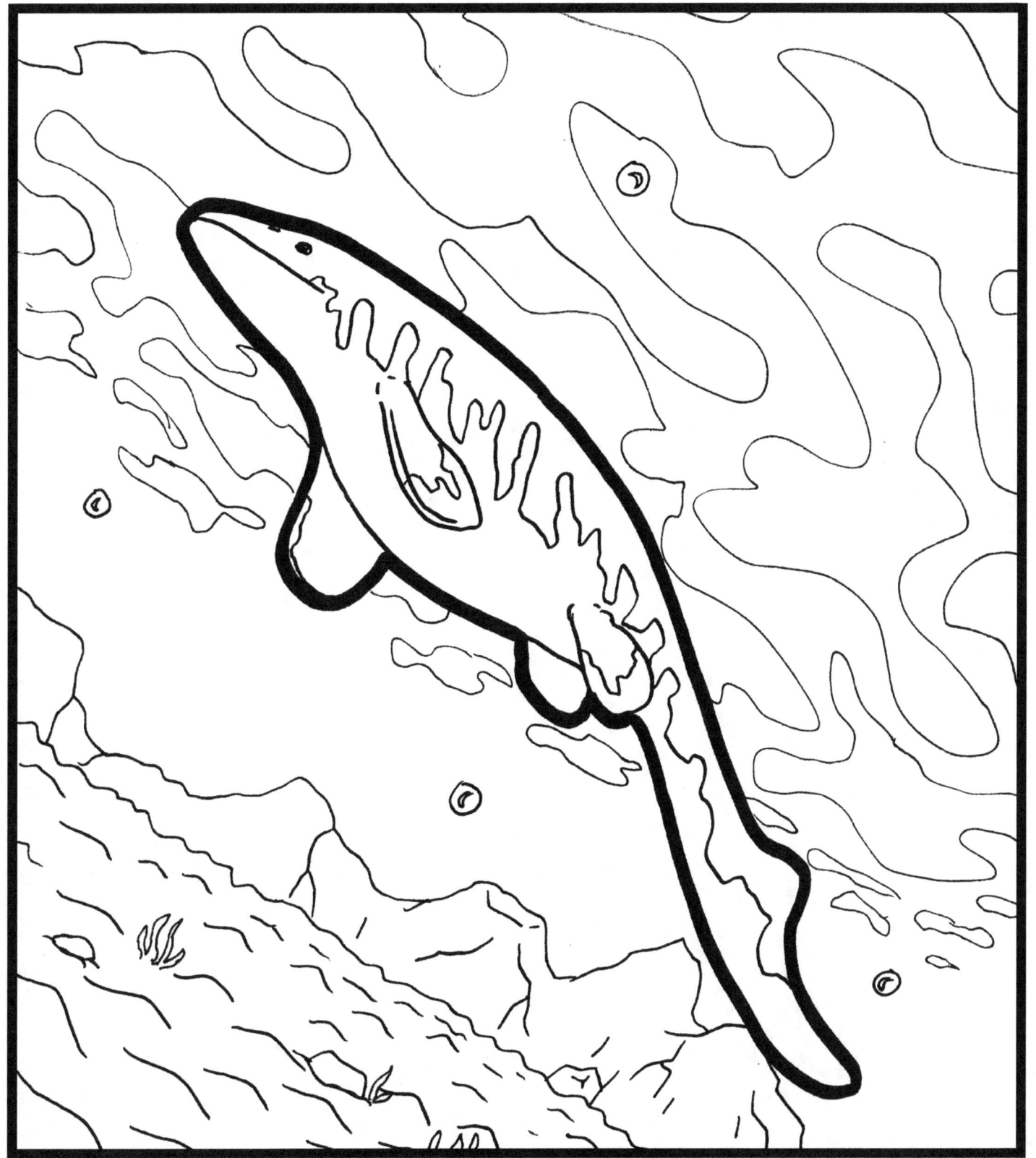

MOSASAURUS

70 - 66 MYA, Late Cretaceous, North America, Western Europe

This large mosasaur had a preference for big, slow-moving prey. It was very heavily built, and was originally mistaken for another species of mosasaur that did prey upon slow, armored animals. It was among the last of the mosasaurids, living up until the end of the Cretaceous.

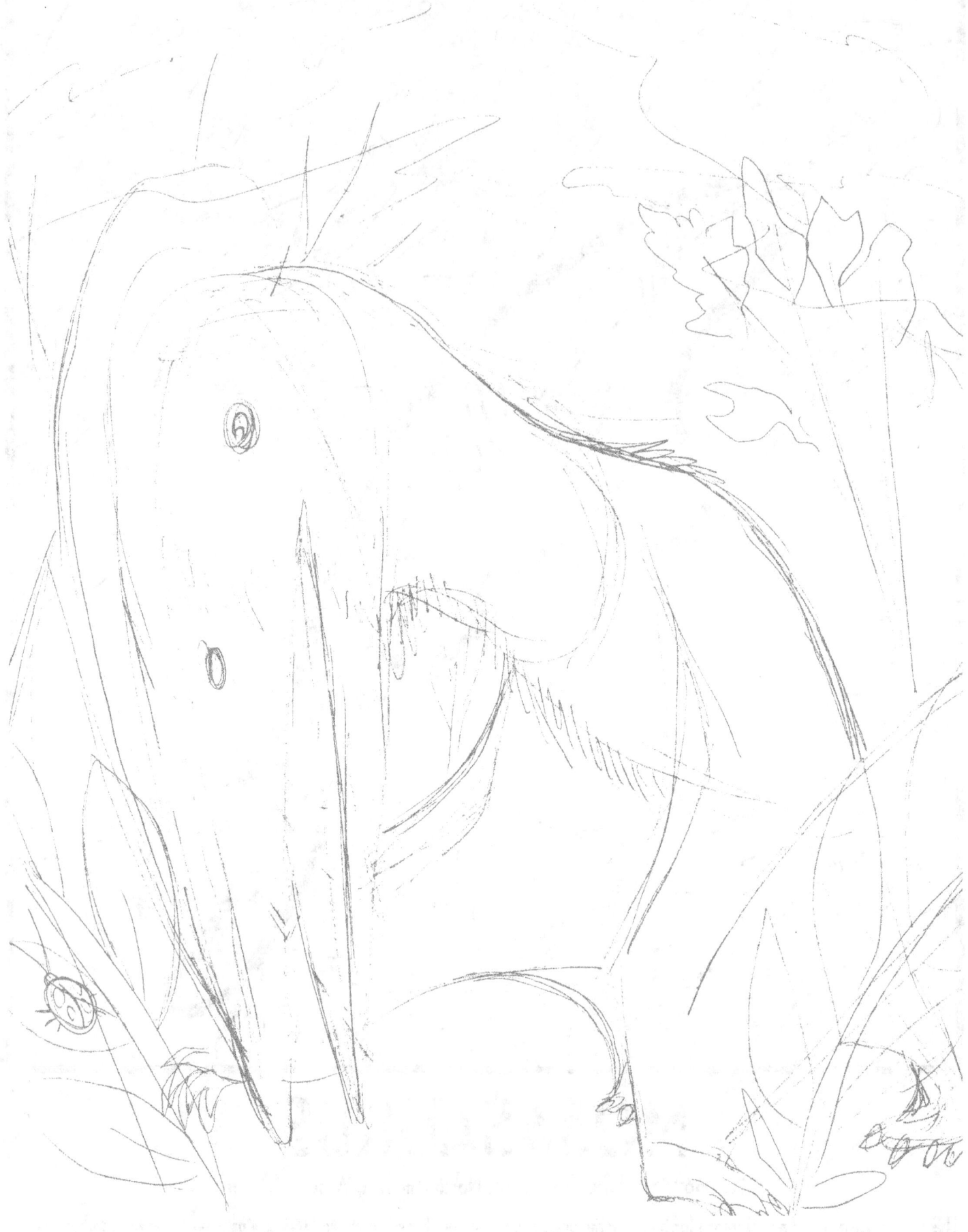

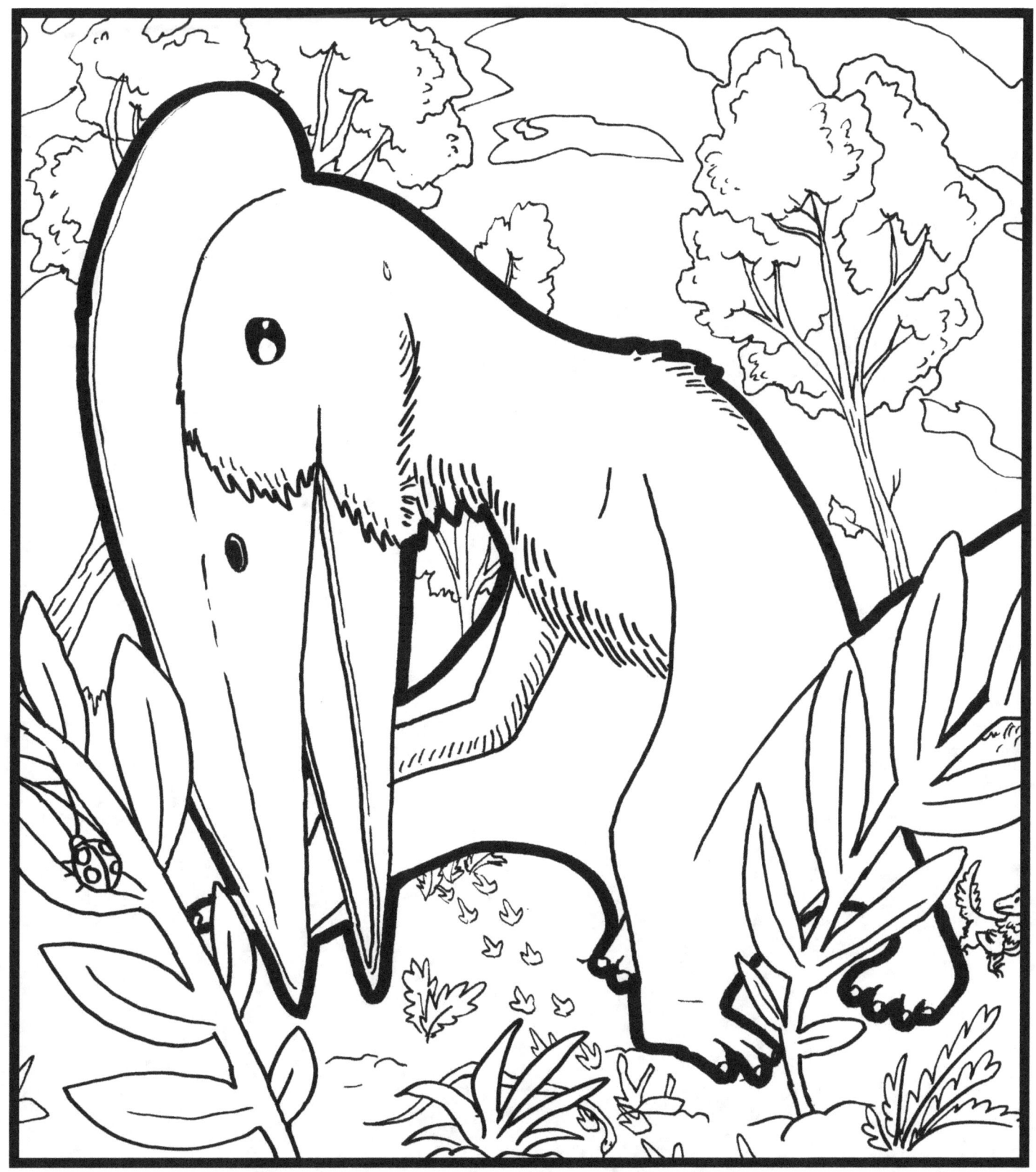

HATZEGOPTERYX

66 MYA, Late Cretaceous, Romania

This giant azdharchid pterosaur lived right up to the end of the Cretaceous. The skull alone was 3 meters long, and the wingspan stretched 12 meters wide. It was a terrestrial foraging predator, though this by far did not mean it was flightless. It played the role of apex predator of Hateg Island, preying upon larger dinosaurs than what other pterosaurs could handle.

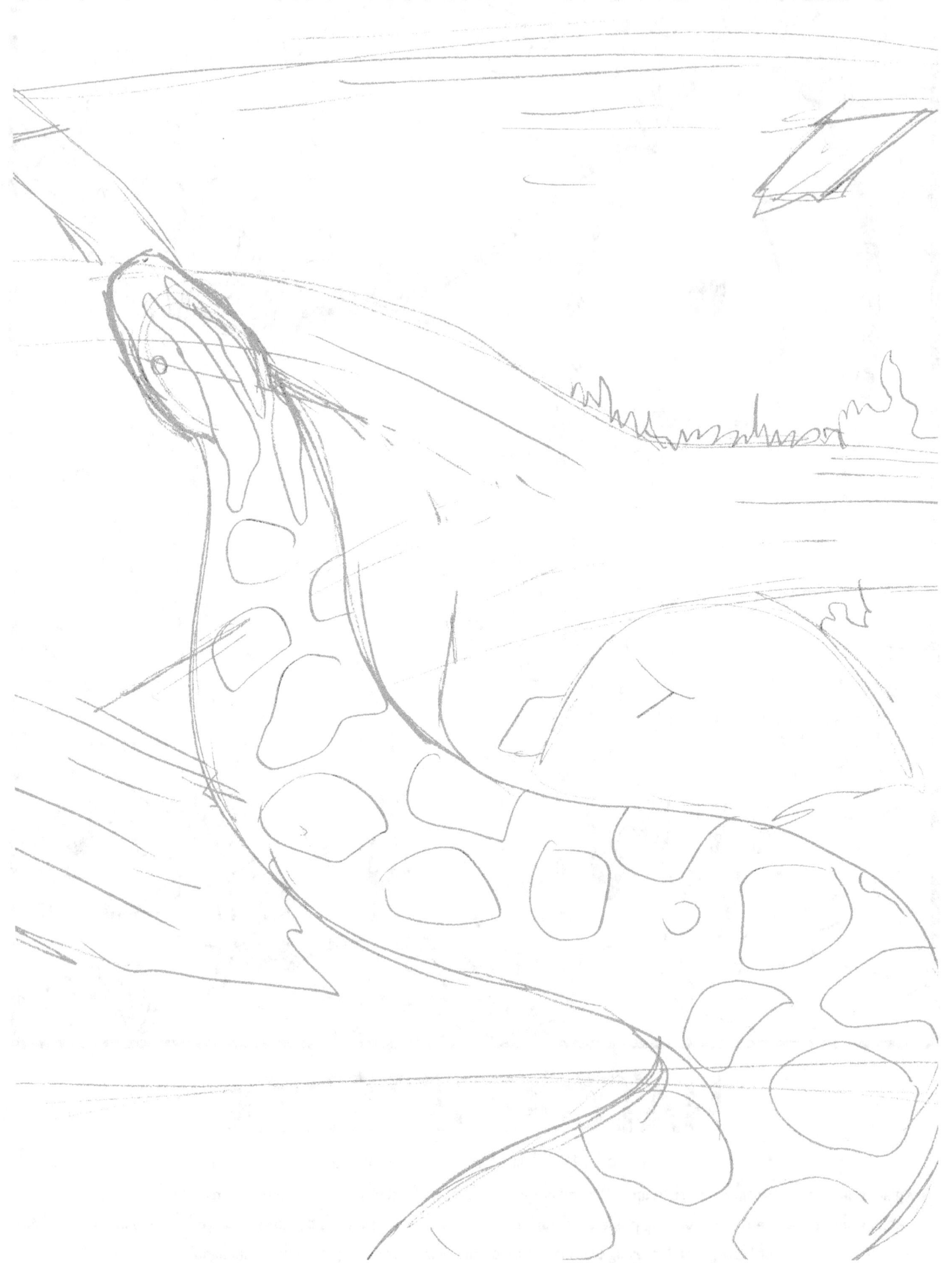

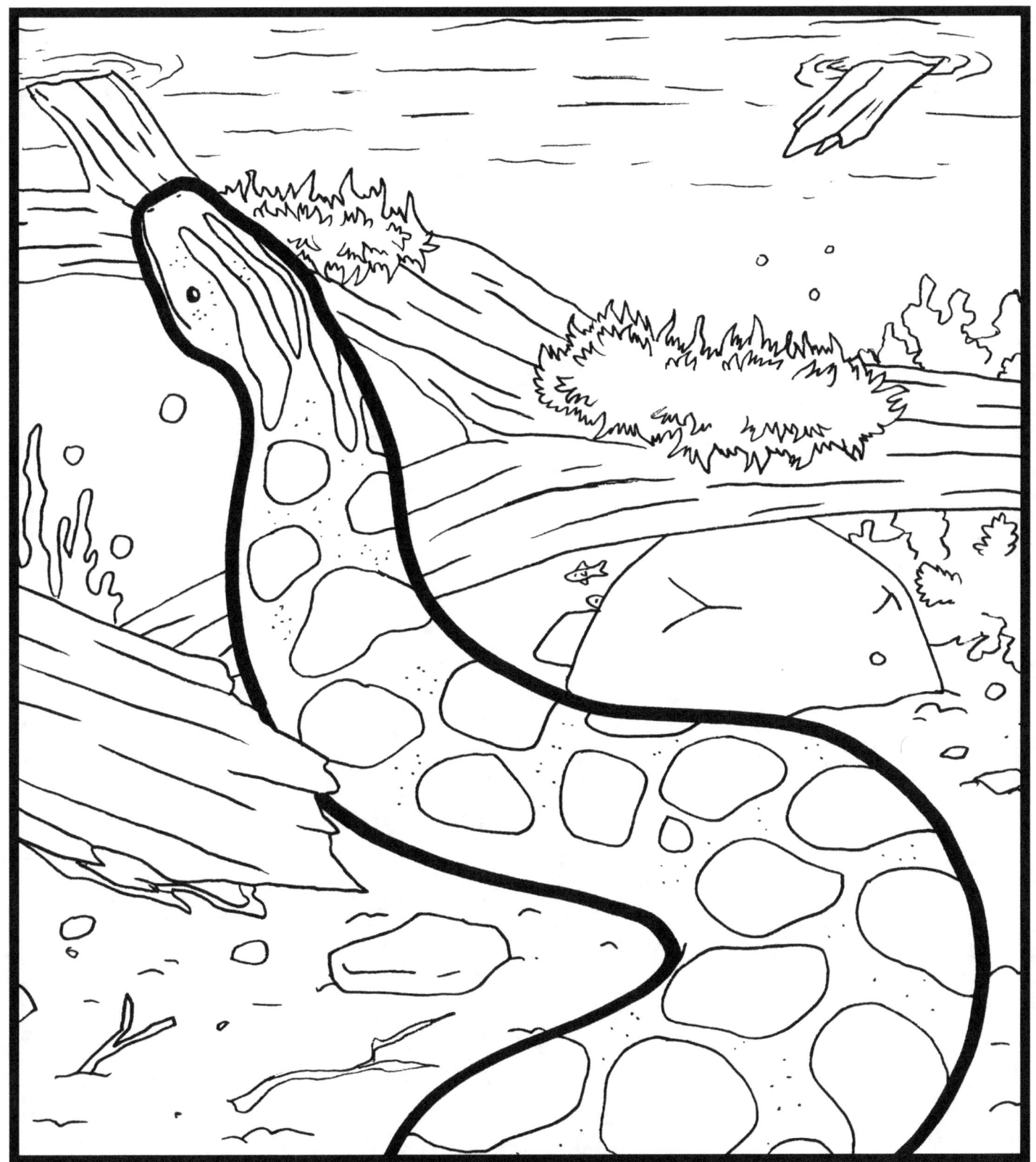

TITANOBOA

60 - 58 MYA, Palaeocene, Columbia

This giant snake thrived in the era that followed the mass extinction that wiped out the non-avian dinosaurs. It spent its days in tropical rivers and other bodies of water, hunting fish. The climate was warmer overall, allowing cold-blooded animals to grow to exceptional sizes.

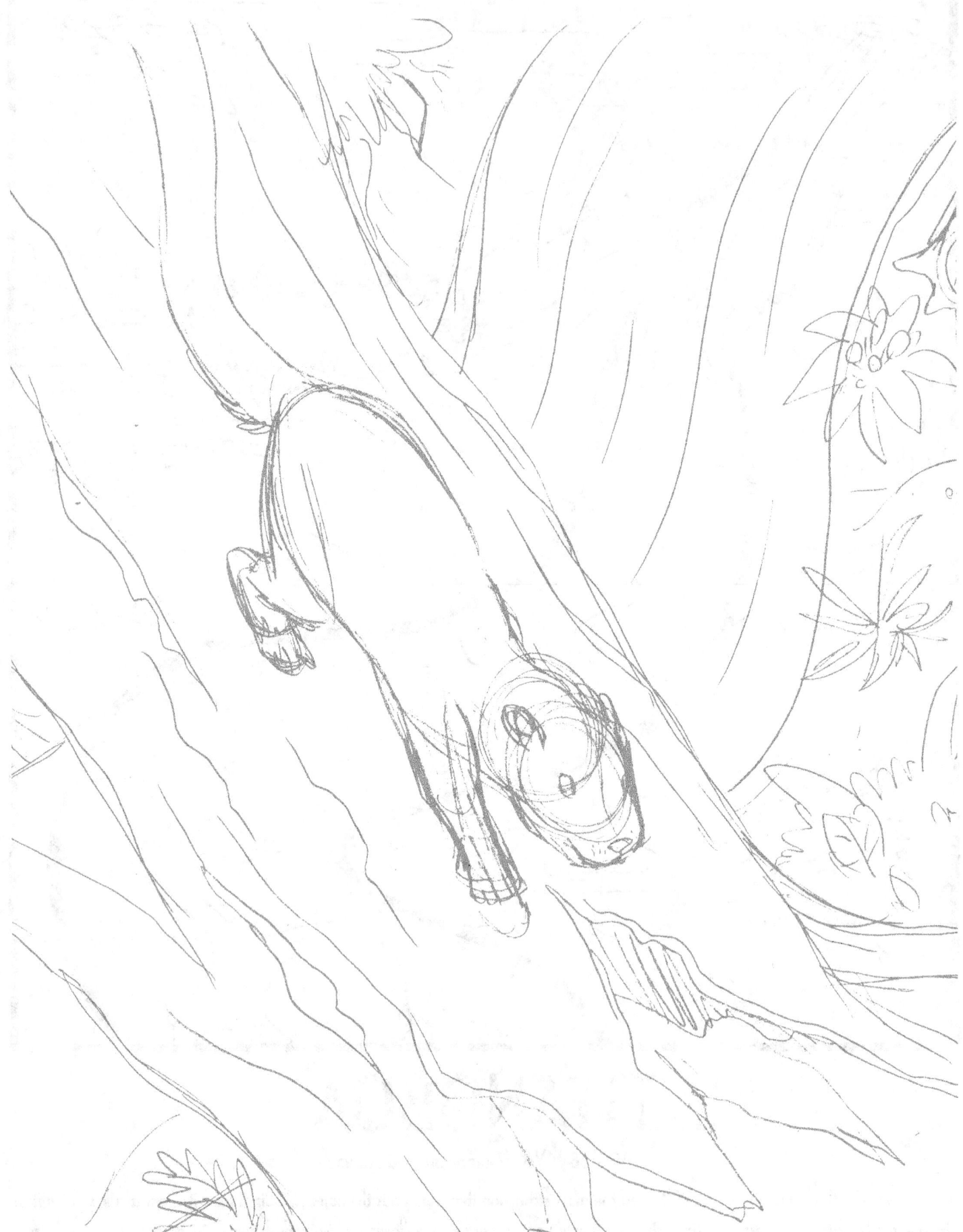

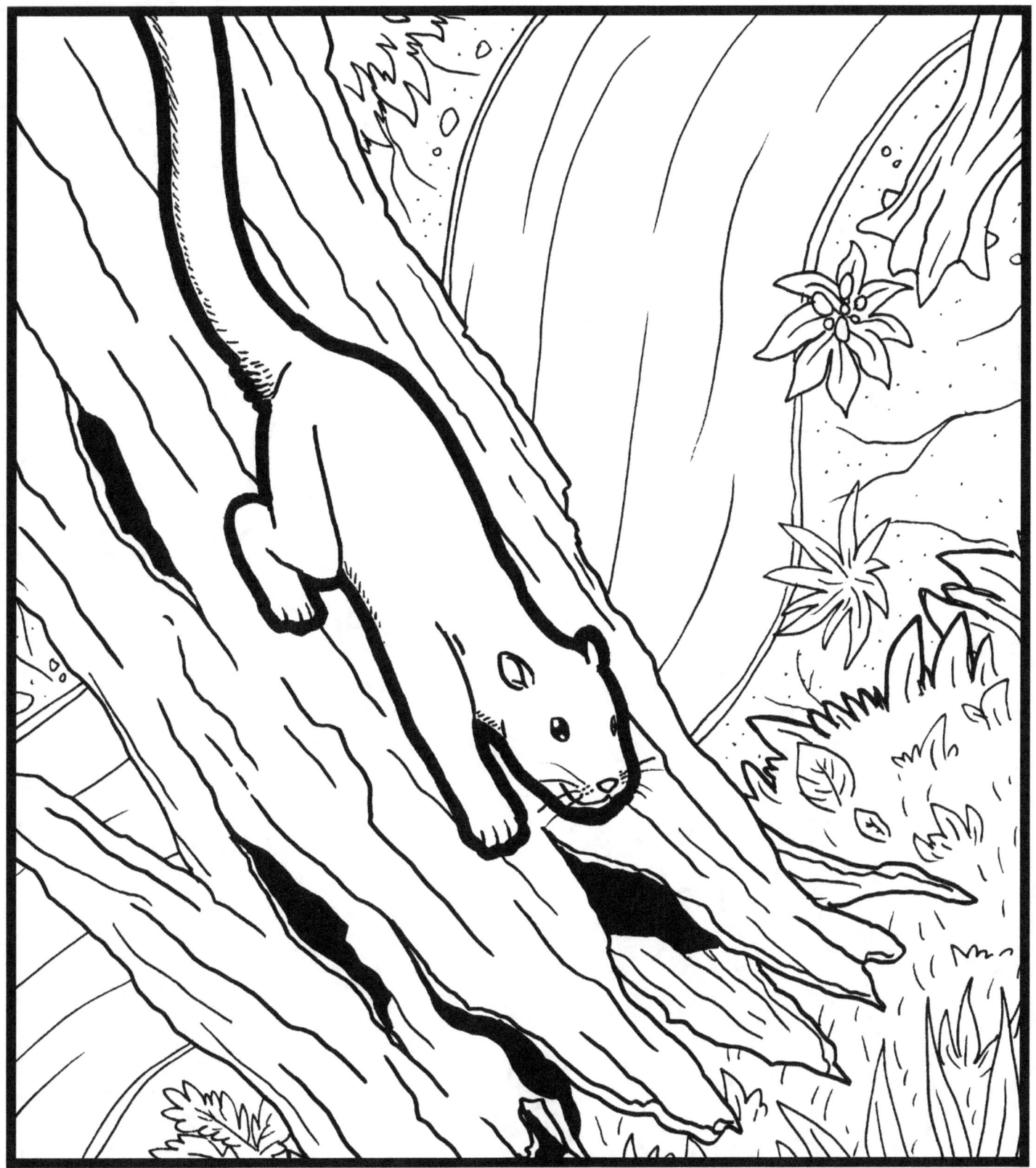

OXYAENA

60 - 50 MYA, Late Paleocene to Early Eocene, United States

Despite the physical appearance resembling modern cats, it was not related to them. Other differences include walking on the entire foot as a plantigrade, as opposed to a cat's digitigrade toe-walking. It was about a meter long and around 20 kilograms heavy, with a long body, and even longer tail.

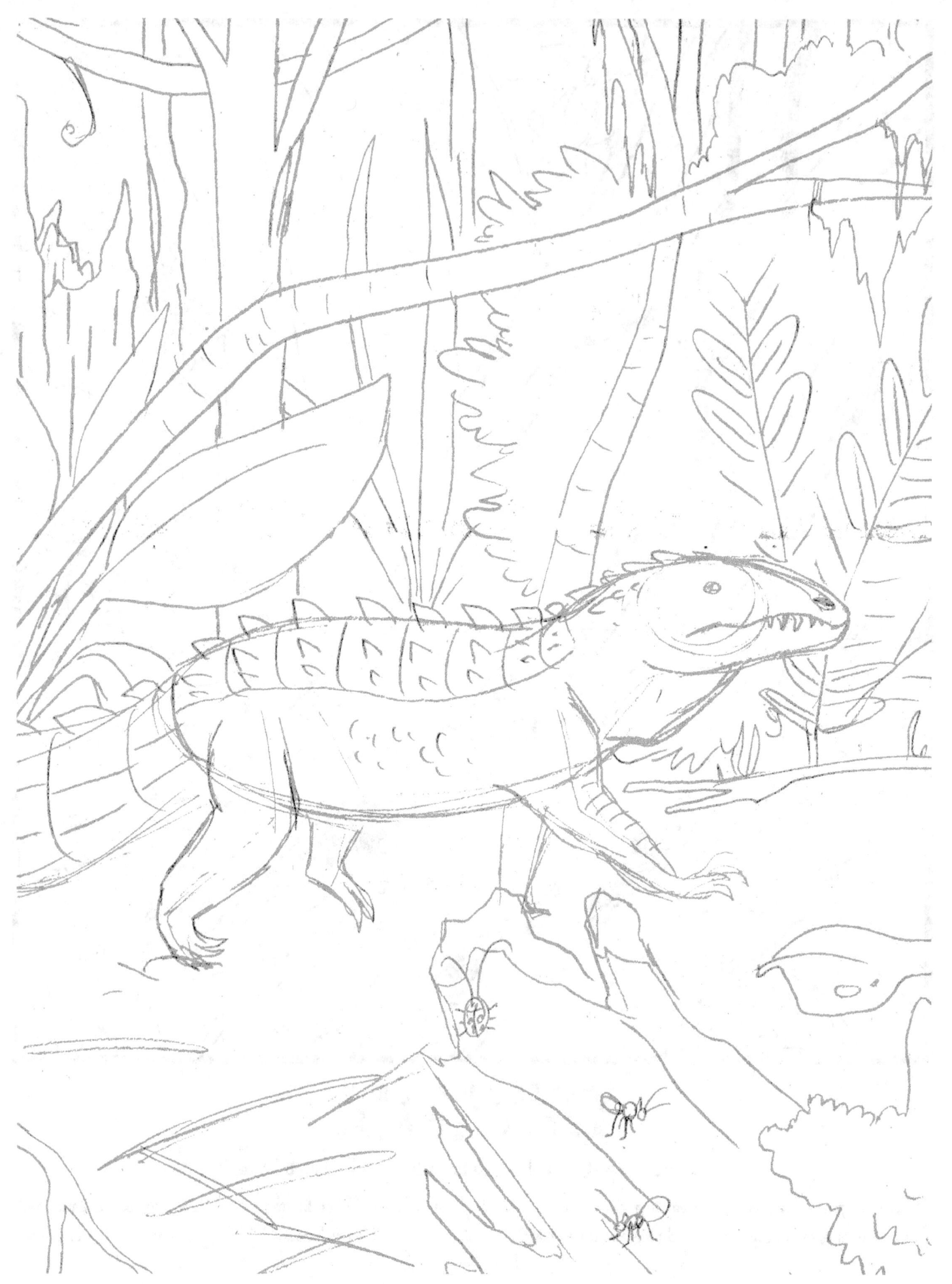

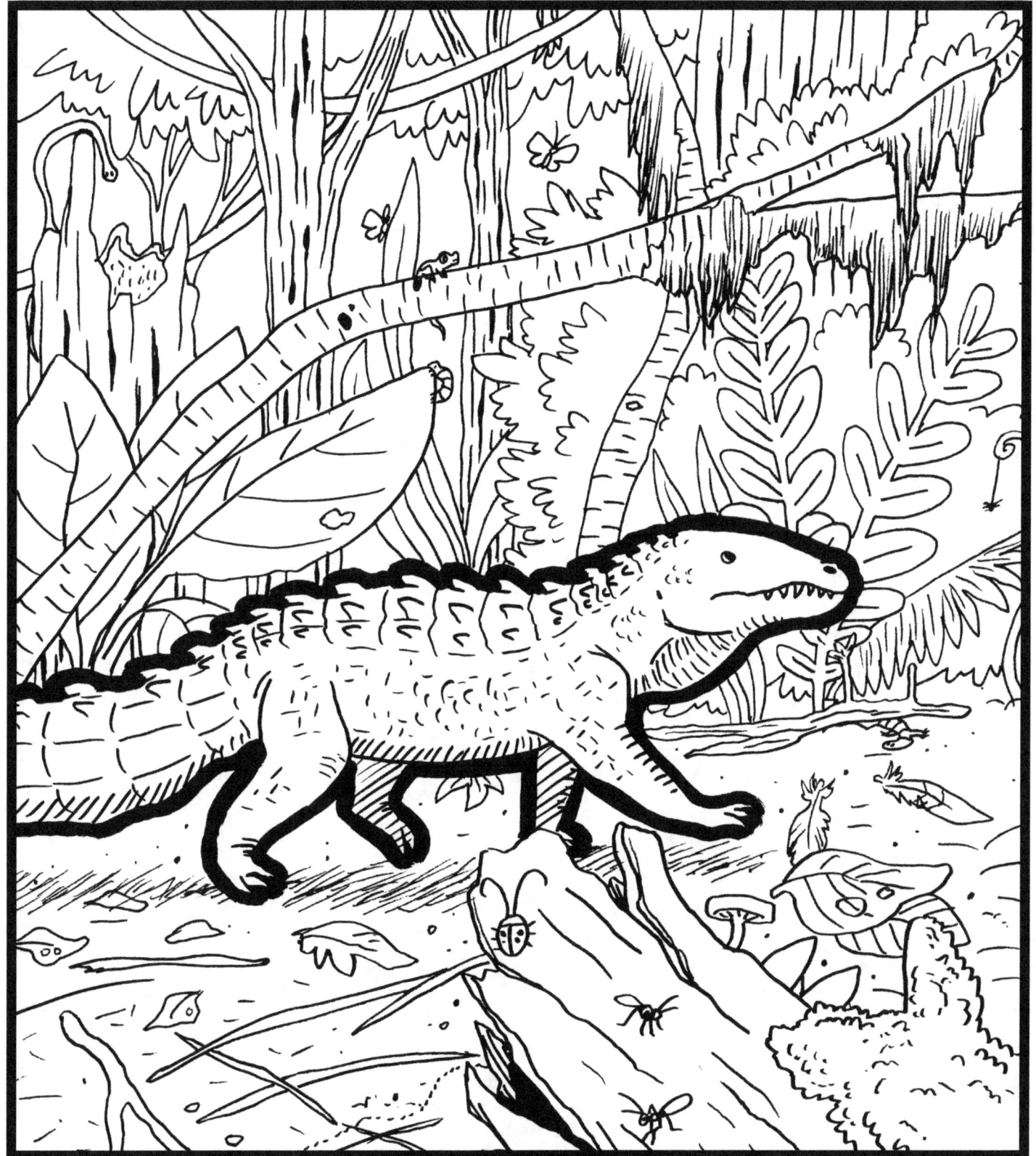

SEBECUS

56 - 50 MYA, Eocene, Argentina, Bolivia

Sebecus had an oddly-shaped skull among crocodyliforms. Instead of the skull being raised up where the eyes are, the top of the head was level. Also unlike the familiar alligators and crocodiles of today, it was entirely terrestrial. It had longer legs than its water-dwelling cousins, and eyes on the sides of the head instead of on top.

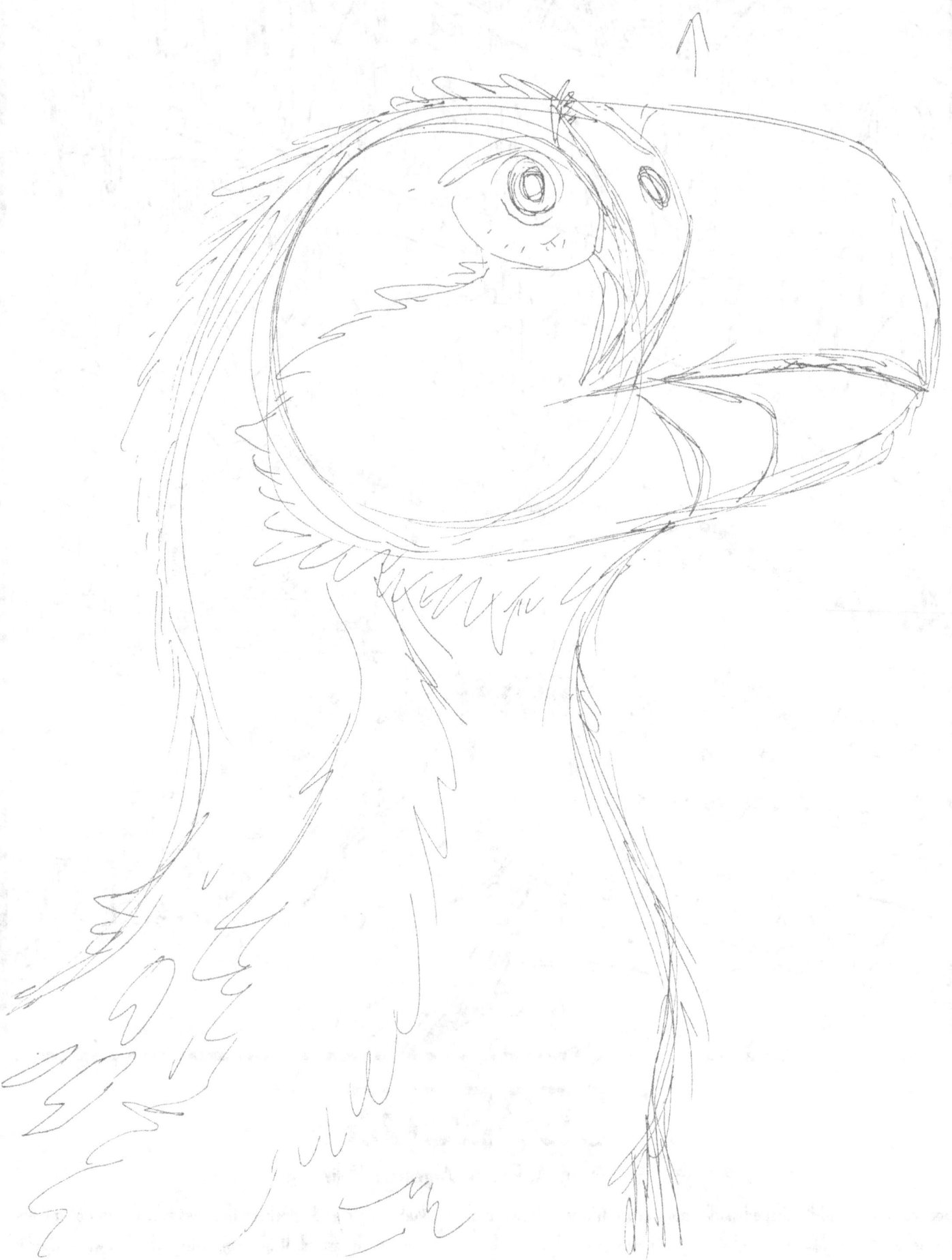

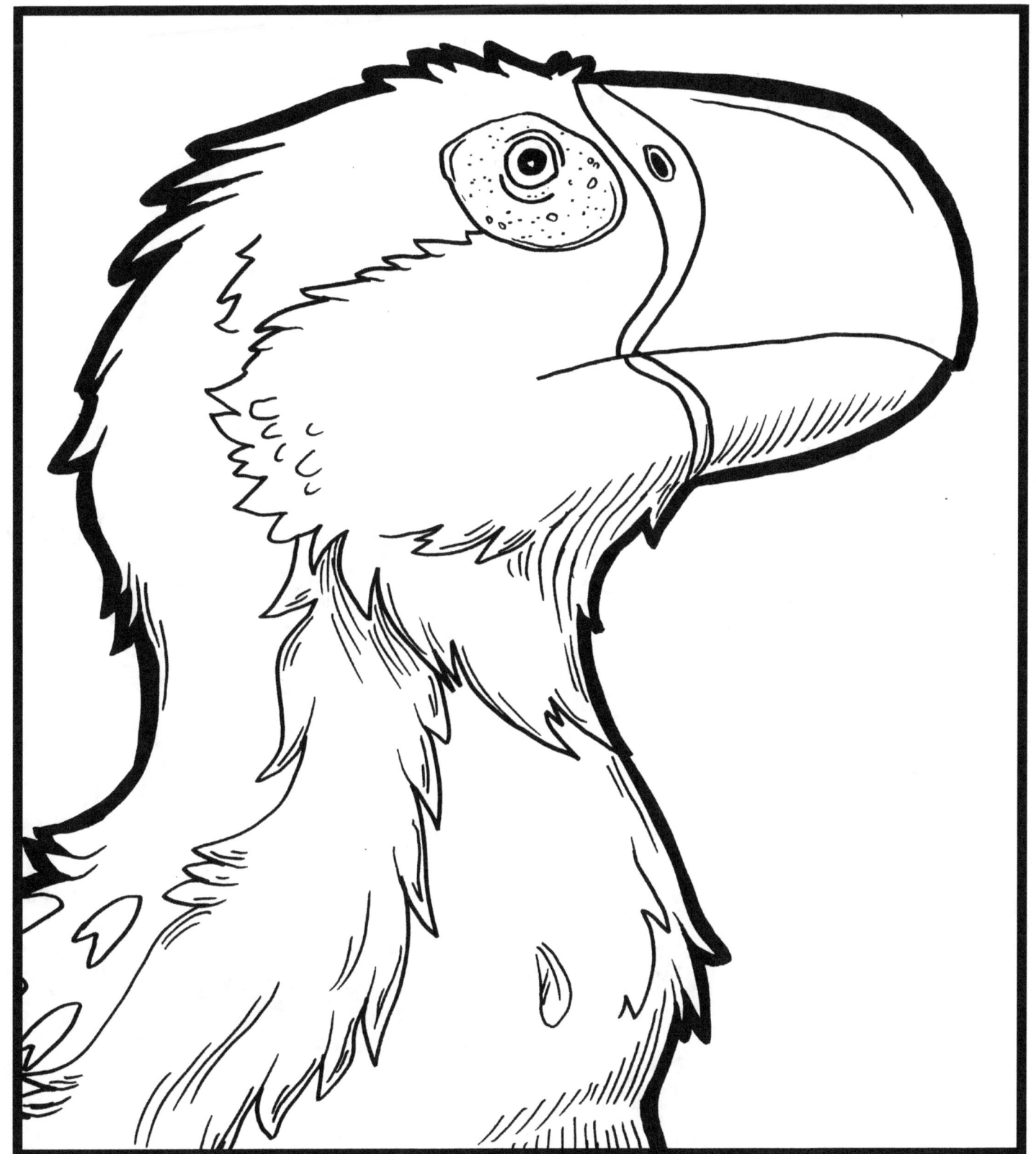

GASTORNIS

56 - 45 MYA, Paleogene, China, North America, Western Europe

This dinosaur was a large, flightless bird. Originally assumed to have used its impressive beak to disembowel small prey, new speculation points to the powerful jaws being used to crack open nuts and seeds, rather than animal bones. Predatory dinosaurs tend to also have curved, pointed claws, which these birds lacked, and the lack of a hooked beak further suggested herbivory.

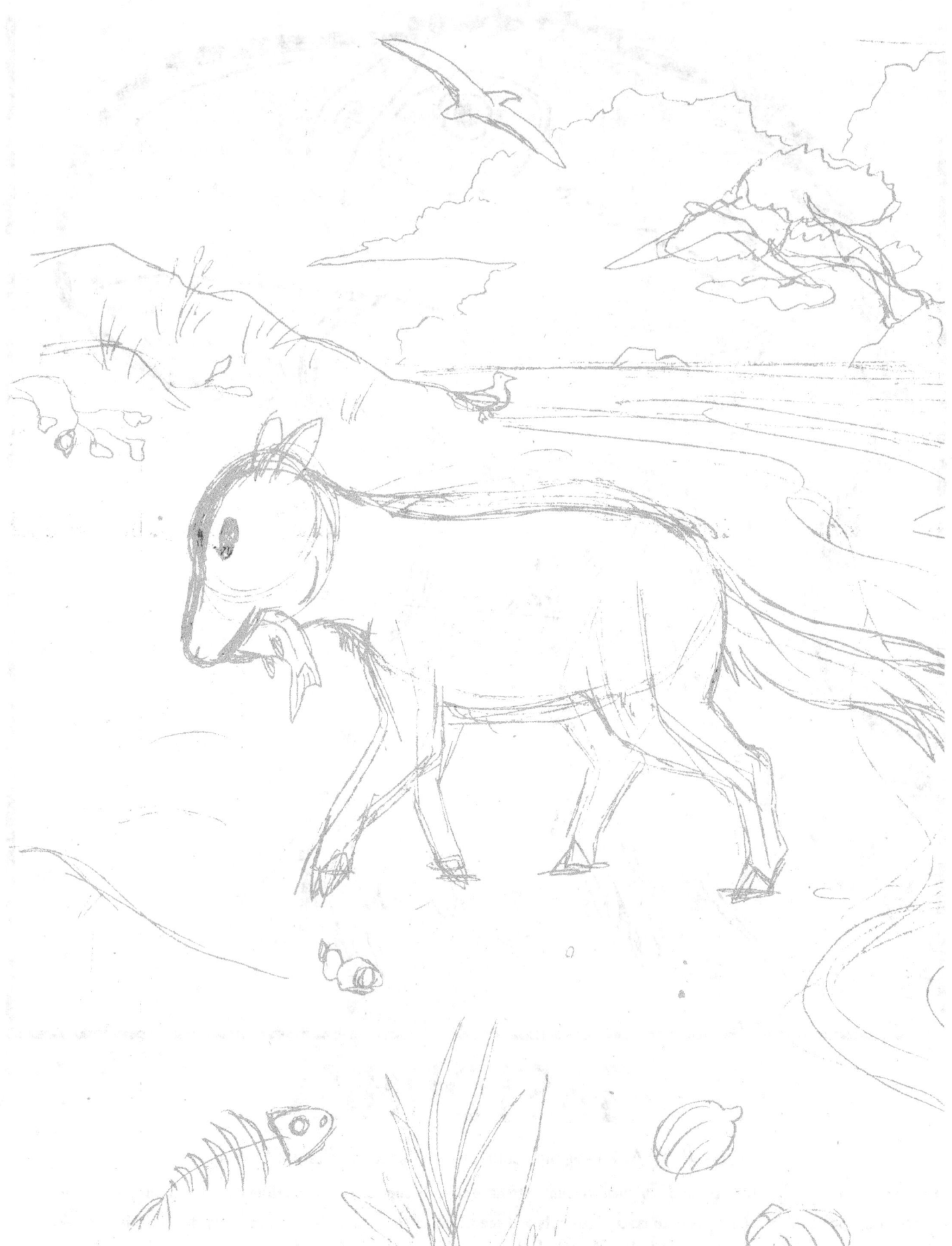

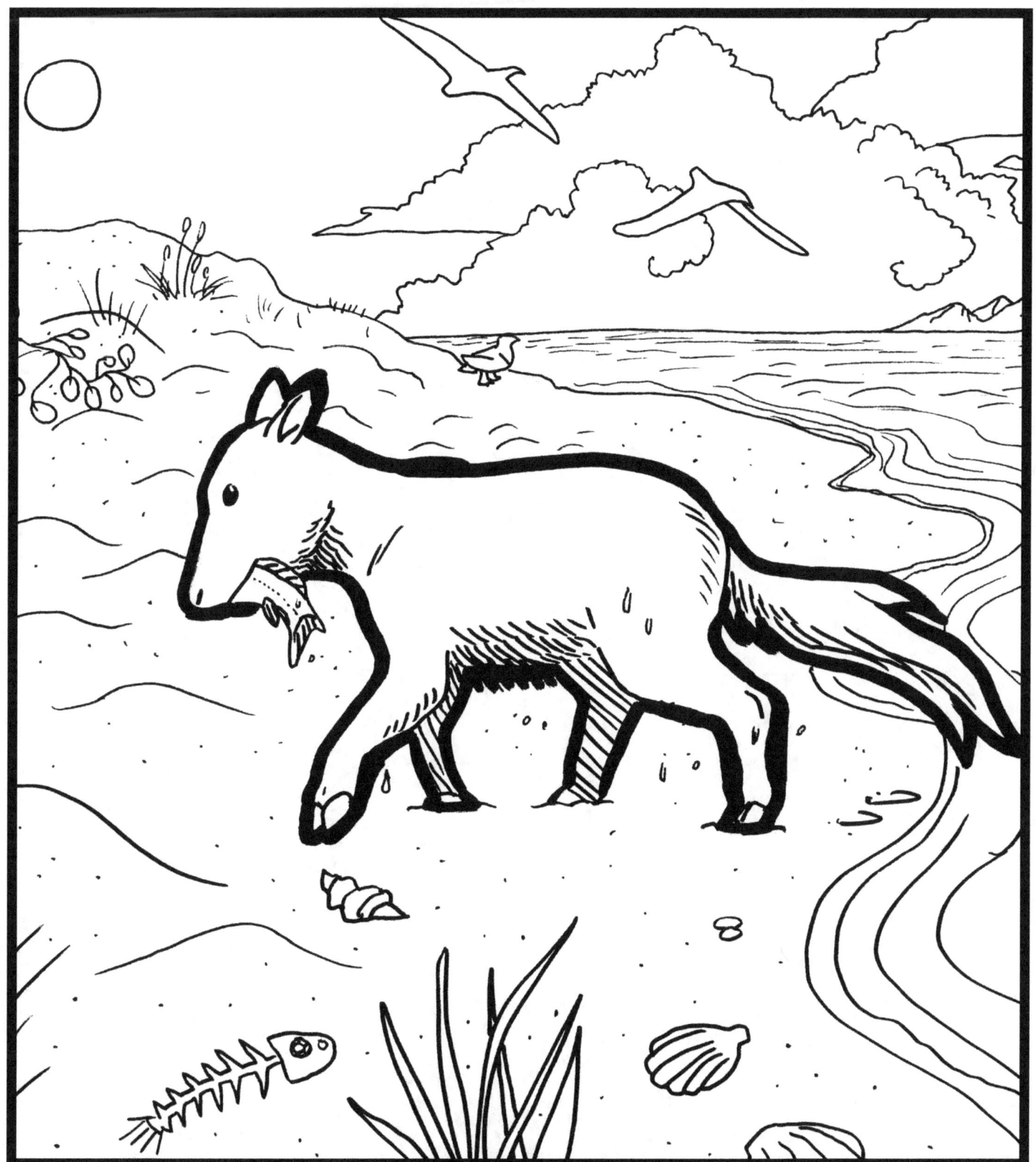

INDOHYUS

56 - 41 MYA, Middle Eocene, India, Pakistan

This omnivorous whale descendant bore little external resemblance to the cetaceans familiar to many of us today. It was a domestic housecat-sized hooved mammal with dense bones that lowered its bouyancy, allowing it to more easily stay underwater. It's name means "India's pig."

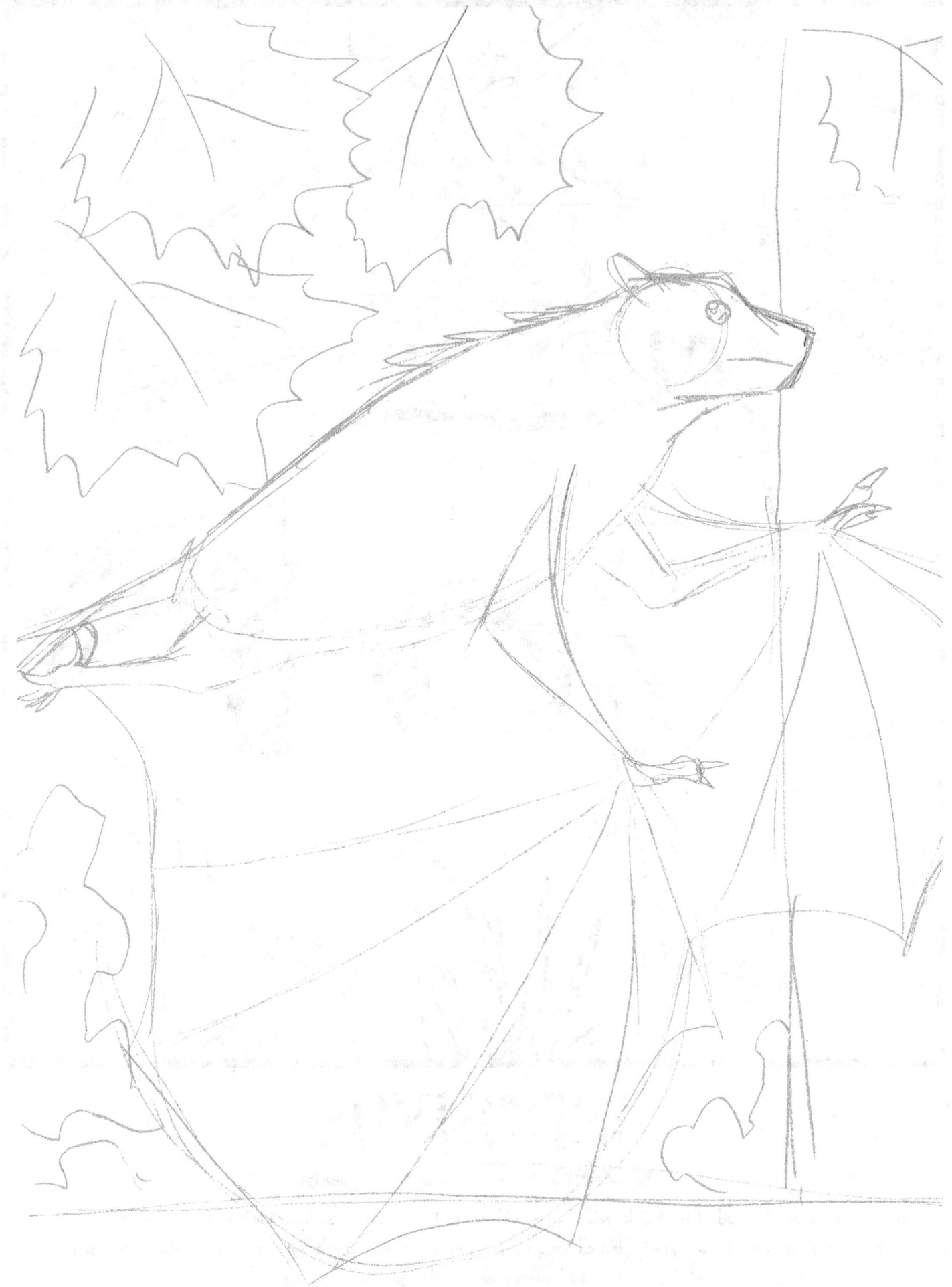

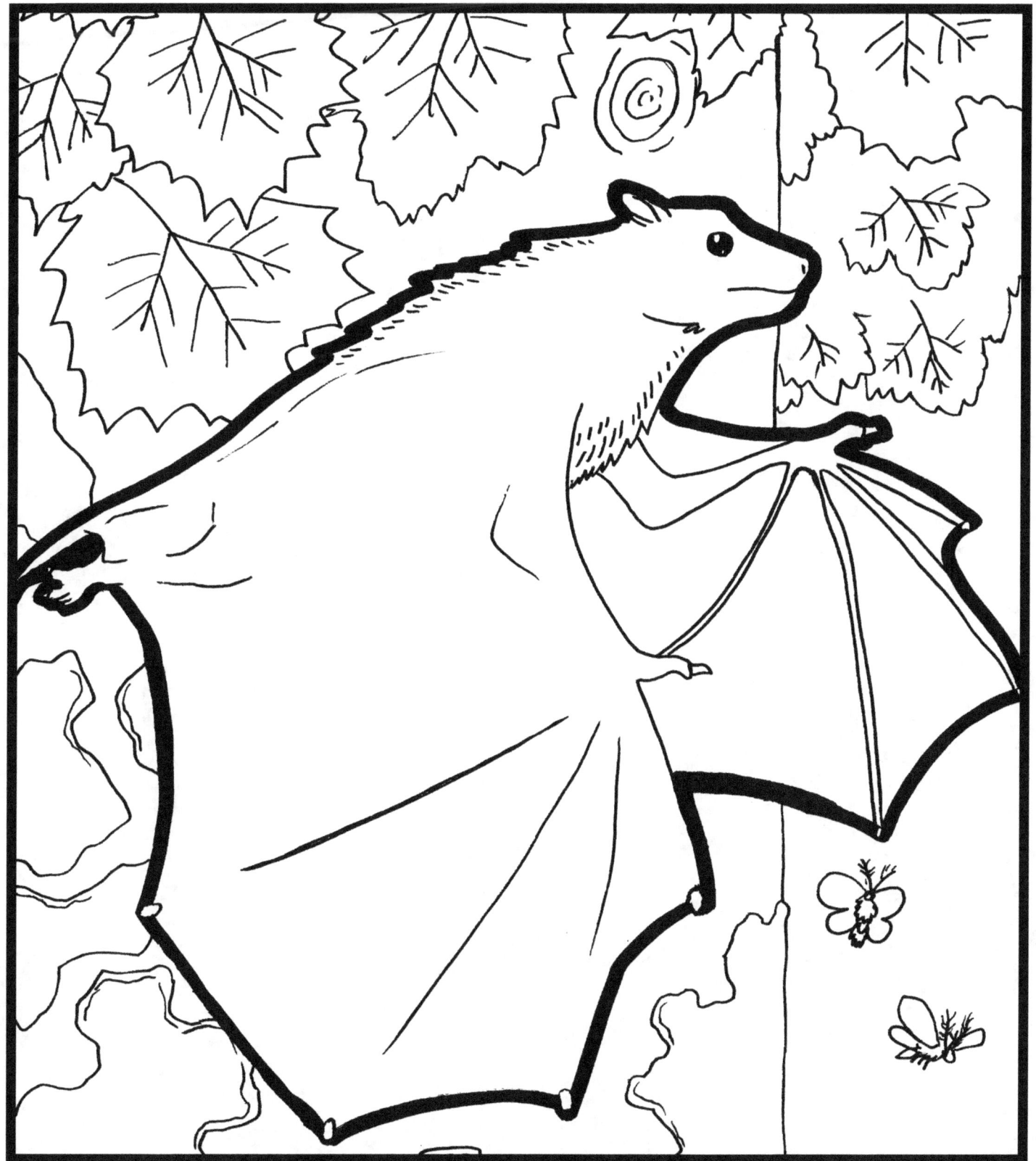

ONYCHONYCTERIS

52.5 MYA, Early Eocene, United States

True to its name, each of this "clawed bat's" wing fingers ended in a claw, while modern bats have one or two claws. It could not echolocate, was probably diurnal (active during the day), and had a longer tail than its current relatives, but it already had many of the adaptations exhibited by bats today.

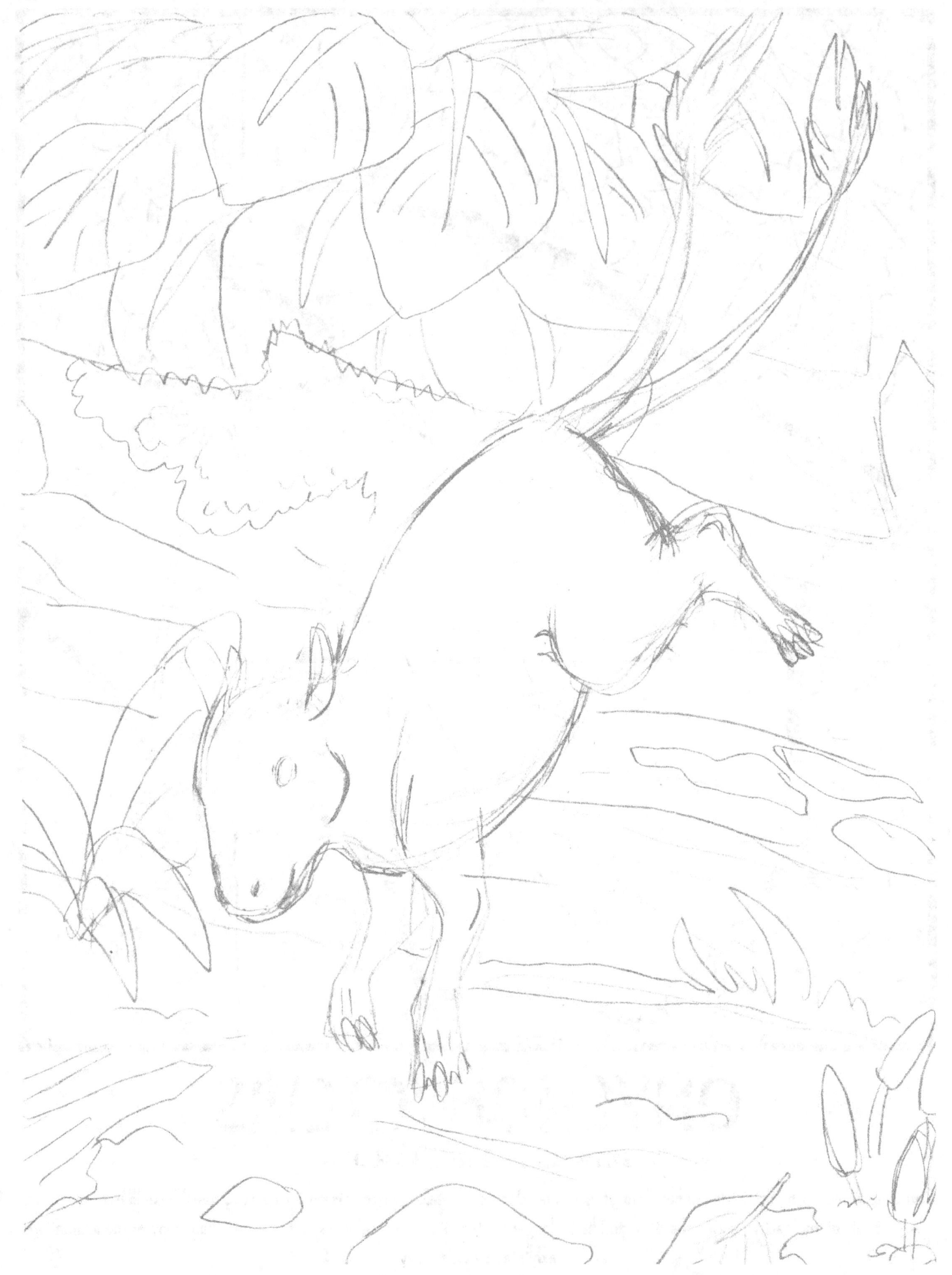

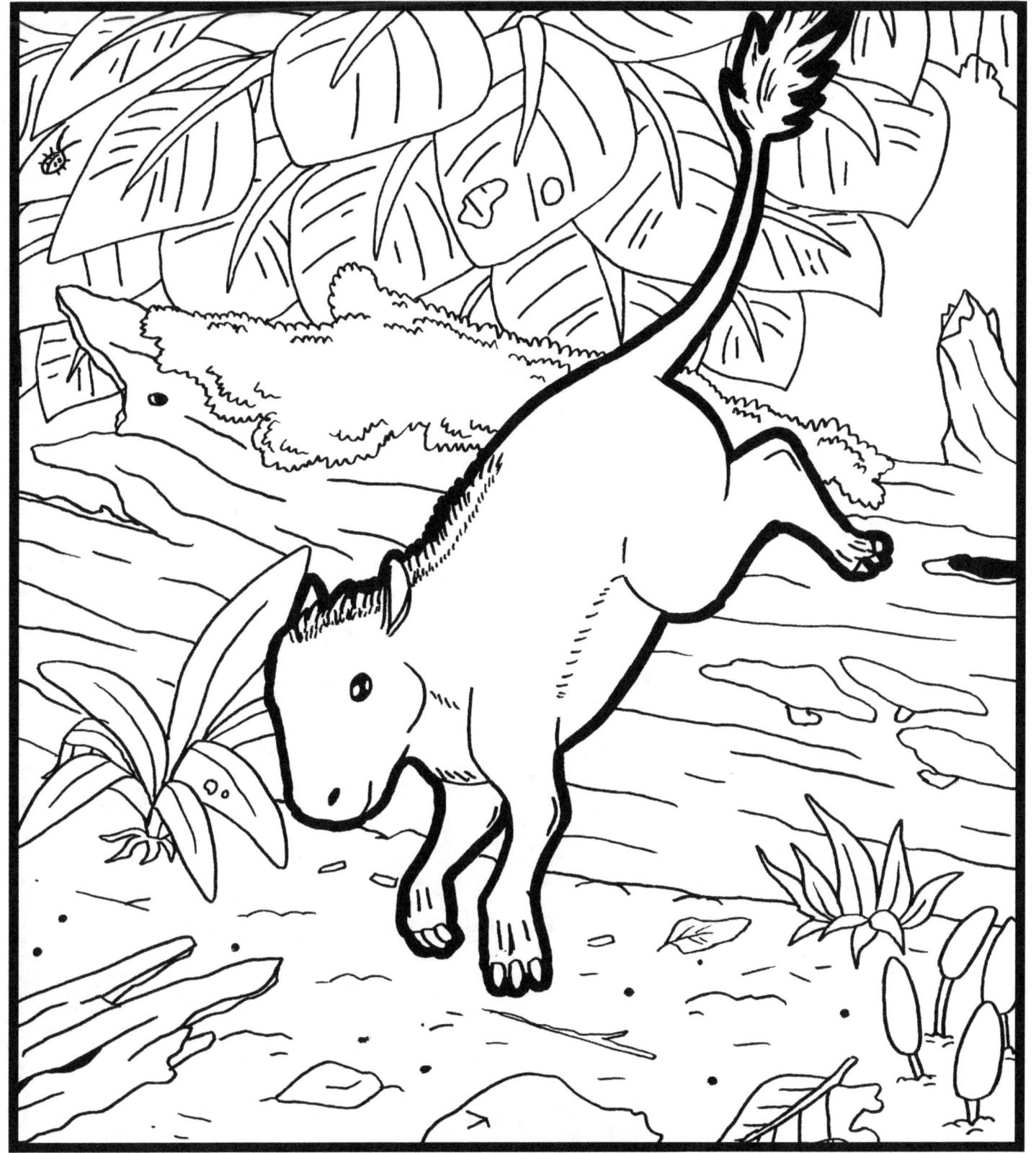

EOHIPPUS

52 MYA, Early Eocene, North America

This "dawn horse" was a very small, very early equid ungulate that dwelled in forests. It still had five toes on each foot (with four touching the ground in the forelegs and three touching the ground in the back legs), and was an omnivore slowly adapting for a diet that included more grasses as the environment it lived in changed.

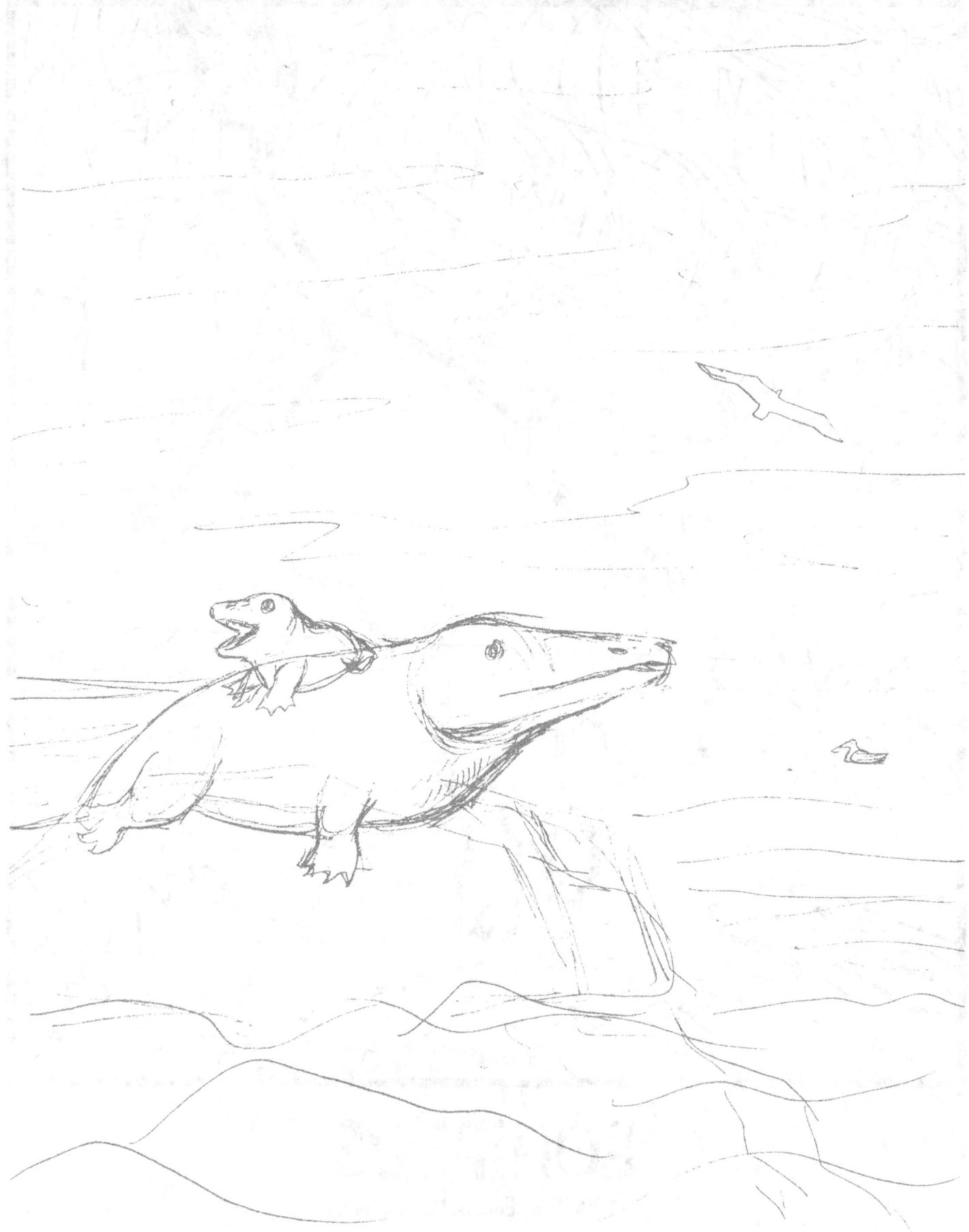

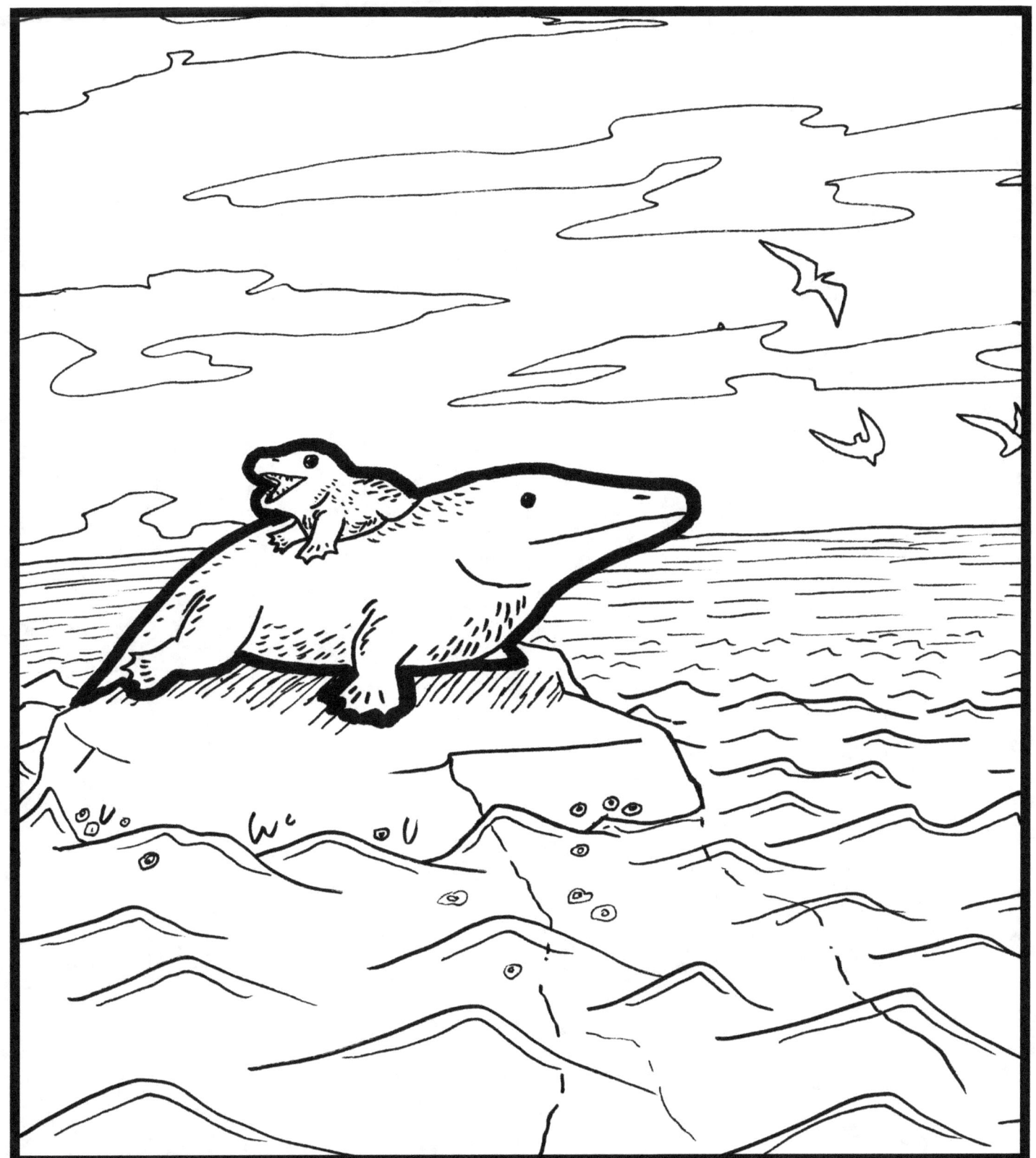

MAIACETUS

47.5 MYA, Middle Eocene, Pakistan

Since the infant was discovered to have exited head-first, these animals probably gave birth on land, where there would be no risk of the infants drowning. It was roughly dolphin-sized, at 2.5 meters long, and weighing between 280 and 390 kilograms. This animal was an intermediate between the land- and sea-dwelling whale ancestors.

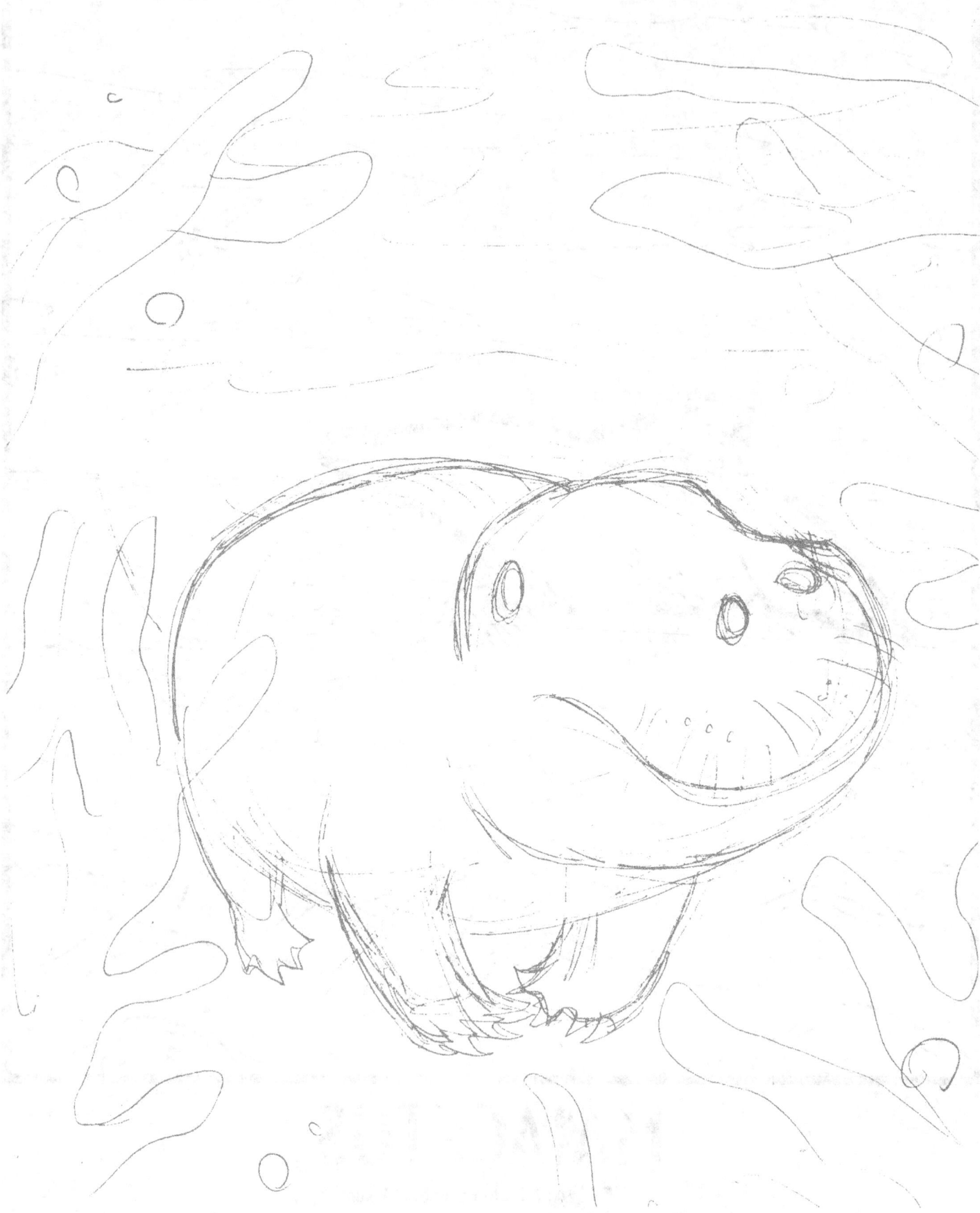

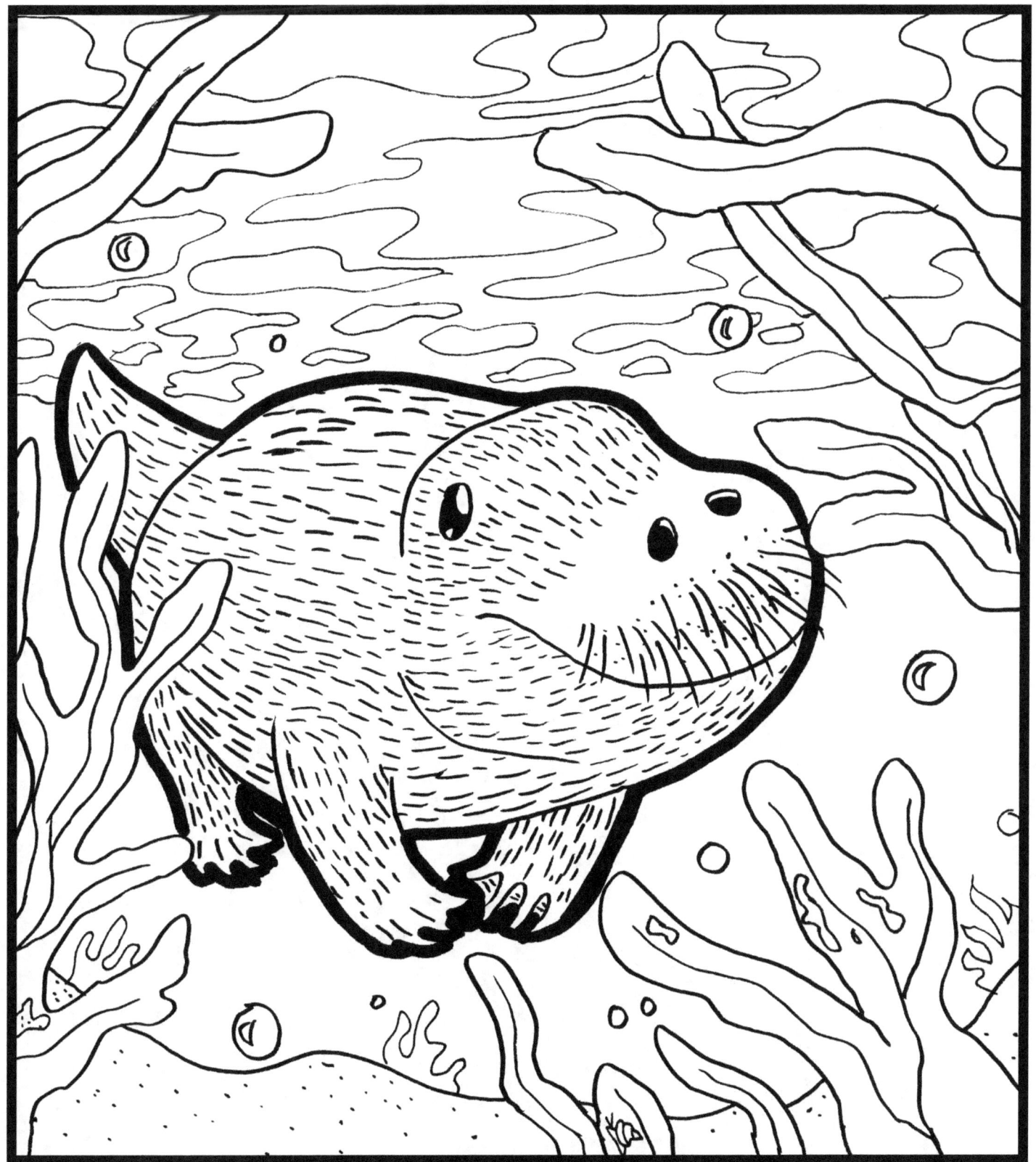

MAKARACETUS

47 - 41 MYA, Middle Eocene, Pakistan

Though it was a carnivore, this early whale had a walrus or tapir-like muscular snout. Modern walruses use their snouts to feed on mollusks, but their heads and teeth are also more suited for this activity. More specimens of this animal will need to be discovered to infer more about its lifestyle.

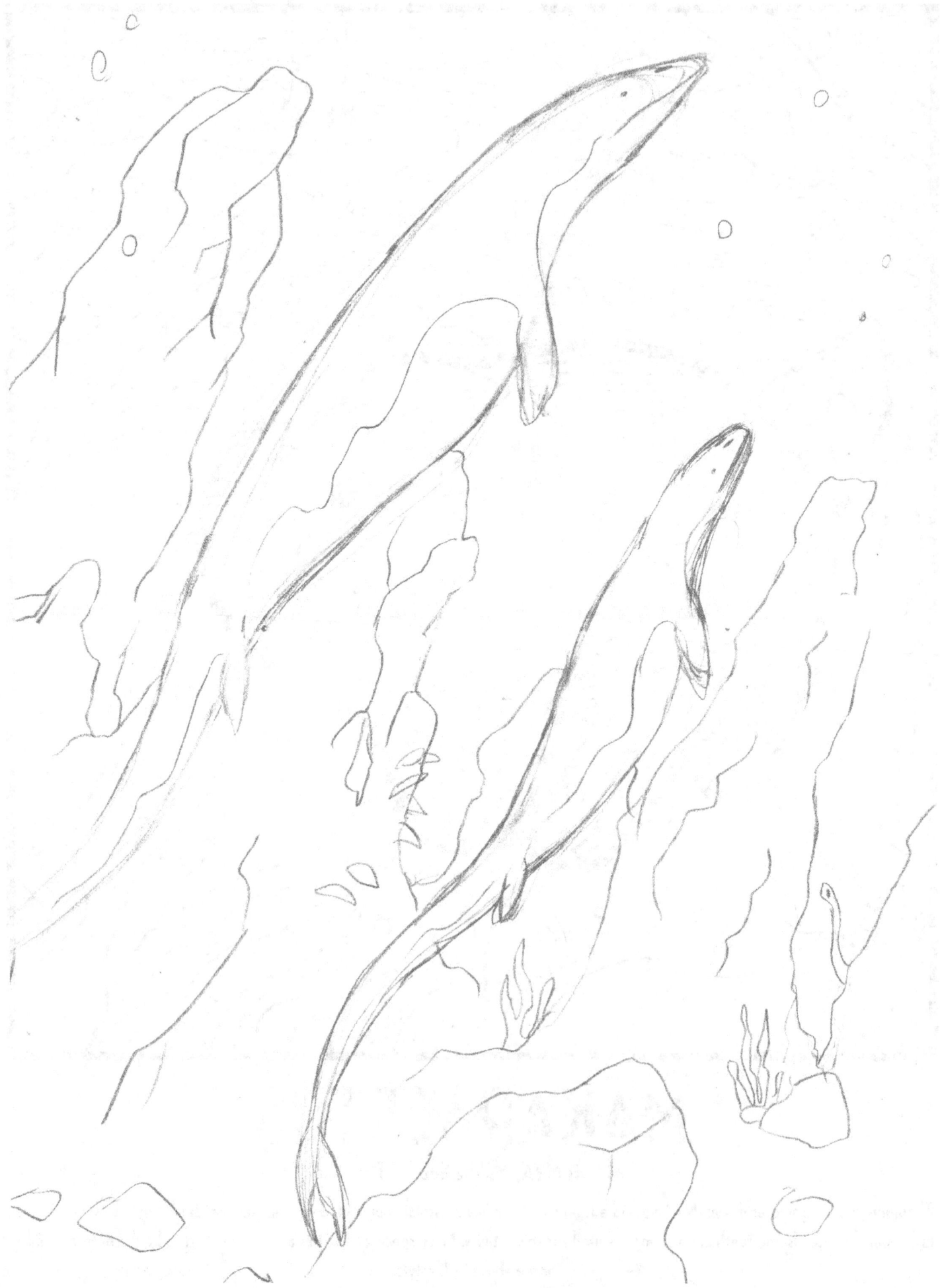

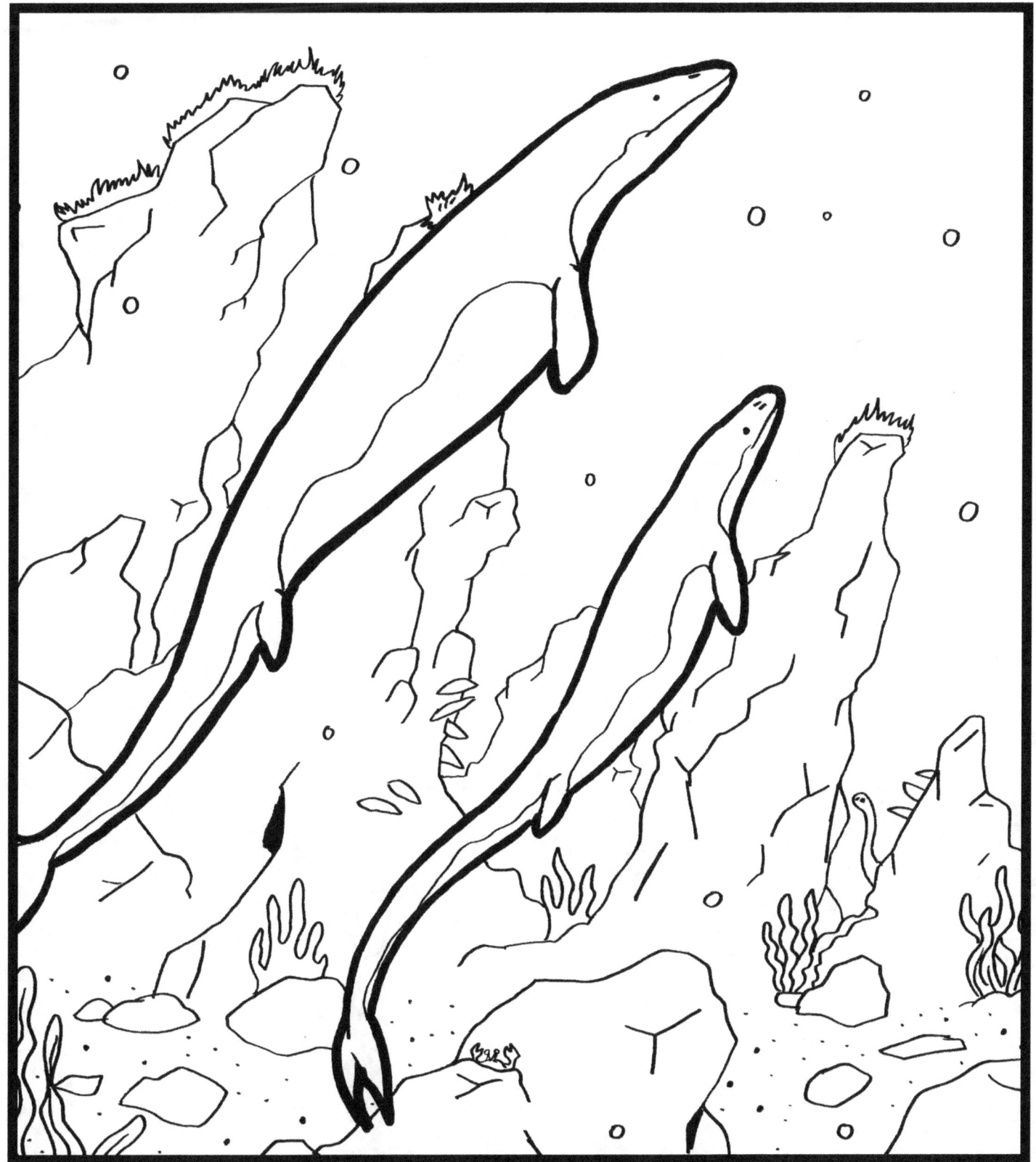

BASILOSAURUS

40 - 34 MYA, Late Eocene, Egypt, Jordan, United States

Originally thought to have been a large marine reptile, this "lizard king" was an ancestor of the whales we have today.
It fed on sharks and other fish, and even other whales. It was not capable of diving as deep as some whales today, or swimming quickly
for long periods of time, but it had a powerful bite.

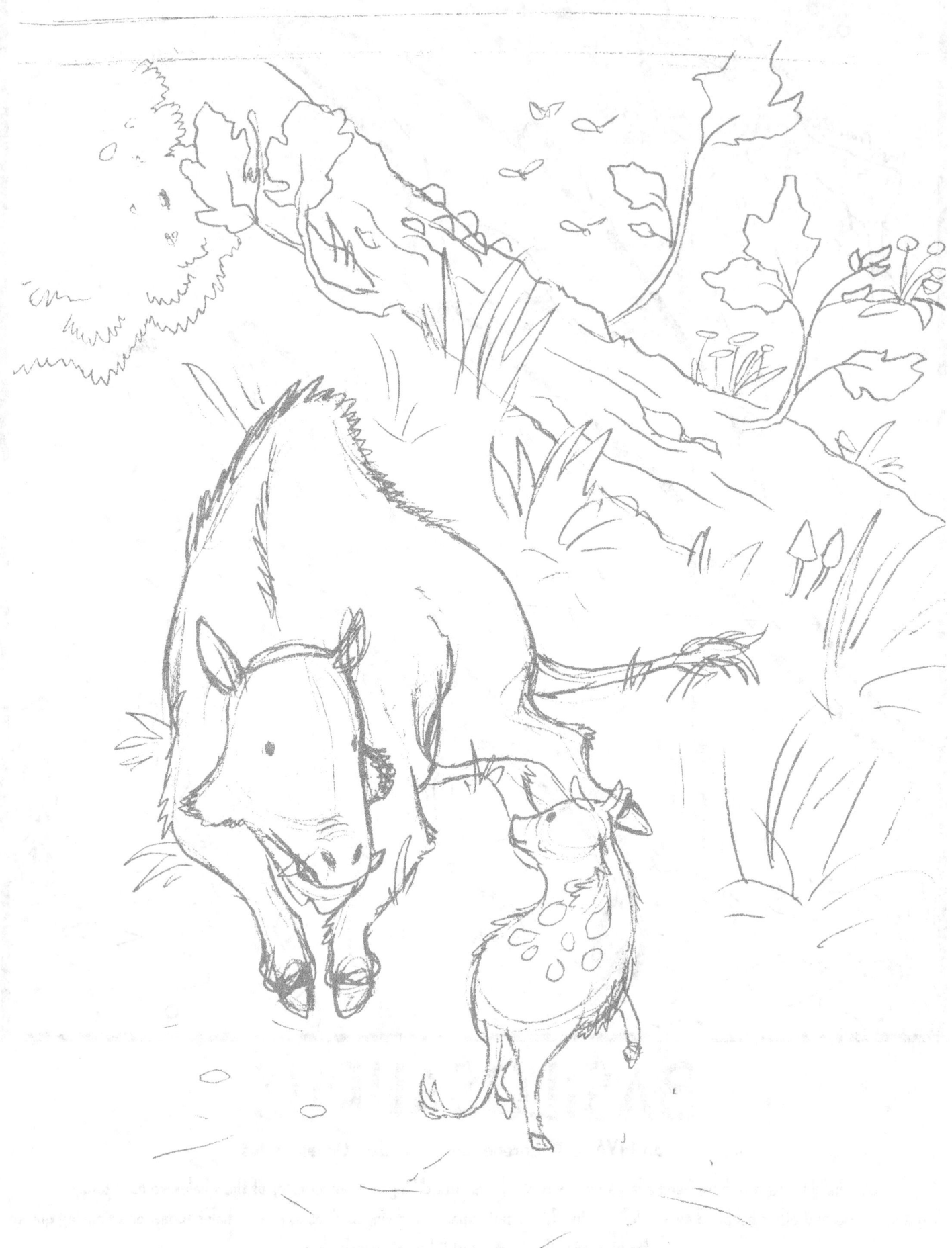

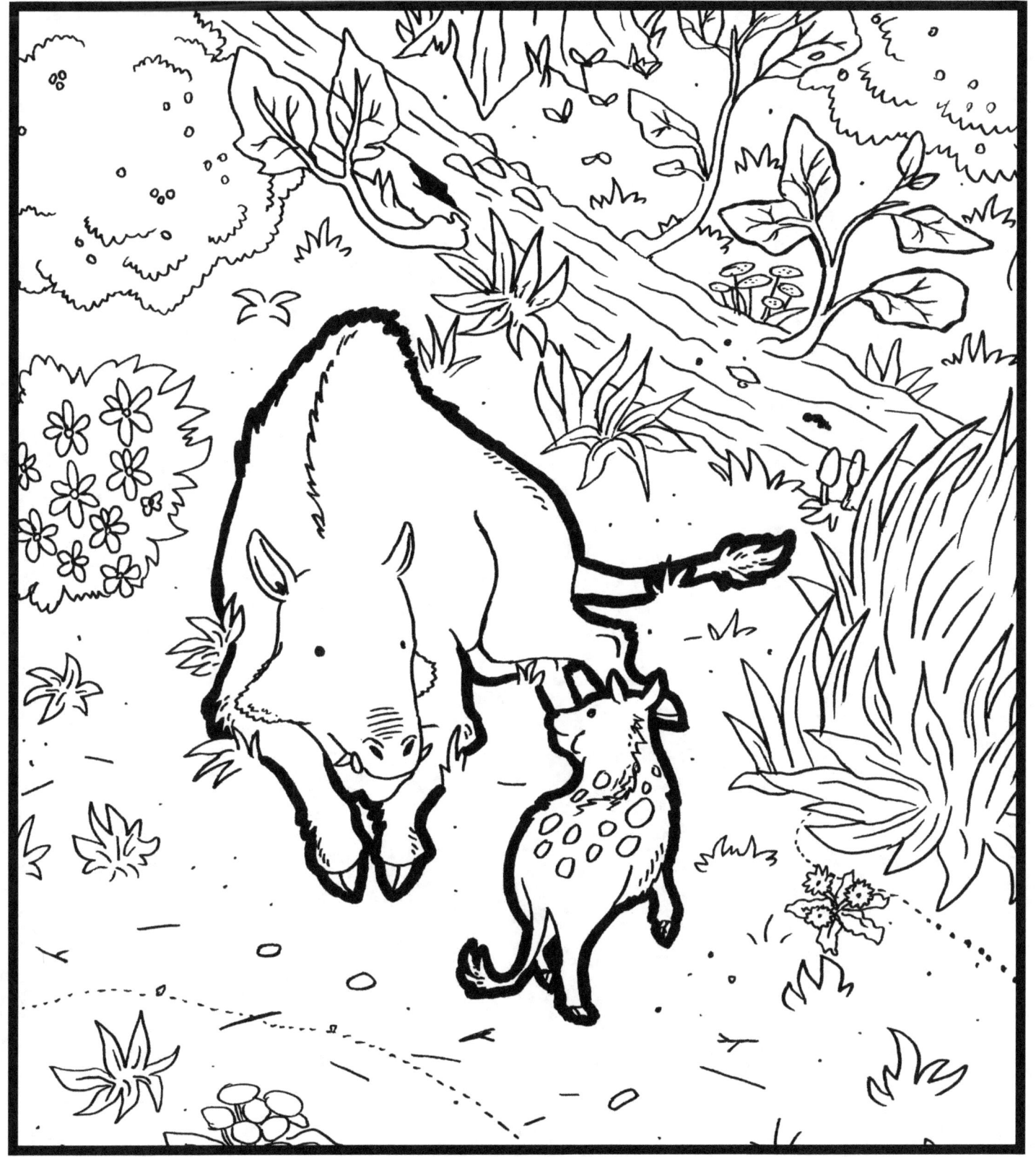

ENTELODON

37 - 28.4 MYA, Late Paleogene, China, Mongolia, Spain, France, Germany, and Romania

Although commonly called a "killer pig," entelodon was an artiodactyl, and was more closely related to whales and hippos than to pigs. It was a large, fast-moving omnivore that was widespread over Eurasia.

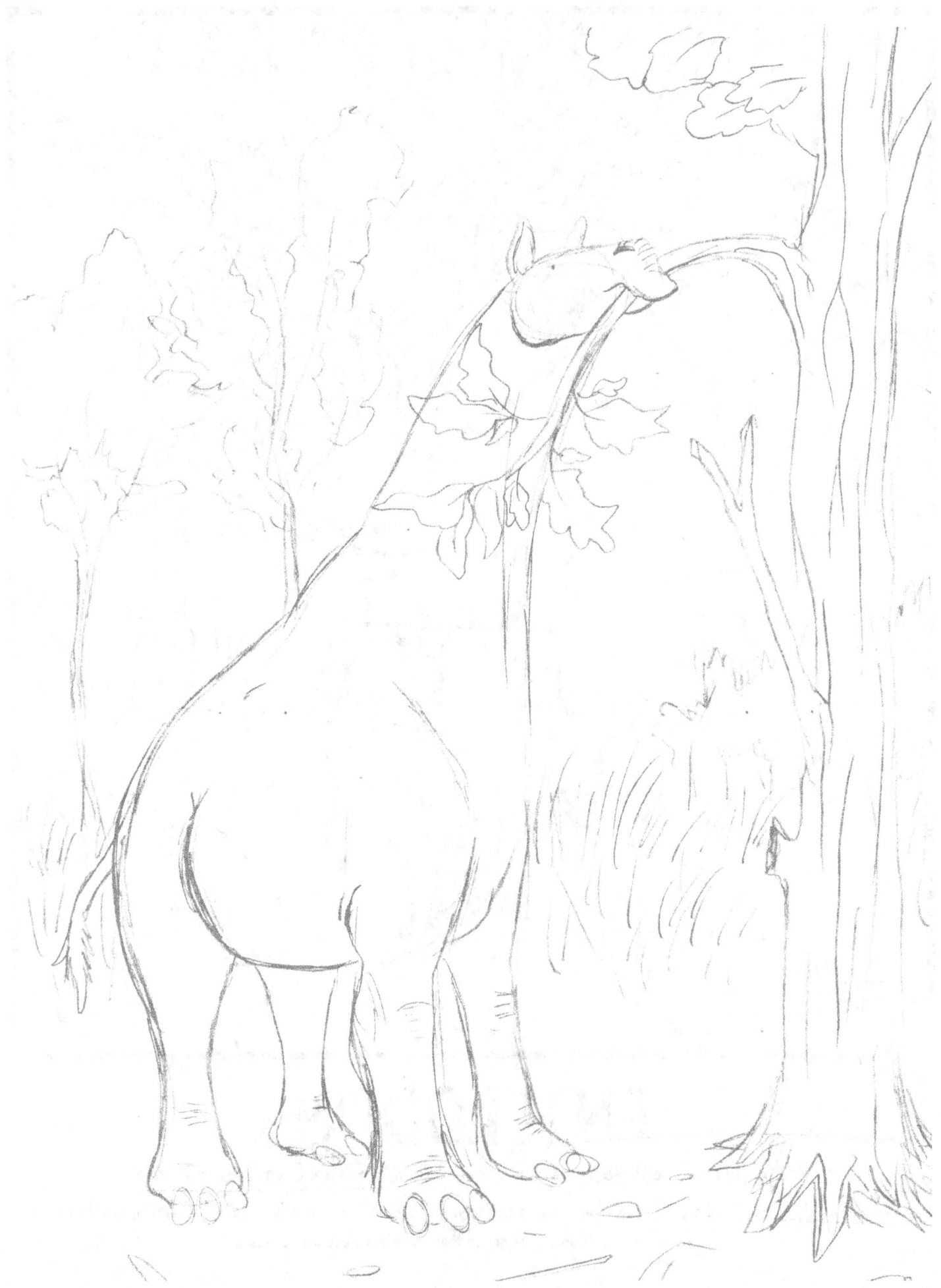

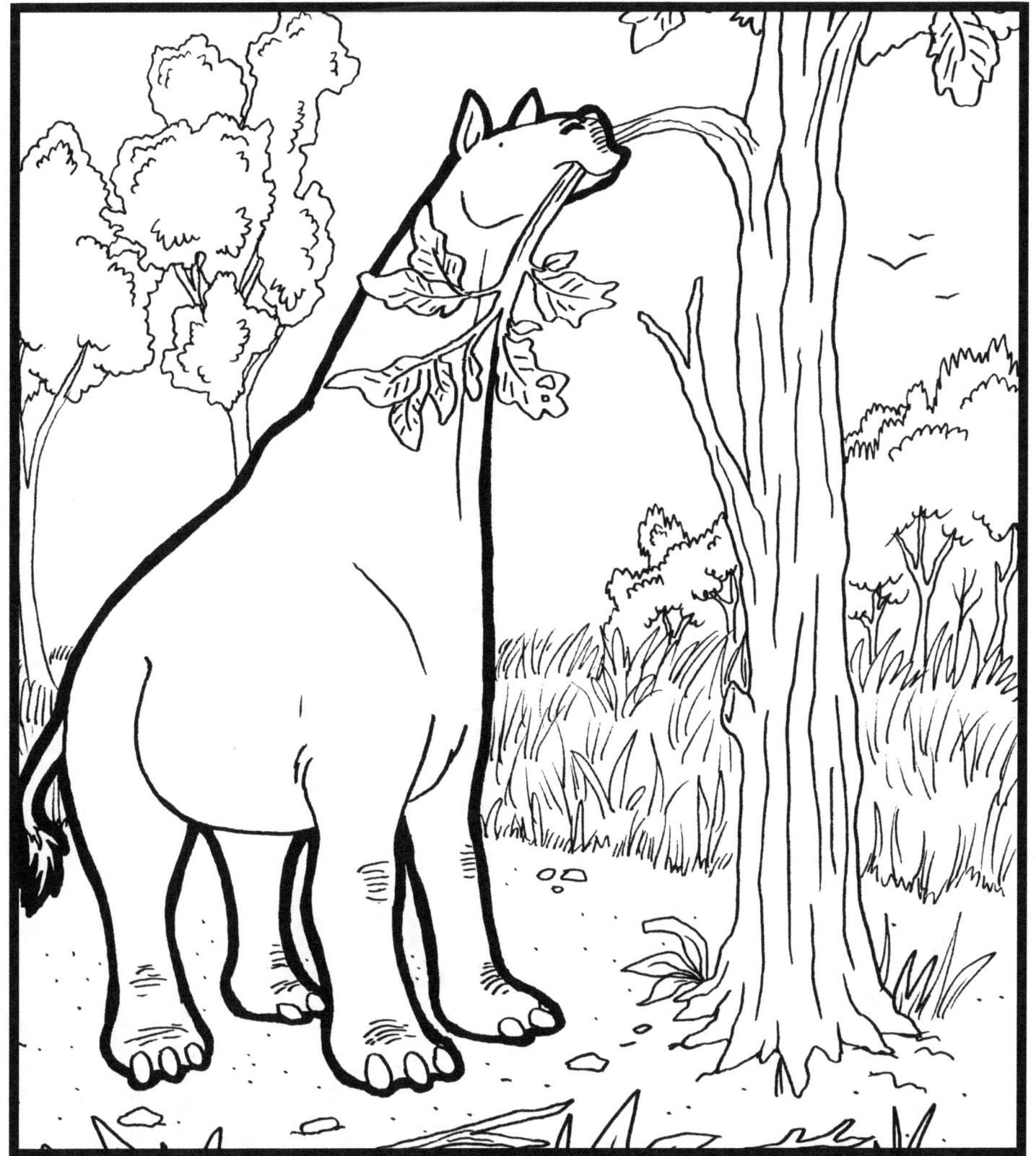

PARACERATHERIUM

34 - 23 MYA, Oligocene, China, Mongolia, India, Pakistan, Kazakhstan, Georgia, Turkey, Romania, Bulgaria, the Balkans

This leaf-eating browser was the largest land mammal to ever have walked the earth, weighing in at 20 tons, and with a length of 7.4 meters. It is said to have been near the limit of the maximum possible size for land mammals. It could not move quickly, but could probably traverse long distances.

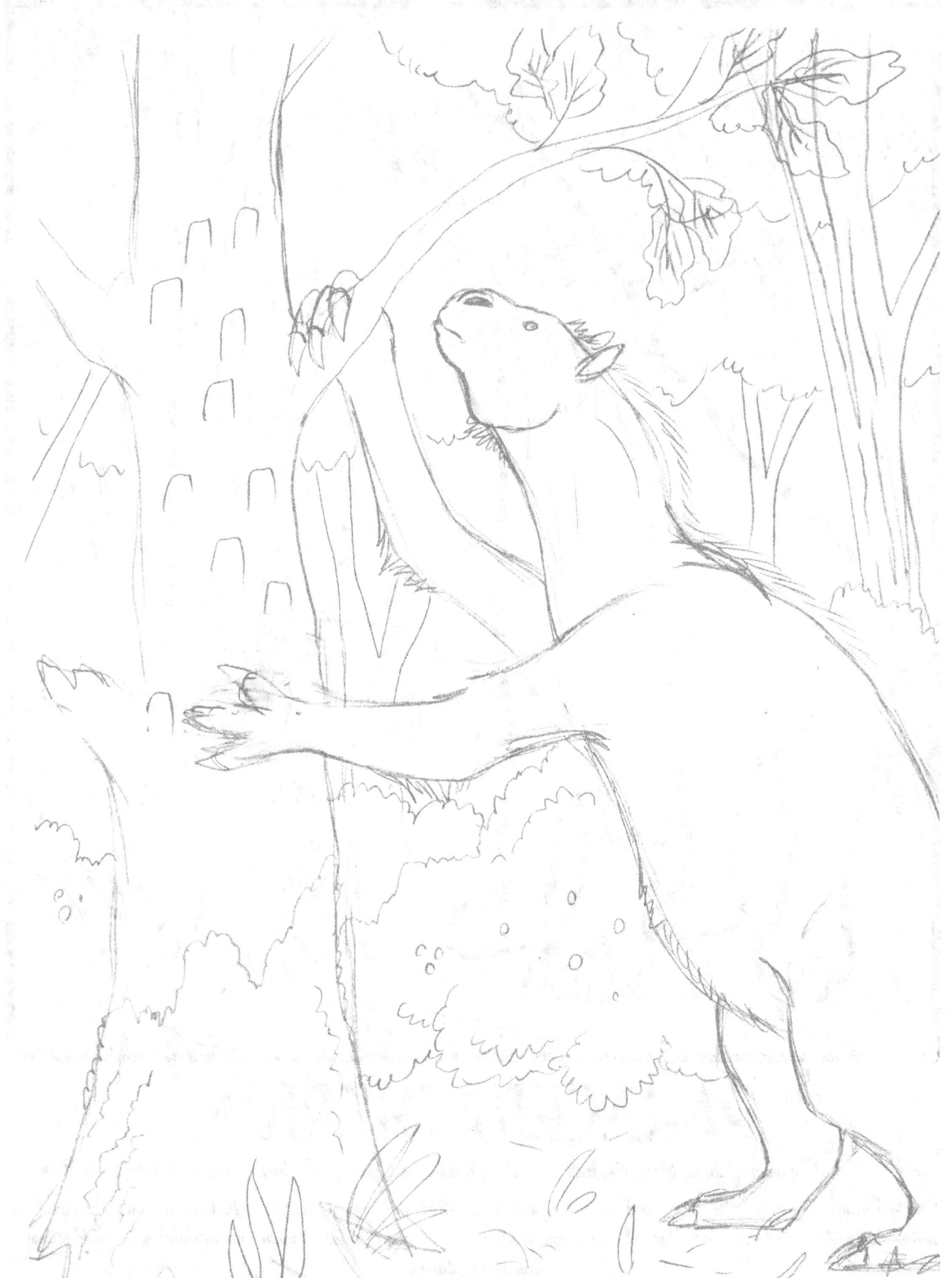

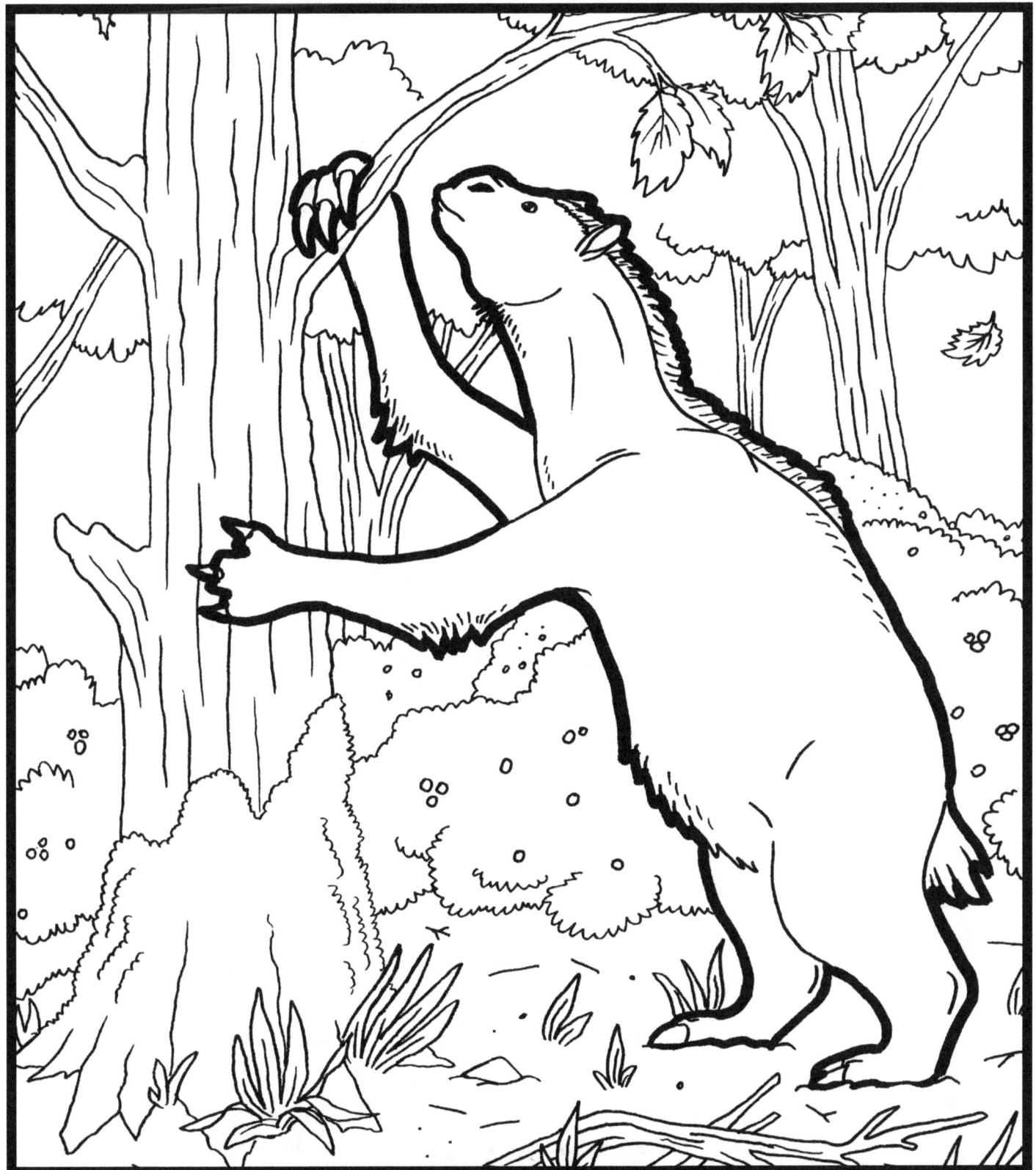

CHALICOTHERIUM

28.6 - 3.4 MYA, Late Oligocene to Early Pliocene, Tajikistan, Germany, China, India

These odd herbivores had well-developed knuckles that were used to support weight while walking. They were once thought to have lived off of tubers dug up from the ground, but new research suggests they used their long arms to grab ahold of branches, and bring them down to where the leaves could be stripped off. The knuckle-walking could also be useful in keeping the claws from wearing down.

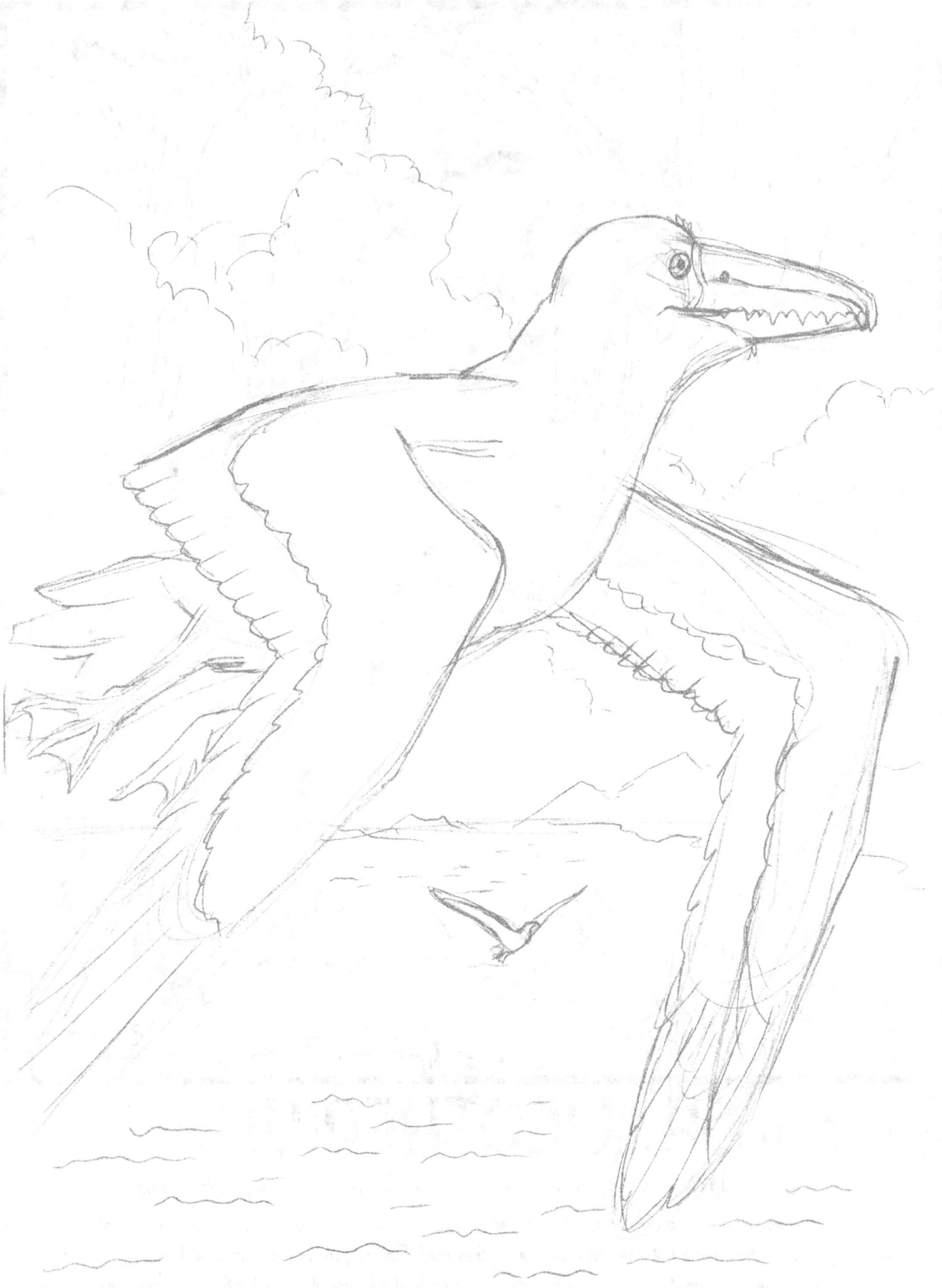

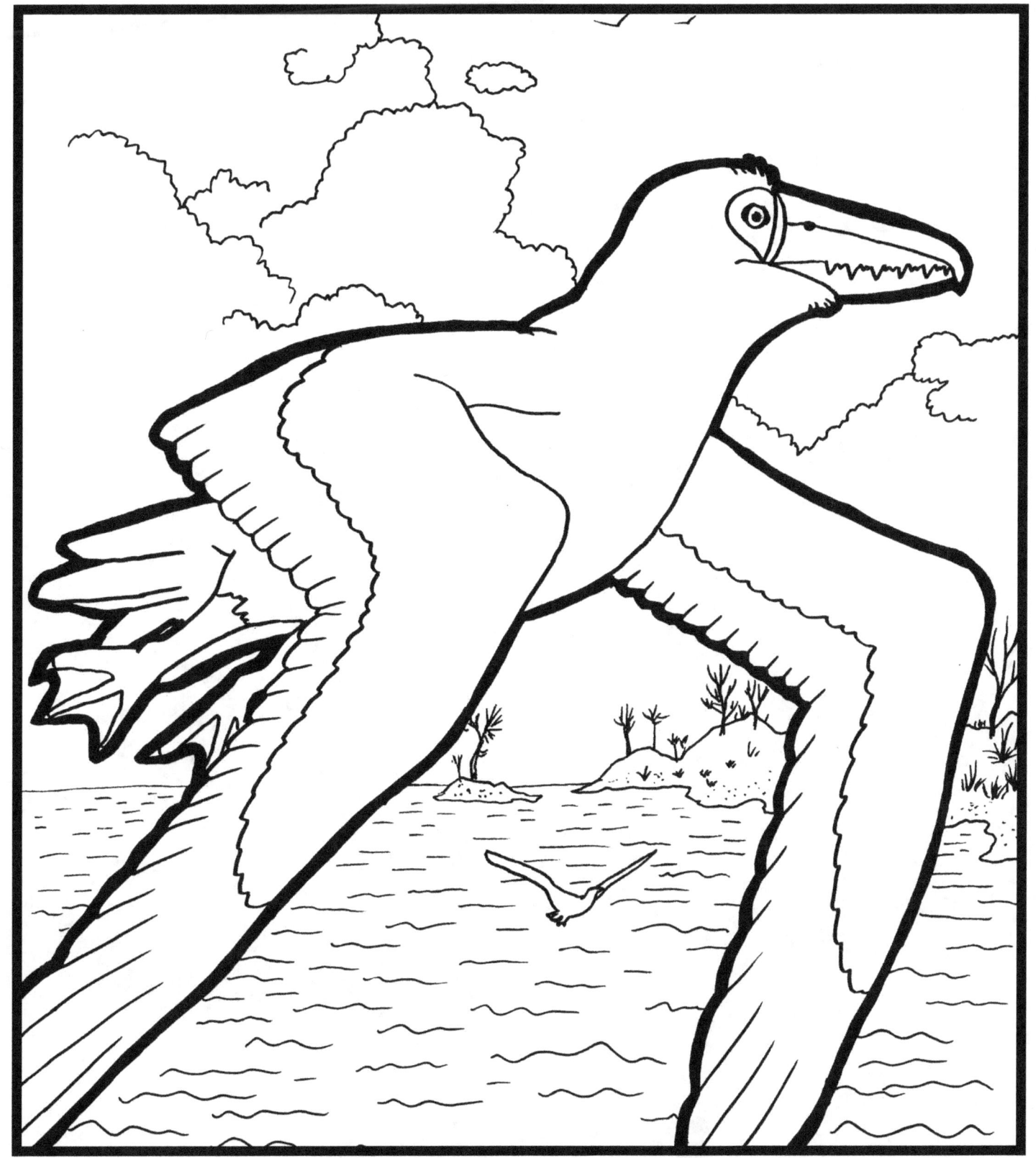

PELAGORNIS

25 - 2.5 MYA, Late Oligocene to Early Pleistocene, United States

This bird had the longest wingspan of any dinosaur, at between 5 and 6 meters long. It probably had a similar sillhouette to a modern-day albatross, but the most striking difference, besides its size, were the pseudoteeth in its beak. These needle-like points were extensions of the beak, and allowed it to better grasp slippery prey, such as fish.

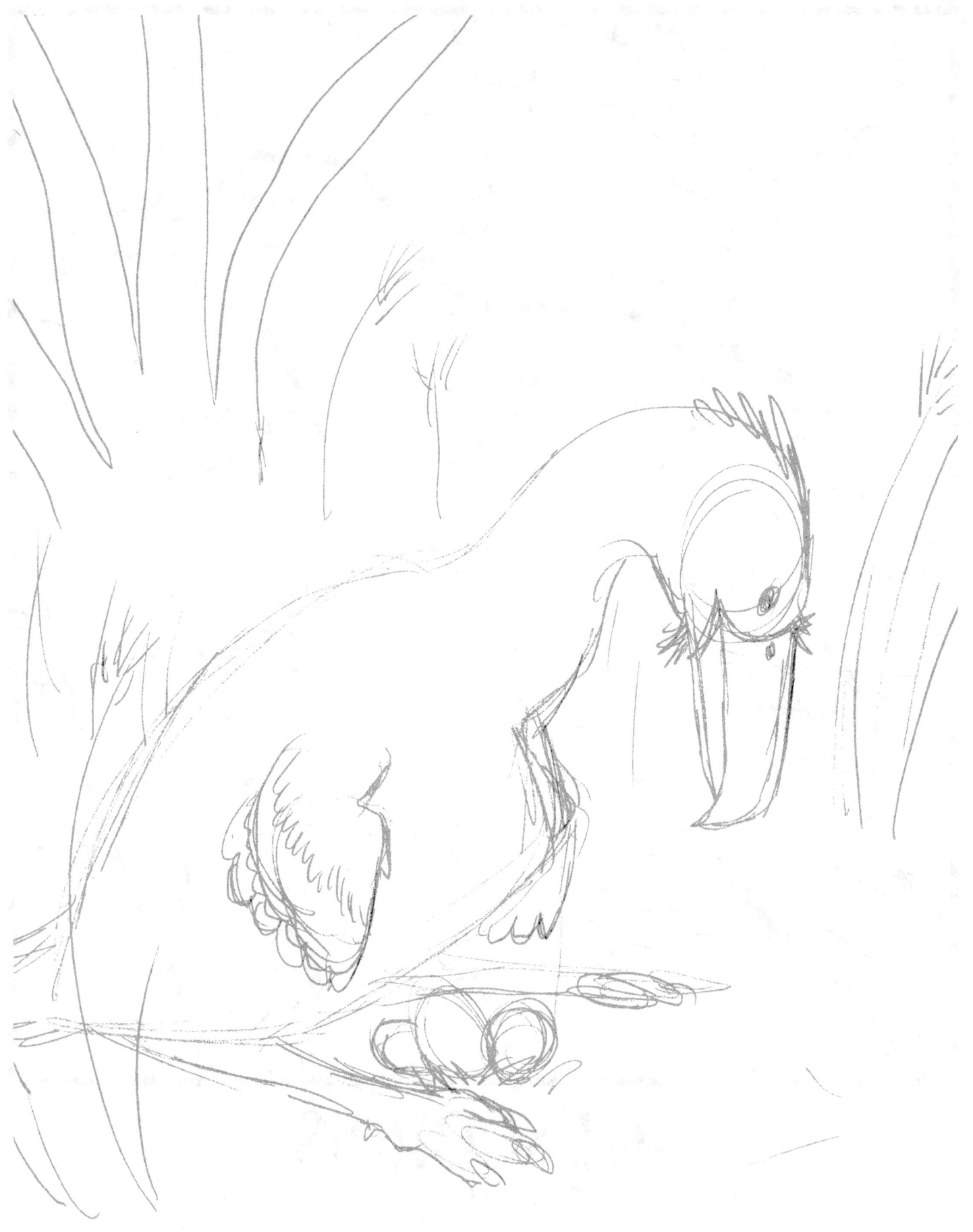

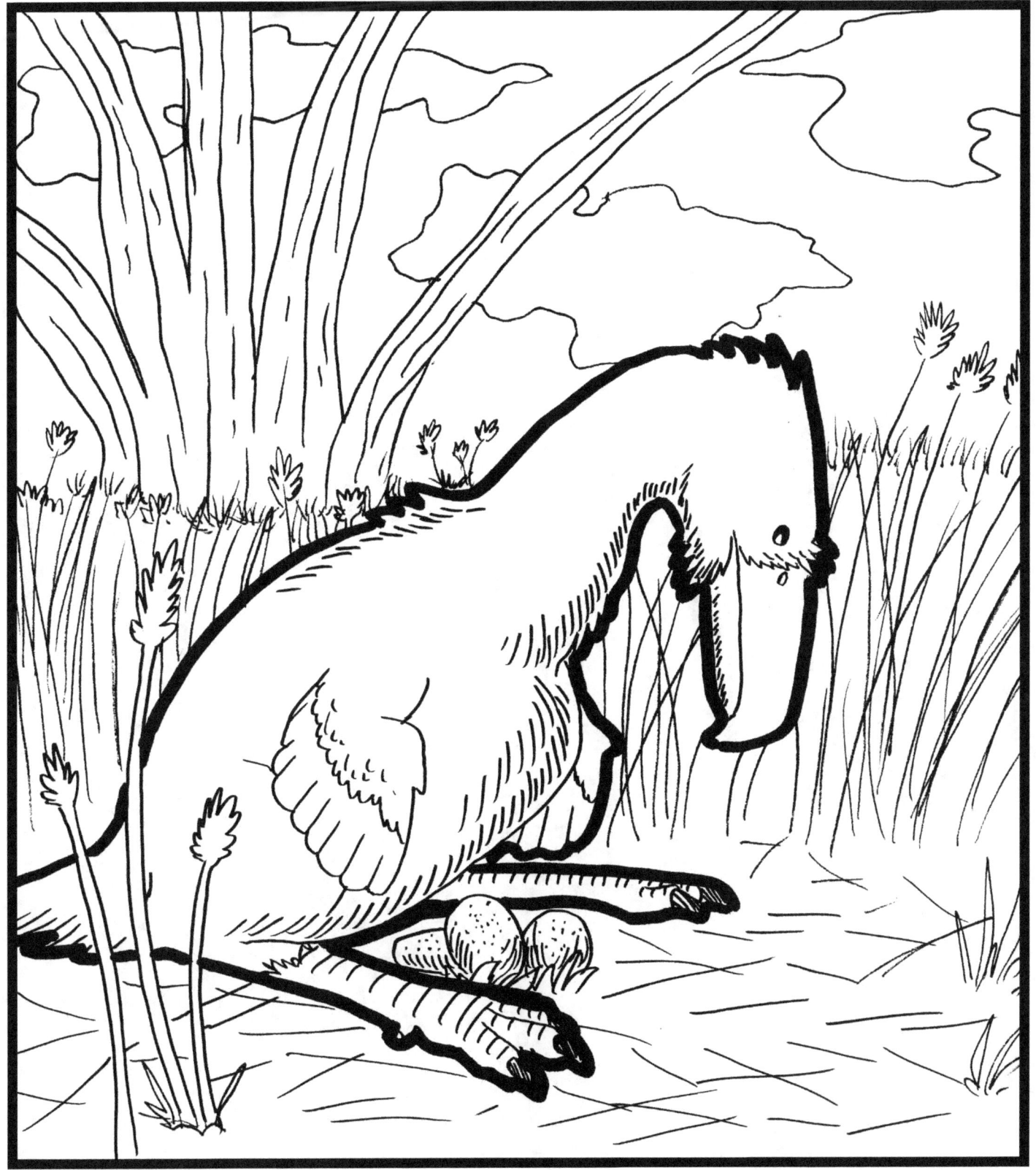

KELENKEN

15 MYA, Miocene, Argentina

At 2.3 to 3 meters tall, this was the largest known predatory bird since the mass-extinction that ended the Cretaceous, and half of its 70 centimeter-long skull was beak. It was a fast runner, and would use its speed to run down small prey that would then be swallowed whole.

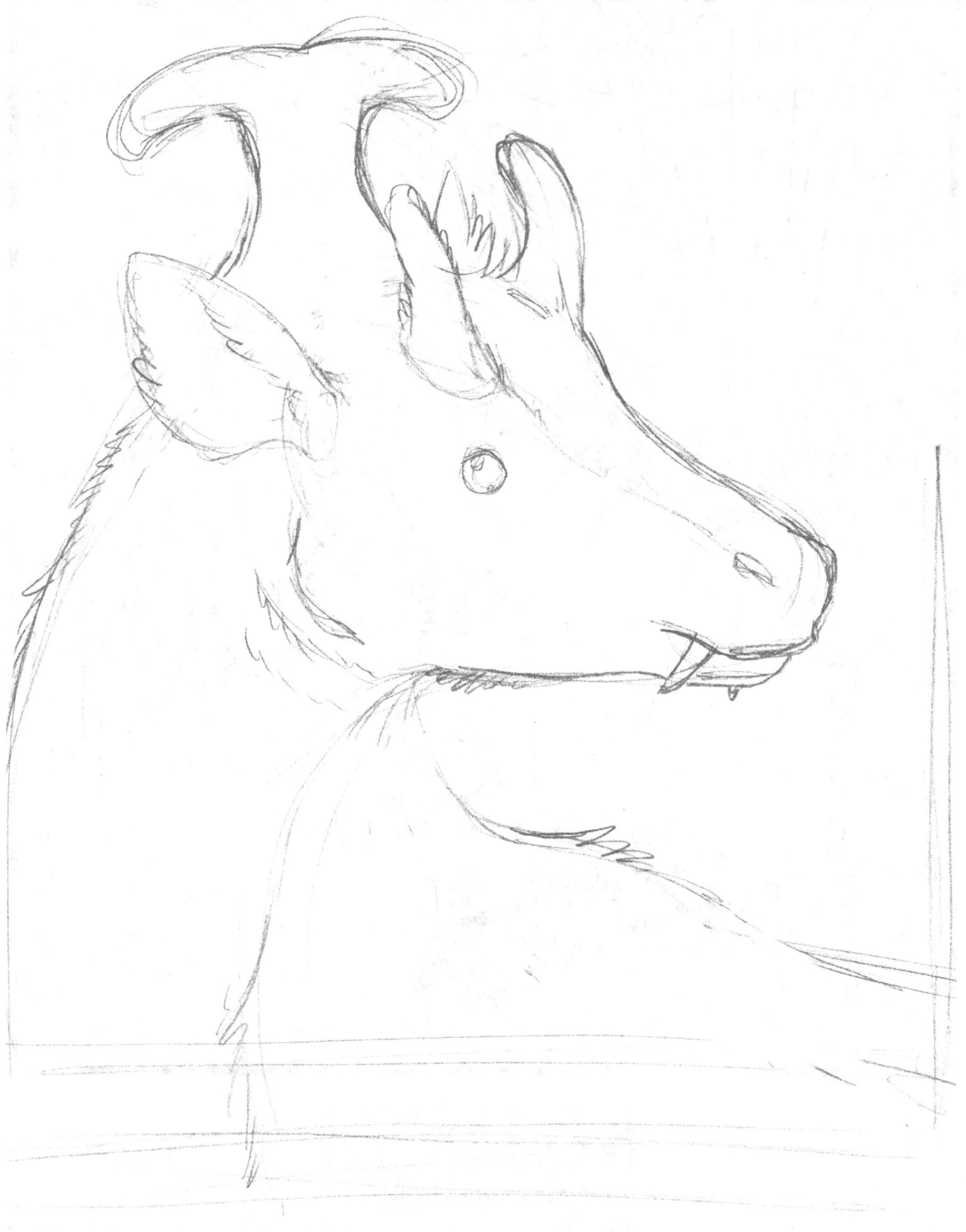

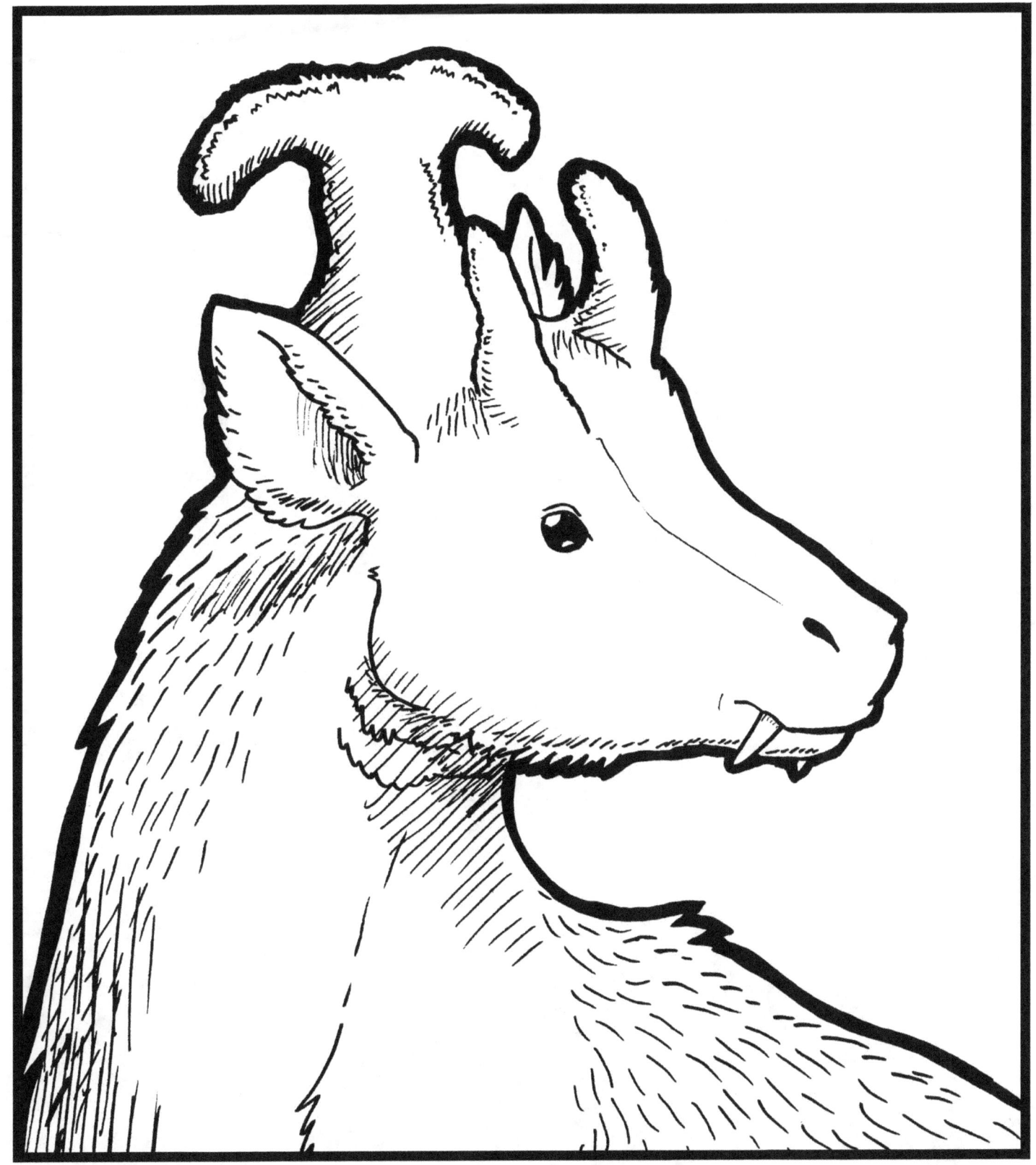

XENOKERYX

15 MYA, Miocene, Spain

This animal that had a T-shaped "strange horn" was a ruminant from Europe, and may have been part of the same clade as modern-day giraffes. Usually, head protrusions such as horns and antlers come in pairs, but this herbivore had three, an odd number.

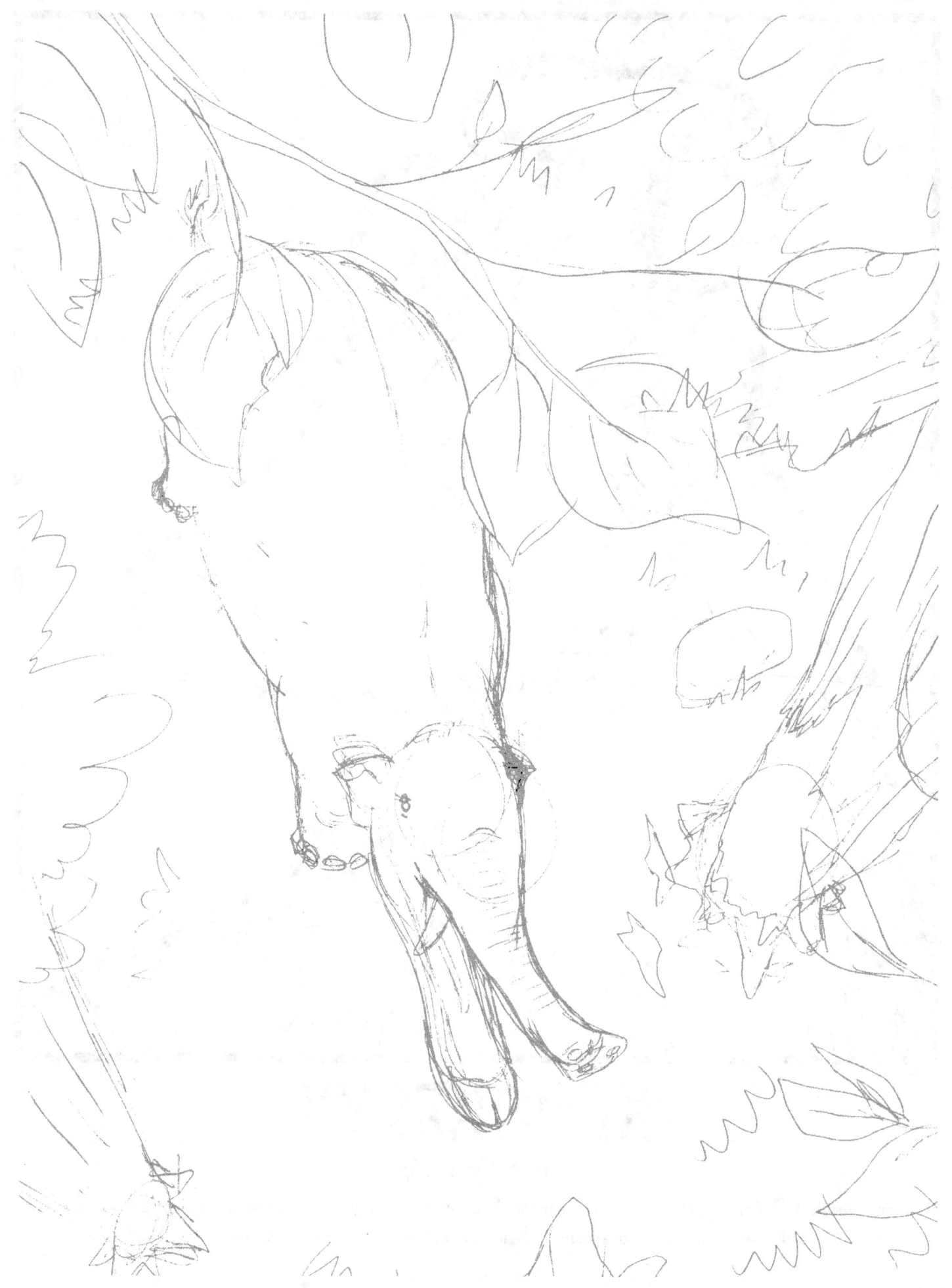

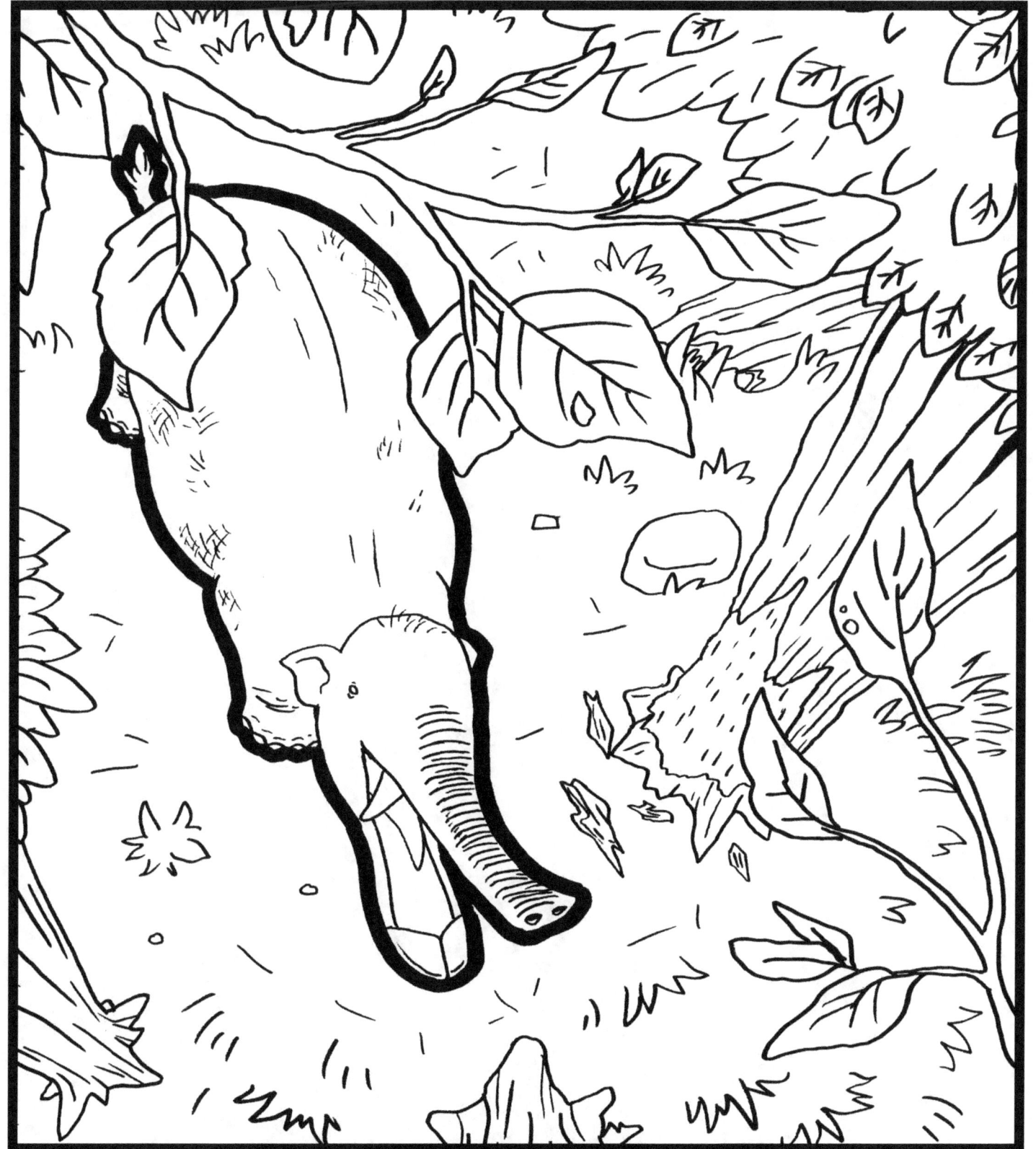

PLATYBELODON

15 - 10 MYA, Miocene, North America, Eastern Europe, Africa, China

About 3 meters long, this elephant relative could be found on several continents.
It used the sharp, flat side of its teeth to strip bark and cut branches from the trunks of trees, but was previously thought to have used the teeth more like a shovel to scoop up soft aquatic vegetation.

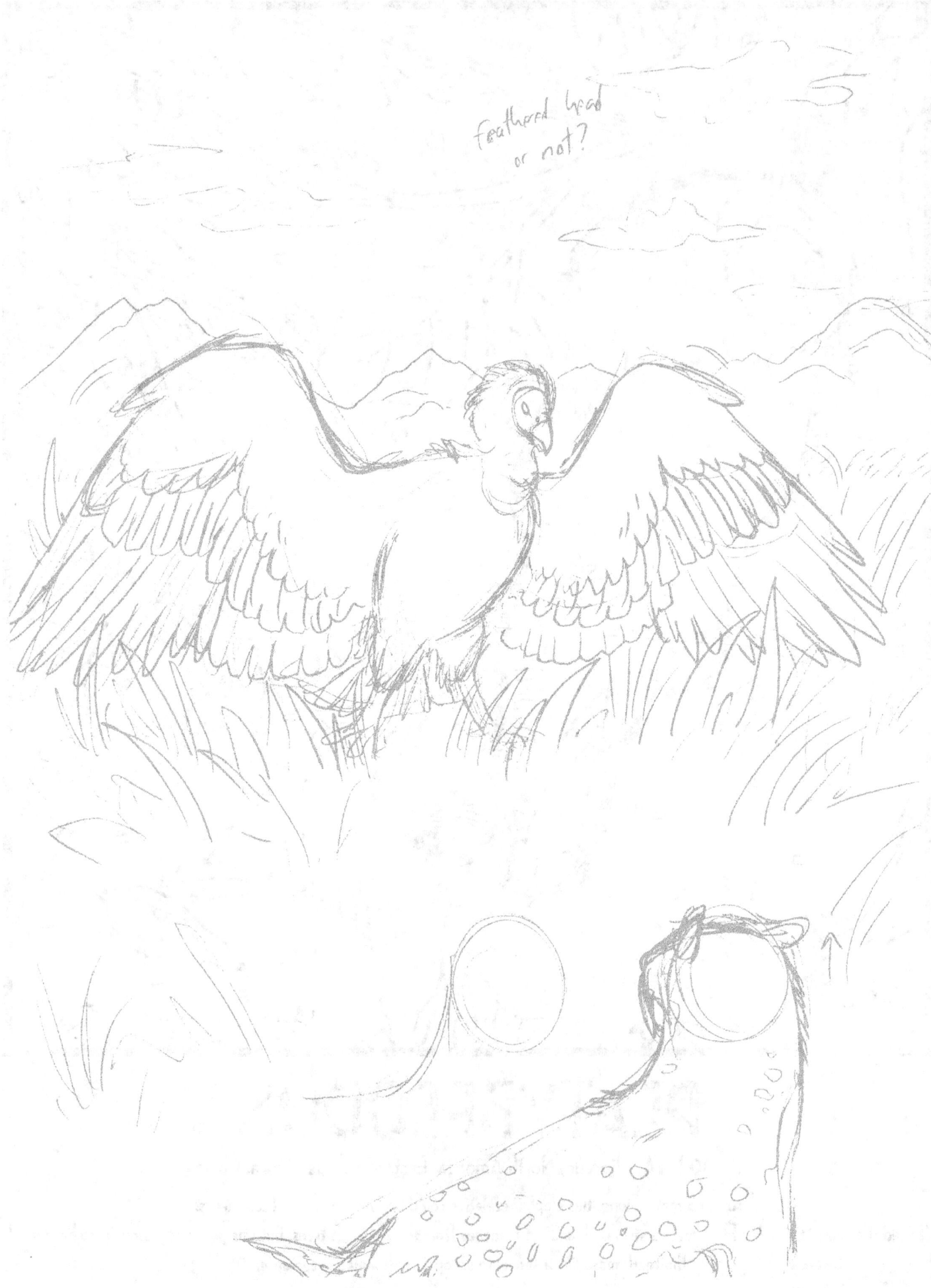

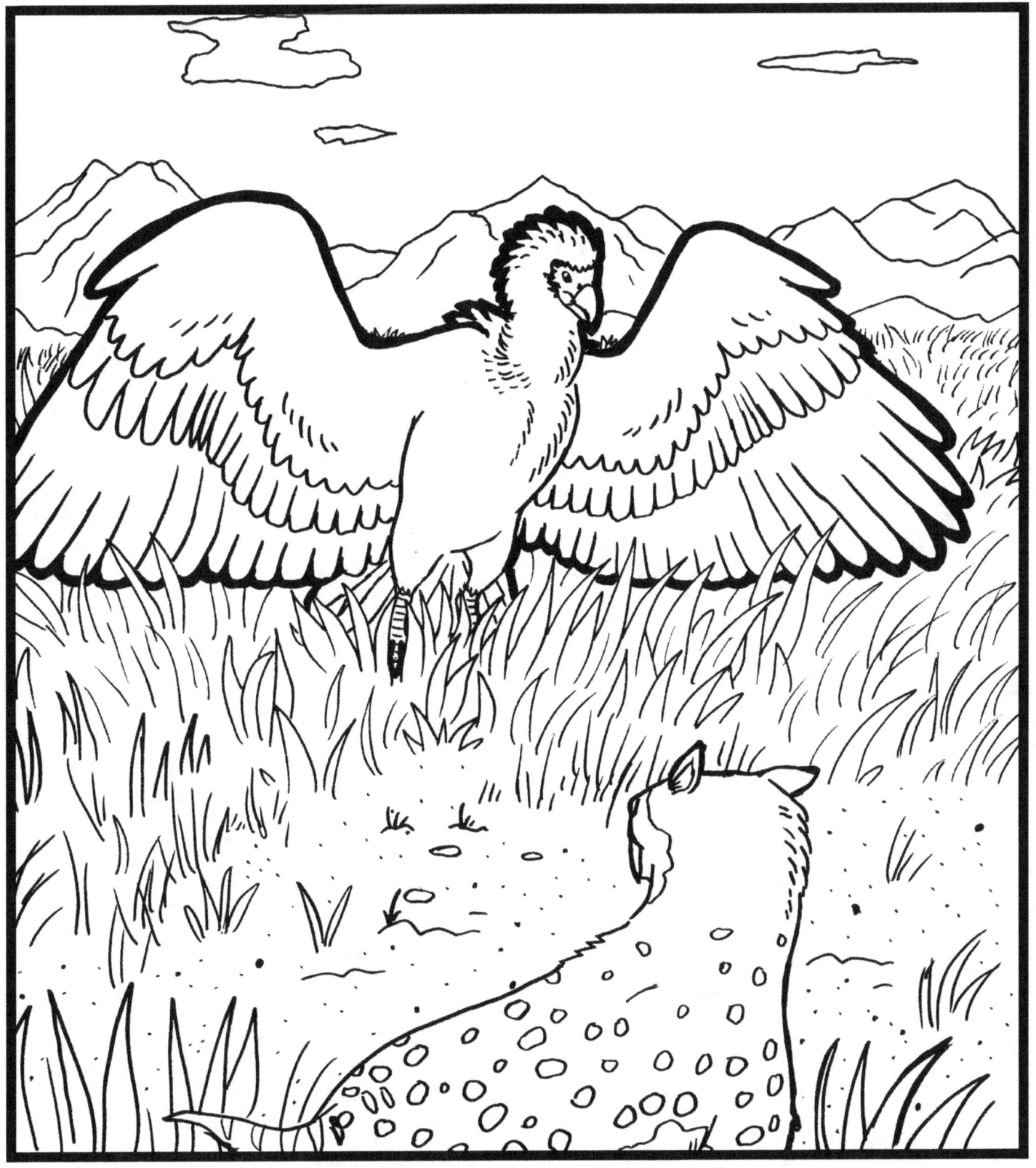

ARGENTAVIS

6 MYA, Late Miocene, Argentina

These birds had huge territories that they patrolled, and both hunted and scavenged for prey. They were large enough to easily scare off predators like the Thylacosmilidae from their catches. They preferred to swallow prey whole, though the beak's hooked end was still useful for tearing apart flesh from larger animals.

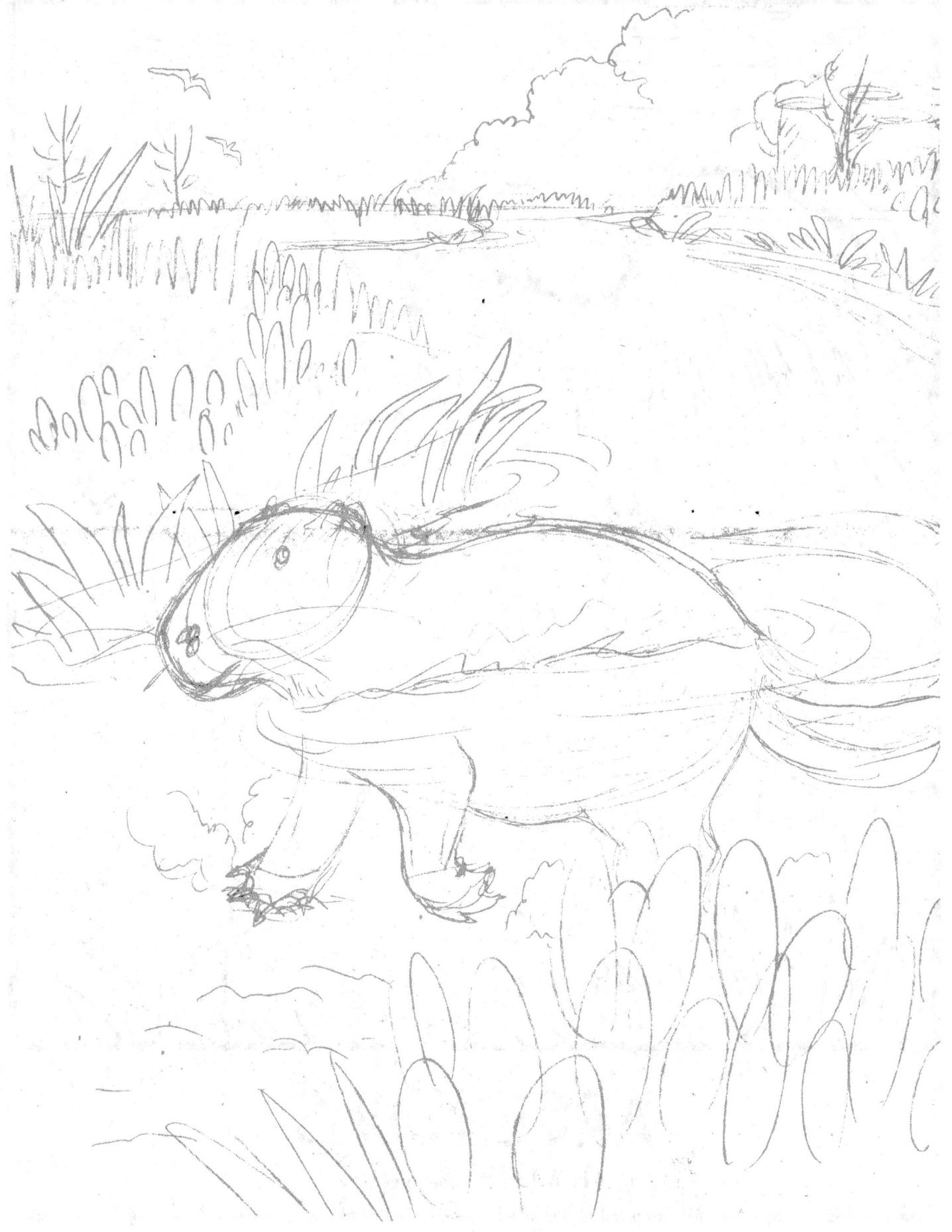

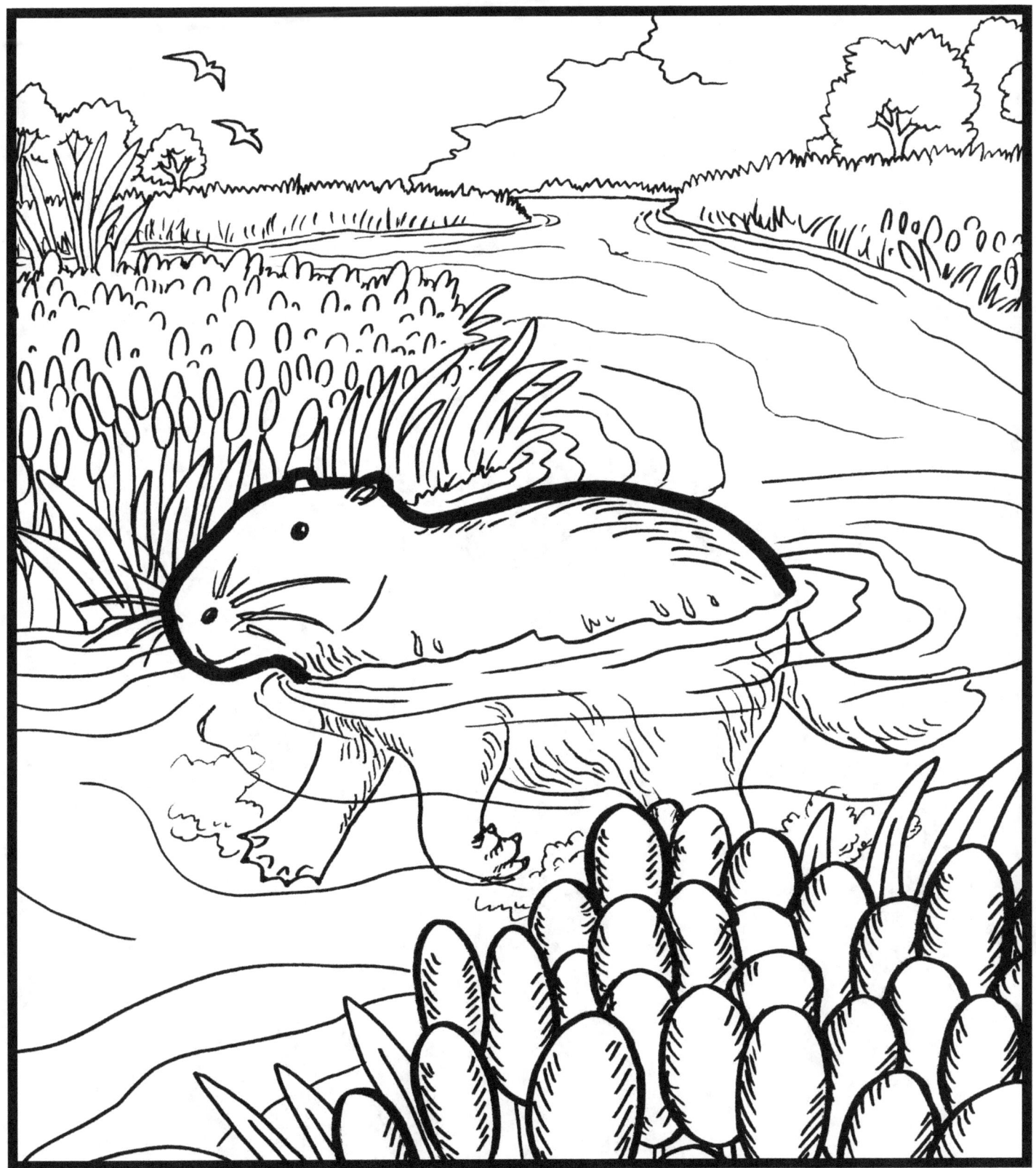

JOSEPHOARTIGASIA

4 - 3 MYA, Pliocene, Uruguay

This cow-sized rodent weighed over 1,000 kilograms was closely related to modern pacaranas. It lived primarily in wetlands, eating grass and other plants. When the climate shifted for drier conditions, the wetlands reduced in size, putting a strain on this animal, and ultimately their inability to adapt quickly enough led to the extinction of this enormous creature.

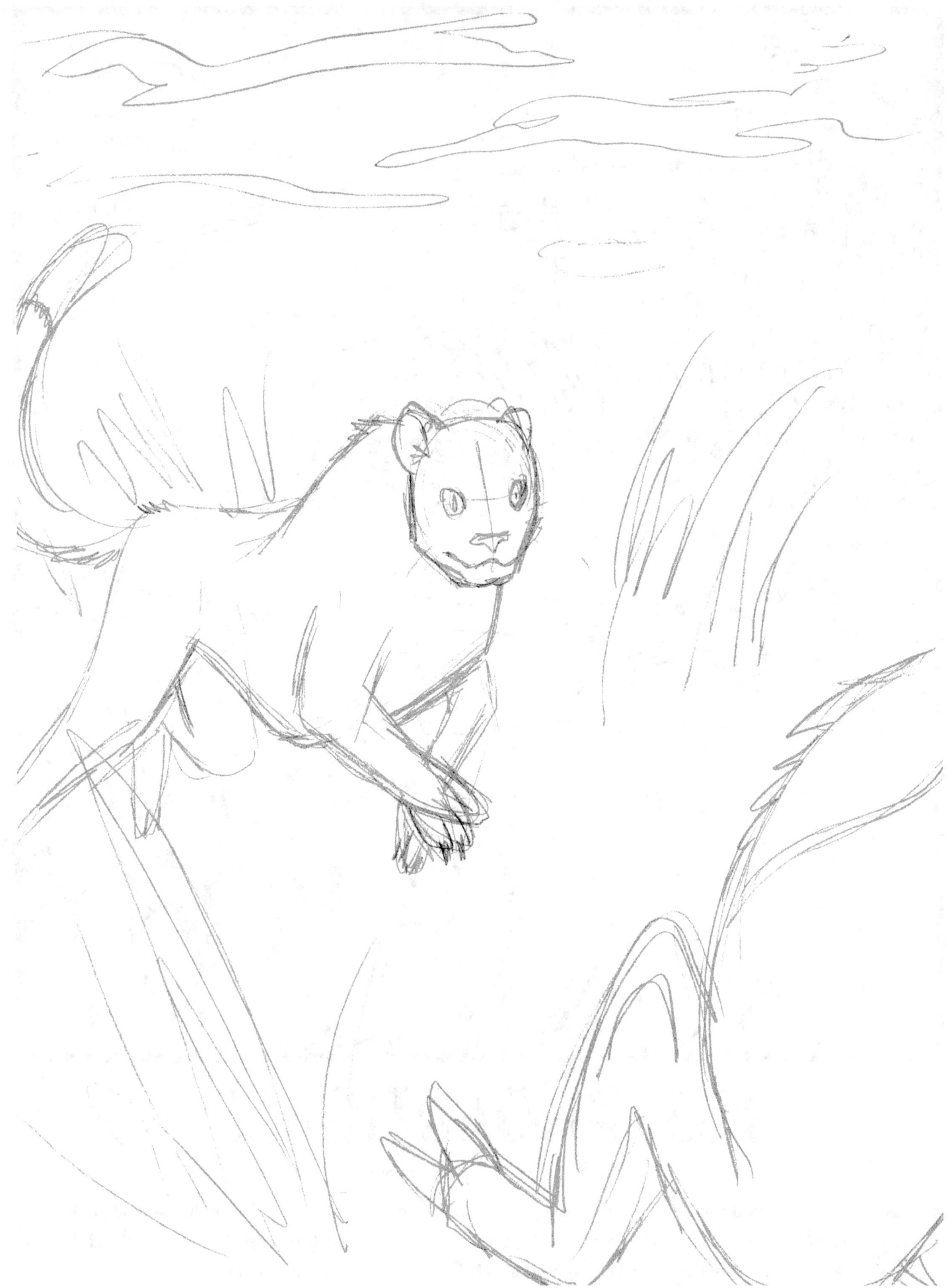

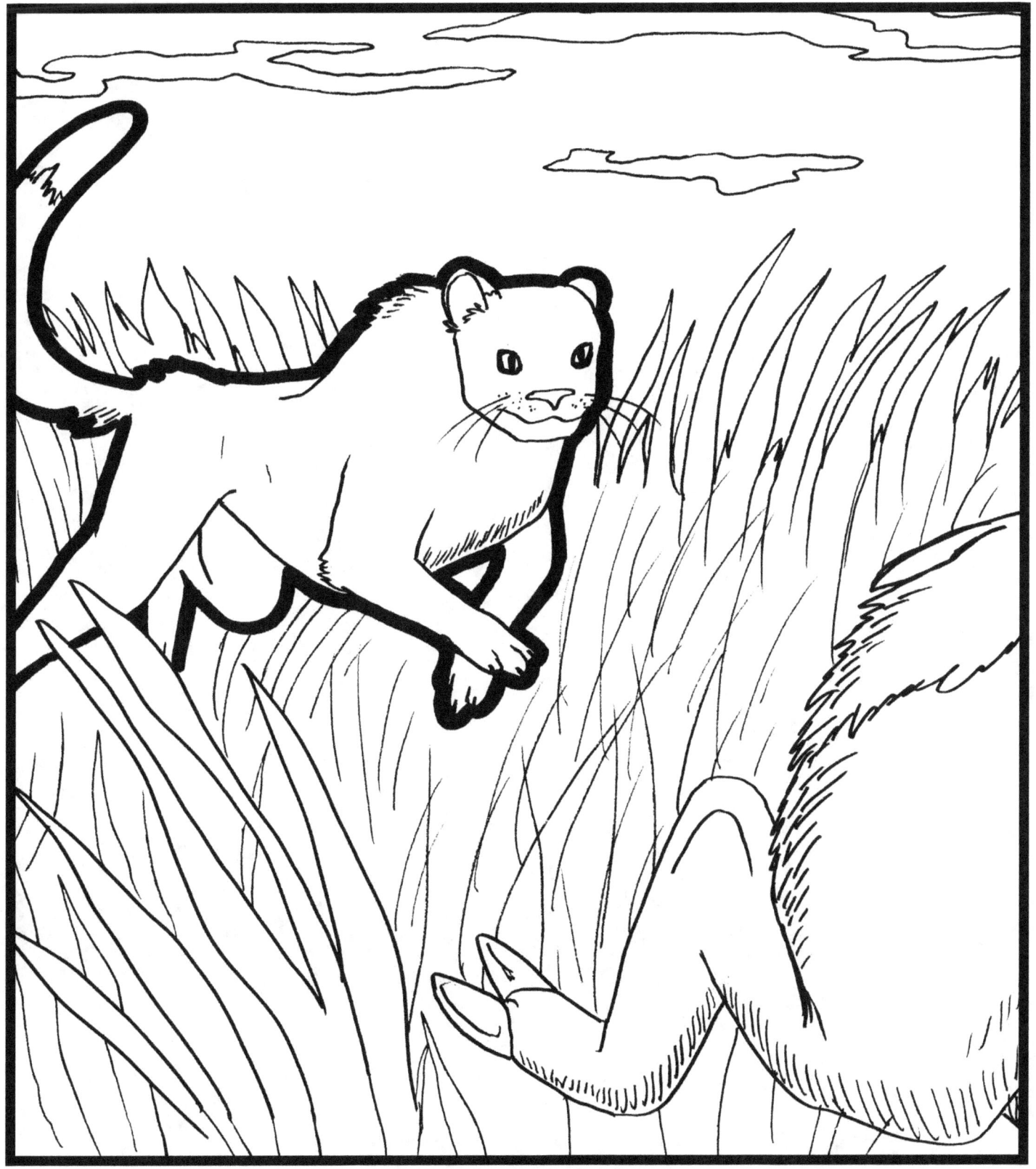

MIRACINONYX

2.6 - 0.01 MYA, Late Pliocene to Late Pleistocene, United States

Even though it is more commonly known as the "American cheetah," and bore many physical similarities to the modern cheetah, this prehistoric cat was a relative of the cougar. It was fast enough to hunt its speedy pronghorn antelope prey, which still roam the wild today.

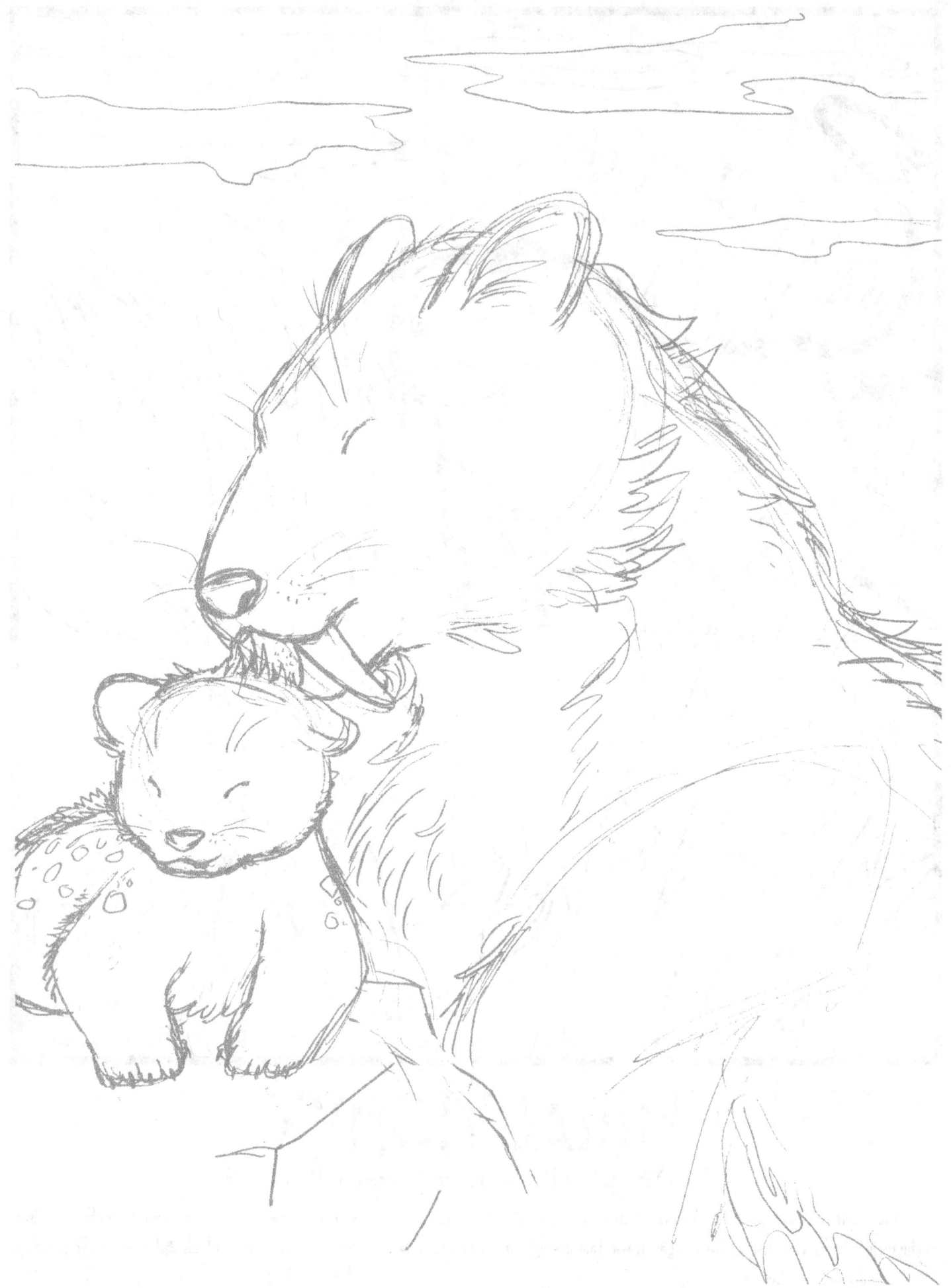

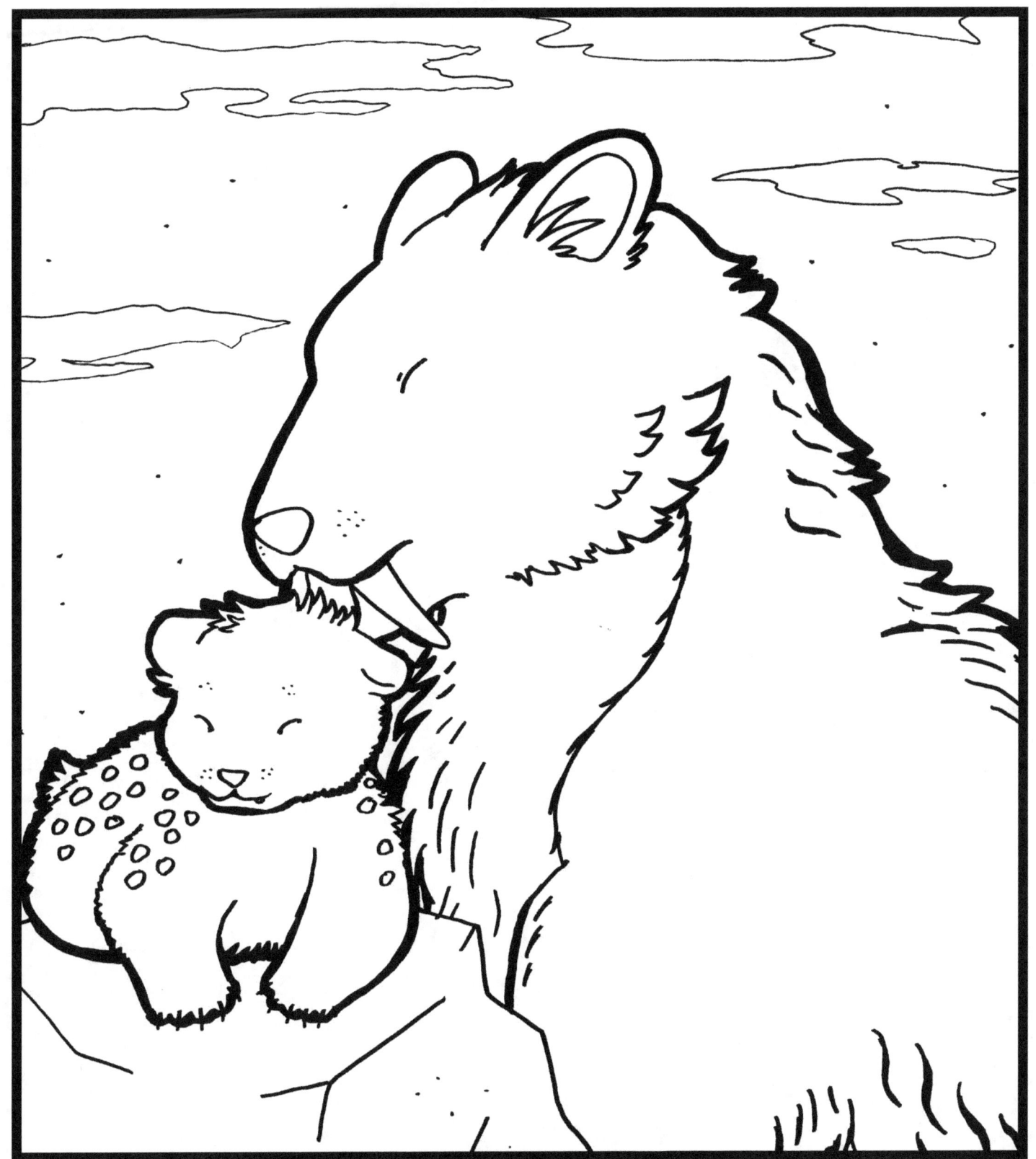

SMILODON

2.5 - 0.01 MYA, Early Pleistocene to Early Holocene, The Americas

This robustly-build cat was a predator of the megafauna that were abundant before the last ice age. The two large canines could pierce deeply into an animal's body, and allow them to take down animals much larger than themselves. It is not known whether this animal lived in groups, or led a more solitary lifestyle, but the current evidence suggests either way of living was possible.

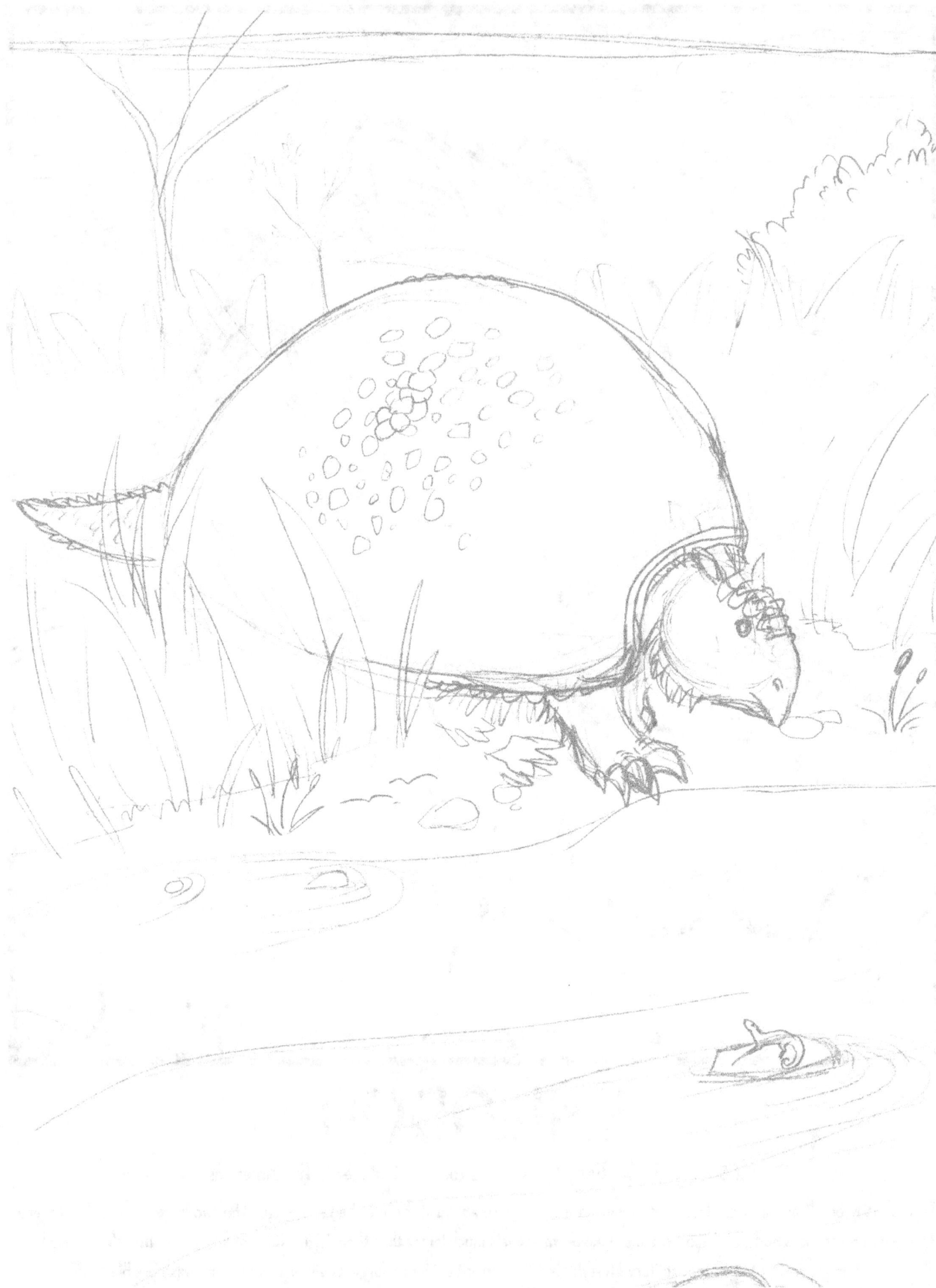

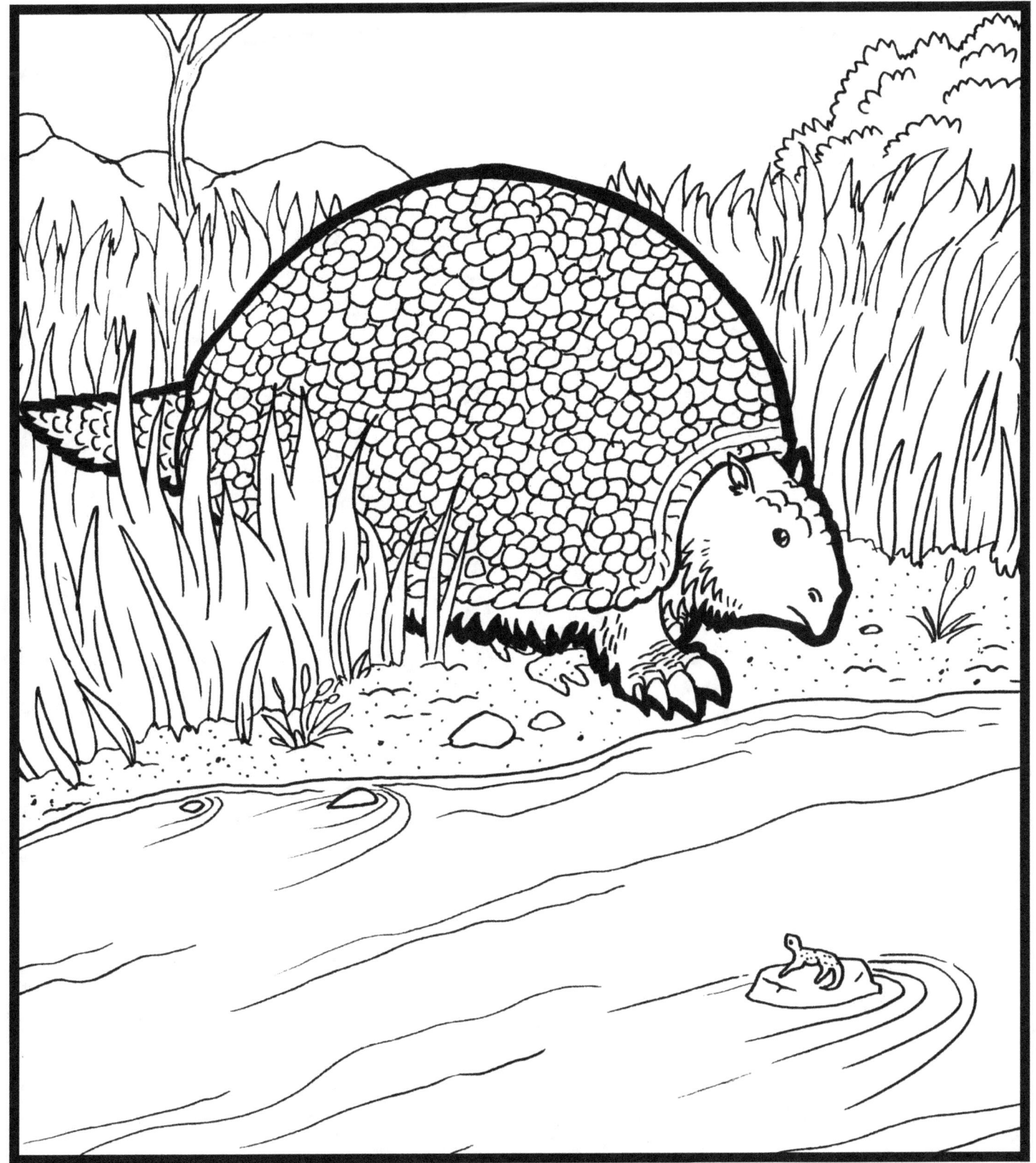

GLYPTODON

2.5 - 0.01 MYA, Pleistocene, North and South America

Like the ankylosaurs and the turtles, bony armor and a similar body shape also arose in this relatively recent mammal.
This animal had very poor vision, and a limited range of movement, but the tough armor probably compensated for its difficulty in spotting
predators from afar and its lack of speed and agility.

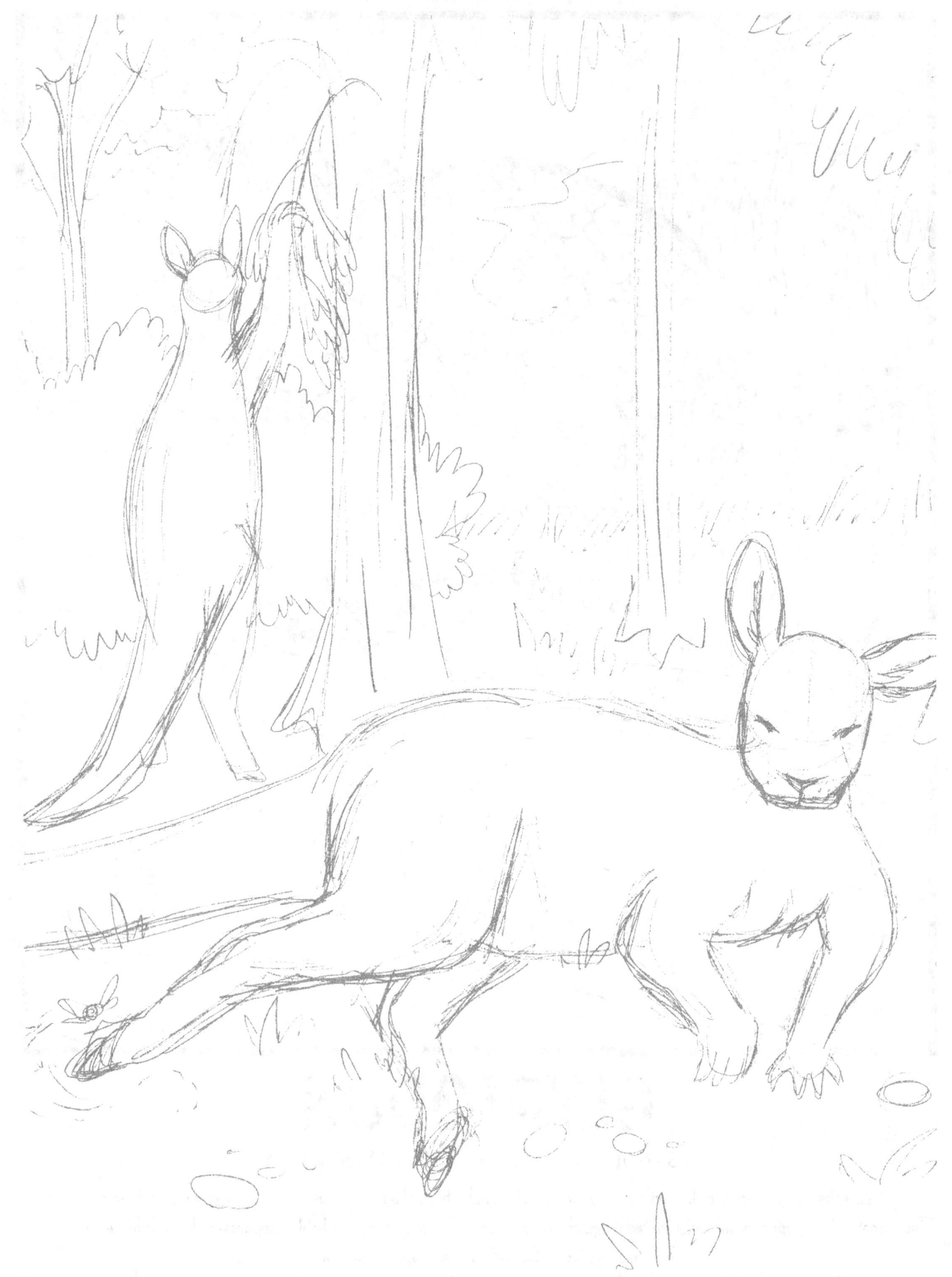

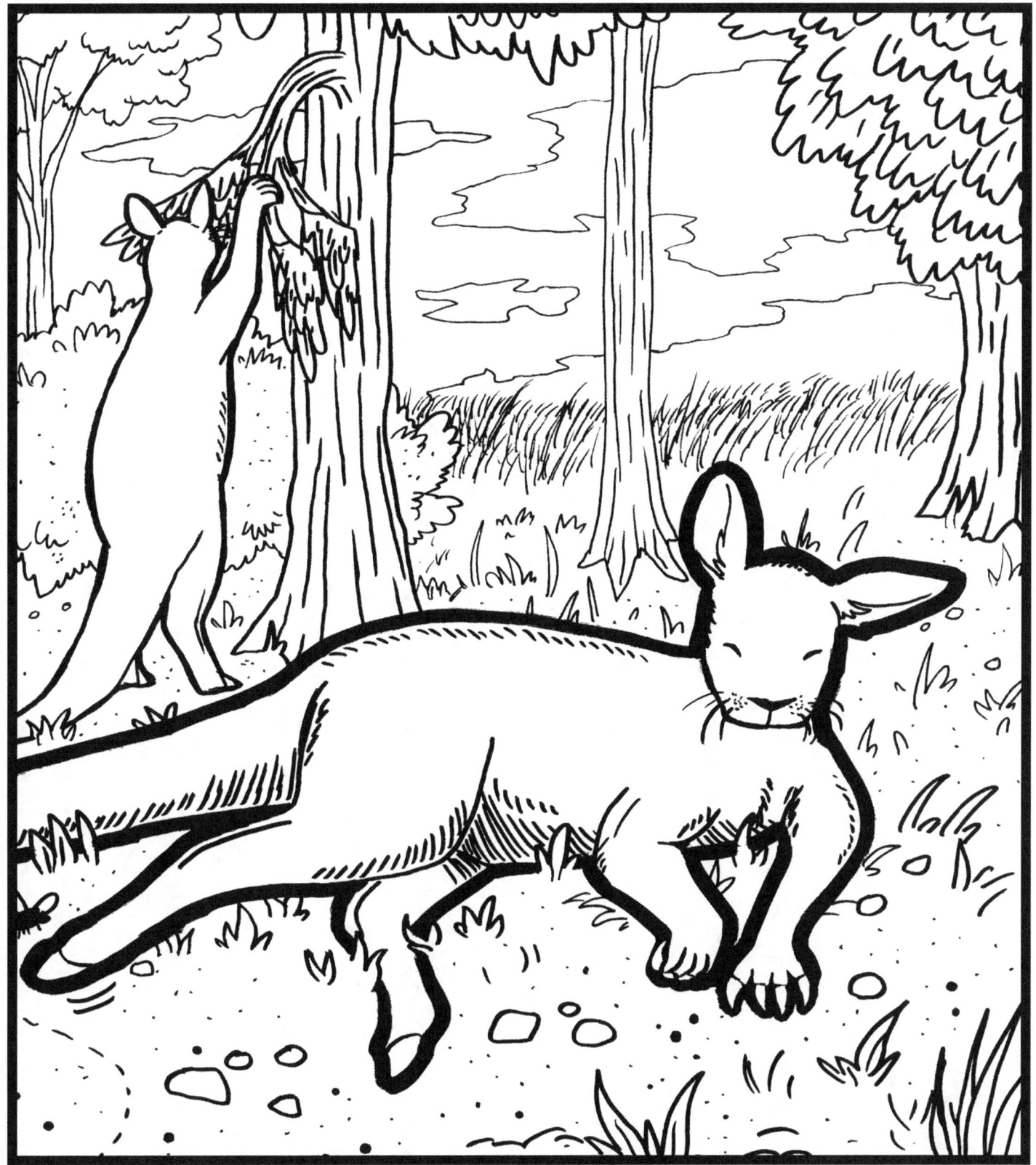

PROCOPTODON

1.6 - 0.03 MYA, Pleistocene, Australia

Also known as a short-faced kangaroo, these browsers walked in a human-like manner, stood 2 meters tall, were heavily-built, and walked on their ungulate-like hooves. The hands ended in long claws that were used for grasping and pulling down branches, and it could get at food that would have been out of reach of smaller kangaroos.

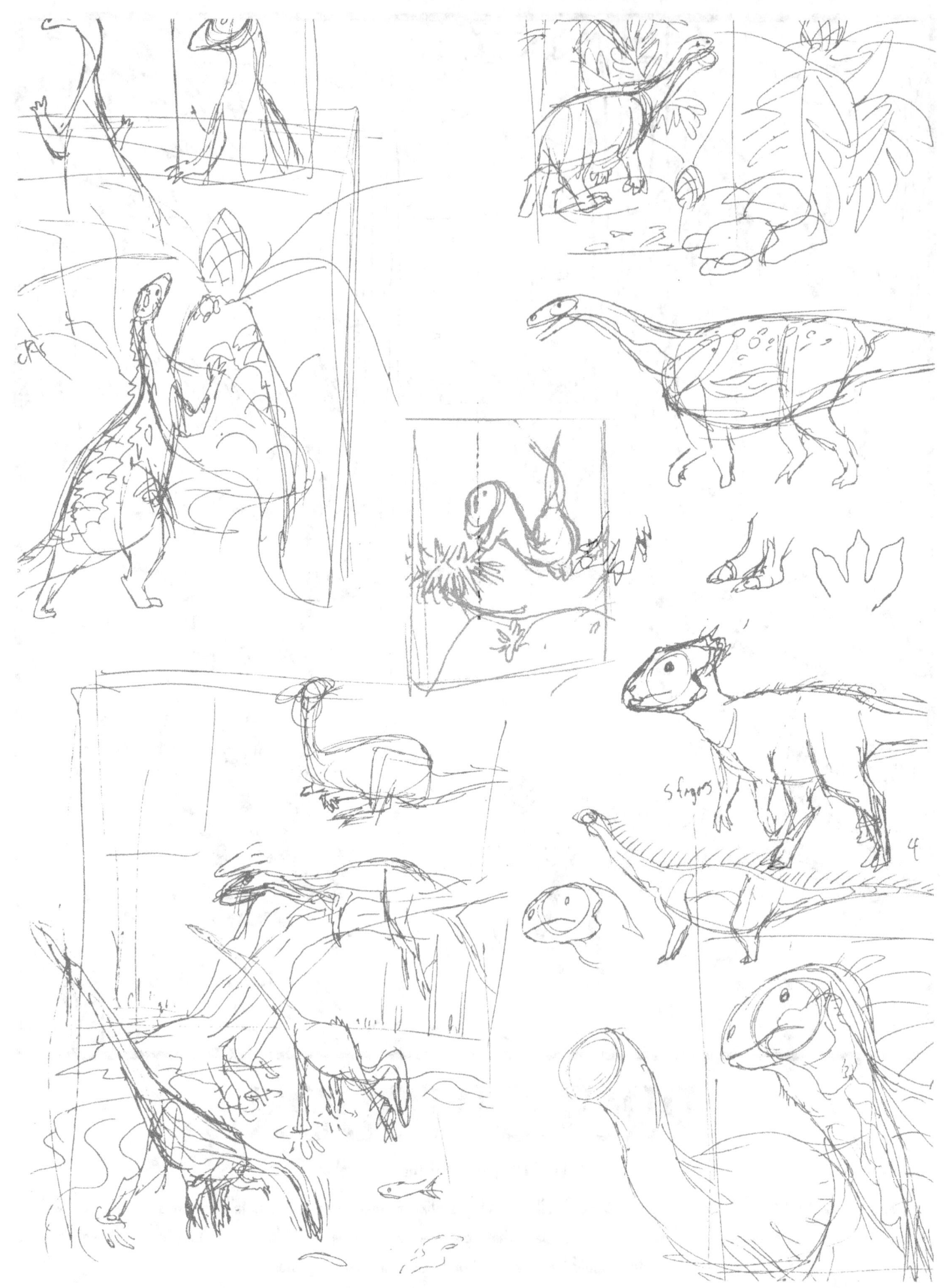

5 fingers

4

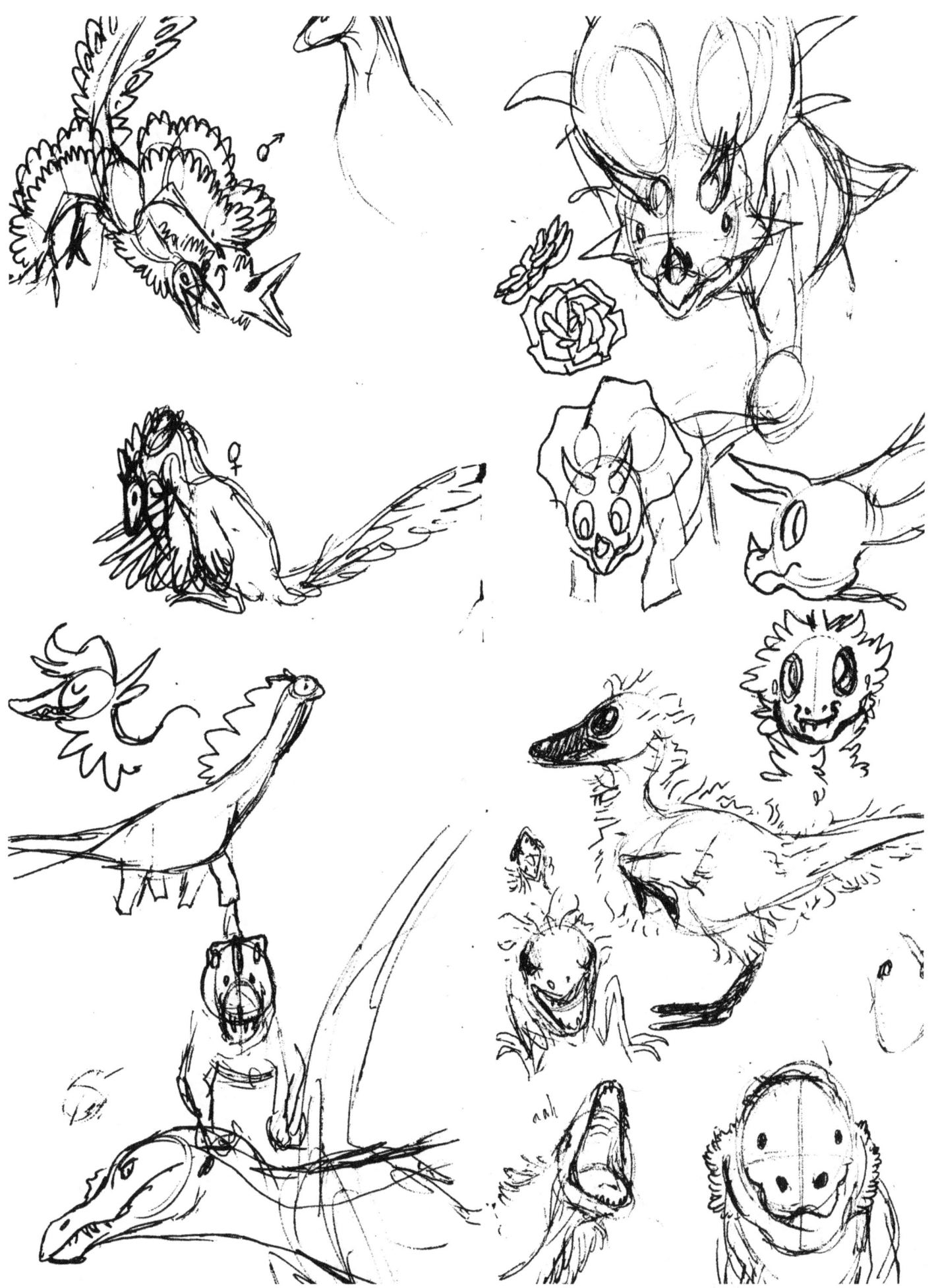